Springer Tracts in Natural Philosophy

Volume 7

AF769493

Edited by C. Truesdell

Co-Editors: L. Collatz · G. Fichera

P. Germain · J. Keller · A. Seeger

The General Problem of the Motion of Coupled Rigid Bodies about a Fixed Point

Eugene Leimanis

Springer-Verlag · Berlin · Heidelberg · New York 1965

Dr. Eugene Leimanis
Professor of Mathematics
The University of British Columbia, Vancouver, B. C.

All rights reserved, especially that of translation into foreign languages
It is also forbidden to reproduce this book, either whole or in part, by photomechanical means
(photostat, microfilm and/or microcard) or any other means
without written permission from the Publishers

© by Springer-Verlag, Berlin · Heidelberg 1965

Library of Congress Catalog Card Number 65-22096

ISBN 978-3-642-88414-6 ISBN 978-3-642-88412-2 (eBook)
DOI 10.1007/978-3-642-88412-2

Titel-Nr. 6735

To the Memory of my Parents

Preface

In the theory of motion of several coupled rigid bodies about a fixed point one can distinguish three basic ramifications.

1. The first, the so-called classical direction of investigations, is concerned with particular cases of integrability of the equations of motion of a single rigid body about a fixed point,[1] and with their geometrical interpretation. This path of thought was predominant until the beginning of the 20th century and its most illustrious representatives are L. Euler (1707—1783), J. L. Lagrange (1736—1813), L. Poinsot (1777—1859), S. V. Kovalevskaya (1850—1891), and others. Chapter I of the present monograph intends to reflect this branch of investigations.

For collateral reading on the general questions dealt with in this chapter the reader is referred to the following textbooks and reports: A. Domogarov [1], F. Klein and A. Sommerfeld [1_1, 1_2, 1_3], A. G. Greenhill [10], A. Gray [1], R. Grammel [4_1], E. J. Routh [2_1, 2_2, 3_1, 3_2], J. B. Scarborough [1], and V. V. Golubev [1, 2].

Chapter II is concerned with the motion of a symmetric as well as an asymmetric self-excited rigid body. A body is said to be self-excited if the torque applied is fixed in the body or moves in a prescribed manner. Prior to the modern age of jet propulsion such a problem seemed to lack a physical meaning. Today, however, various devices with internal reactions are commonly used to influence rotational motions; for example, devices for the steering of space vehicles. Therefore the problem of motion of a self-excited rigid body about a fixed point is now meaningful and important.

Chapter III considers the motion of an externally excited rigid body. While the earlier literature on the motion of rigid bodies is mainly concerned with torque-free and heavy bodies, in the more recent literature several authors such as W. Braunbek [1] and

---

[1] Any rotating body having freedom in one or more planes at right angles to the plane of rotation is called a gyro or gyroscope. A gyro having complete freedom in three planes at right angles to each other is called a free gyro. Mechanically complete freedom of a wheel in three planes can be realized by mounting it in a system of gimbals. However, a rotating ball held in the air would also be a gyroscope; in fact, the Earth itself is a gyroscope.

R. WIEBELITZ [1] have discussed the motion of a rigid body subject to periodic torque vectors. Such rotating bodies are of interest in astronomy and atomic physics. In astronomy our main concern is the perturbation of the Earth's rotation about its axis under the influence of forces arising from the planetary system (precession and nutation of the Earth's axis). For additional reading on this subject the reader is referred to a report by E. W. WOOLARD [1] and papers by S. D. POISSON [2]. In atomic physics electrons and nuclei in high frequency magnetic fields represent atomic gyroscopes subject to periodic torques. Investigations of this type have been stimulated by the necessity of finding a mechanical model for the phenomenon of nuclear induction. For additional information see F. BLOCH, W. W. HANSEN and M. PACKARD [1], F. BLOCH and A. SIEGERT [1], R. K. WANGSNESS and F. BLOCH [1], and F. KIRCHNER [1].

2. A turn in the direction of research took place in the second decade of the present century when in 1909 the gyroscopic compass and in 1917 the gyro horizon and rate gyro (turn indicator) were constructed. Since the above dates these instruments have been used for guidance and control of ships and aircraft, and today for guidance and control of missiles and spacecraft. As a consequence the applied theory of gyroscopes came into existence and matured rapidly while the classical theory receded into the past. Efforts were now made to investigate the motion and stability of particular gyroscopic devices, to study the effect of motion of the supporting member of such devices, the effect of friction at the bearings and that of the flexibility of the rotor-shaft, and so on. This was the beginning of the second ramification in the general theory of the motion of coupled rigid bodies about a fixed point. The most illustrious representatives of this path of thought are L. FOUCAULT (1819–1878), A. N. KRYLOV (1863–1945), M. SCHULER (1882), R. GRAMMEL (1889–1964), C. S. DRAPER (1901), and others.

For additional reading we refer the reader to the following textbooks and reports: E. S. FERRY [1], R. GRAMMEL [4_2], A. N. KRYLOV and YU. A. KRUTKOV [1], A. L. RAWLINGS [1], M. DAVIDSON [1], K. I. T. RICHARDSON [1], C. S. DRAPER, W. WRIGLEY, and L. R. GROHE [1], E. J. SIFF and C. L. EMMERICH [1], B. V. BULGAKOV [3, 4], R. N. ARNOLD and L. MAUNDER [1], P. SAVET [1], and H. ZIEGLER [2]. Part 4 of F. KLEIN and A. SOMMERFELD's treatise [1] on technical applications of the gyroscope is now rather out of date.

3. A third ramification in the theory of motion of coupled rigid bodies about a fixed point was initiated by Lord KELVIN (Sir William THOMSON) and P. G. TAIT [1]. In the eighties of the last century they were concerned with the classification of the various types of forces. According to their terminology, forces which depend on the generalized

velocities $\dot{q}_k$, and the work of which in any real infinitesimal displacement of the system is equal to zero, are called gyroscopic forces. It should be noted, however, that this term has a conditional meaning in the sense that it applies to actual forces applied to the system as well as to certain terms in the equations of motion which are liable to be interpreted as forces. If the gyroscopic forces are denoted by $g_{ik}\dot{q}_k$ (summation over k), then the matrix of the coefficients g_{ik} (depending upon the coordinates q_k) must be skew-symmetric.

Linear terms with respect to the generalized velocities $\dot{q}_k$ with a skew-symmetric matrix appear, for example, in the nonlinear equations of motion of systems containing gyroscopes as well as in the equations of motion of holonomic systems with cyclical coordinates, or moving subject to nonstationary constraints. Further, they appear in the equations for the perturbations of systems subject to stationary constraints, and in the equations of motion of nonholonomic systems in terms of quasi-coordinates. Hence a general theory of systems moving subject to gyroscopic forces is of interest not only for gyrodynamics but also for various types of mechanical and electrical systems containing no gyroscopes at all.

Investigating the effect of gyroscopic forces on the motion of a given system, it is sometimes convenient to assume that the gyroscopic forces depend on a certain parameter H. The introduction of such a parameter H permits us to study the solutions of the equations of motion as functions of this parameter, and to determine some properties of the system in terms of H. For large values of the parameter H it is natural to raise the question concerning a possible simplification of the equations of motion in order to make the integration easier.

The physical reason for introducing a parameter H is also obvious. Namely, for large values of the velocity $\dot{\Phi}$ of the proper rotation of a gyroscope (in comparison with the velocities of precession $\dot{\psi}$ and nutation $\dot{\Theta}$) $H = Cn$ (Cn being the constant of integration corresponding to the cyclic coordinate Φ in the equations of motion of a gyroscope) is approximately equal to the modulus I of the angular momentum $\vec{I}$ of the gyroscope, i.e. $I \approx H = Cn$, and the gyroscope exhibits certain properties which do not occur for small values of $\dot{\Phi}$. This simple example shows the necessity and desirability for investigating gyroscopic systems in terms of a parameter.

It is also well known that in certain cases, discarding terms containing second derivatives and products of first derivatives, equations result which provide acceptable solutions for practical purposes. For example, the equations of motion of a gyroscopic compass admit such a simplification and the solution of the simplified equations describes with great accuracy the actual motion. In other cases, however, such a simplification of the differential equations of motion leads to completely unacceptable solutions. For example, the equations of motion of the FOUCAULT gyroscope, being a system with two degrees of freedom, do not admit such a simplification.

A book which summarizes the results obtained following the third path of thought sketched above is that by D. R. MERKIN [1]. It also contains certain general theorems due to its author concerning the stability of the equilibrium of a system subject to gyroscopic forces and the effect of gyroscopic forces on the motion of a nonconservative system. Also of some practical importance are MERKIN's investigations concerning the conditions under which the equations of a fast rotating gyroscope can be simplified in the sense explained above.

Concerning problems of gyroscopic stabilization the reader is referred to I. I. METELICYN [1].

Two particular cases of several-body systems which are of great practical importance are discussed in Chapters IV and V respectively, namely the gyrostat and the gyroscope in a CARDAN suspension.

4. In connection with the investigation of the motion of the Earth's artificial satellites some old problems of classical celestial mechanics and gyrodynamics became again actual and, in addition, new problems arose. Such problems are, to name just a few:

(i) Separation of the general motion of mutually attracting rigid bodies into translations of their mass centers and rotations about the latter.

(ii) Rotation of an artificial satellite about its mass center.

(iii) Motion of a rigid body about a fixed point in a central NEWTONian force field.

(iv) Motion of a rigid body with fluid-filled cavities about a fixed point.

(v) Motion of a gyroscope with variable mass or moments of inertia.

(vi) Application of gyroscopics to inertial guidance systems.

A rapidly expanding literature of the above problems exists today which is already so extensive that a proper survey would be an undertaking beyond the scope of this monograph. Therefore we shall limit ourselves to the discussion of problem (iii) in Chapter VI and problems (i) and (ii) in Chapter VII.

5. The time EULER spent in Berlin (1741—1766) was rich with discoveries in the fields of celestial mechanics, mechanics of rigid bodies and mechanics of fluids. Although EULER was concerned with the dynamics of rigid bodies from the beginning of his scientific activities his main results, which culminated in the now classical equations of

motion of a rigid body about a fixed point, were obtained by him only during the Berlin period of his life and published in the Mémoires de l'Académie Royale des Sciences et Belles-Lettres de Berlin, vols. 5—16 (1751—1767), the papers being written between 1749—1760.

In particular, in 1758 EULER developed the theory of moments of inertia of a rigid body, proved the existence of three mutually orthogonal axes, called the principal axes of inertia of the body, and obtained the equations for the rotational motion of the body relative to the body-fixed coordinate trihedral, the axes of which are directed along the principal inertia axes of the body. The concept of the ellipsoid of inertia, however, was introduced later by L. POINSOT. The above papers of 1758 were published in 1765 together with his fundamental treatise "Theoria Motus Corporum Solidorum seu Rigidorum", of which the 10th and 15th chapters again contained the derivation of his equations.

My report [1] "On some recent advances in the dynamics of rigid bodies and celestial mechanics" appeared in 1958. The present monograph is an attempt not only to account for the present state of the field which it covers but also of its growth during the last two hundred years. Completeness of the monograph is not claimed.

Acknowledgements. This work was supported in part by the Mathematics Division of the United States Air Force Office of Scientific Research (Grant AFOSR 483—64) while the author was on leave from the University of British Columbia.

My colleague Dr. W. H. SIMONS and former student Dr. R. LEE read the manuscript and made some valuable comments. Dr. R. LEE also prepared the figures, and Mrs. Cathie EASTO typed the manuscript. My Ph. D. students Graham ZELMER, B. Sc. (Hons. Math.), B. Sc. (Eng.) and B. S. LALLI, B. A. (Hons.), M. A., and graduate student L. Barry MULLETT, B. Sc., helped me with proof-reading.

To my University, the above agency, J. Springer-Verlag and its technical staff, and all the people named above I wish to express my deep appreciation.

E. LEIMANIS

July 15, 1965

Department of Mathematics
The University of British Columbia
Vancouver 8, B. C., Canada

Contents

Part I: Single rigid body

PART I

Single rigid body

Chapter I

Heavy rigid body

A. General solution of the Euler and Poisson equations

§ 1. The Euler angles

1.1. Definition. According to a theorem of EULER [1] the general displacement of a rigid body with one fixed point is a rotation about some axis through this point. The position of a rigid body is completely determined by locating a rectangular coordinate trihedral fixed in the rigid body relative to a rectangular coordinate trihedral fixed in space. If the fixed point is taken as a common origin of the body and space trihedrals, then the orientation of the body in space can be described in terms of the direction cosines of the body axes relative to the space axes. Among the nine direction cosines only three are independent. Therefore we must use some set of three independent functions of the direction cosines to specify the position of the rigid body. A number of such sets of independent variables have been described in the literature, the most important and useful being the EULER [2] angles.

Let O be a fixed point of a rigid body about which the rotation of the body takes place, and let $O\,X\,Y\,Z$ be a right-hand rectangular trihedral fixed in space (Fig. 1). Let $O\,x\,y\,z$ be a right-hand rectangular trihedral fixed in the body and moving with it. Furthermore, let the coordinate planes $X\,O\,Y$ and $x\,O\,y$ intersect along the line of nodes $O\,N$ which is perpendicular to the plane through the axes $O\,Z$ and $O\,z$. Choose the orientation along $O\,N$ in such a way that the trihedral $O\,N\,Z\,z$ is right-handed. Denote the angles $Z\,O\,z$, $X\,O\,N$ and $N\,O\,x$ by Θ, ψ, Φ respectively. These are known as the EULER angles. The angle $\Theta\,(0 \leqq \Theta < \pi)$ is called the angle of nutation, the angle $\psi\,(0 \leqq \psi < 2\pi)$

the angle of precession and the angle Φ $(0 \leqq \Phi < 2\pi)$ the angle of proper rotation. If the Euler angles Θ, ψ, Φ are known as functions of the time t, then the position of the trihedral $O\,x\,y\,z$ with respect to the trihedral $O\,X\,Y\,Z$ is defined. In other words, the motion of the rigid body about the fixed point is known.

1.2. The direction cosines of $O\,x$, $O\,y$, $O\,z$ as functions of the Euler angles. The trihedral $O\,x\,y\,z$, the position of which is defined by the Euler angles Θ, ψ, Φ, can be derived from the trihedral $O\,X\,Y\,Z$ by means of the following three successive counter-clockwise rotations: (i) a rotation through the angle ψ about the $O\,Z$-axis, obtaining the trihedral $O\,X_1\,Y_1\,Z_1$, the $O\,X_1$-axis of which coincides with the line of nodes $O\,N$; (ii) a rotation through the angle Θ about the $O\,X_1$-axis, obtaining the trihedral $O\,x_1\,y_1\,z_1$, the $O\,y_1$-axis of which lies in the plane $Z_1\,O\,z_1$ and makes the angle Θ with the $O\,Y_1$-axis; (iii) a rotation through the angle Φ about the $O\,z_1$-axis until the $O\,x_1$-axis coincides with the $O\,x$-axis and the $O\,y_1$-axis with the $O\,y$-axis.

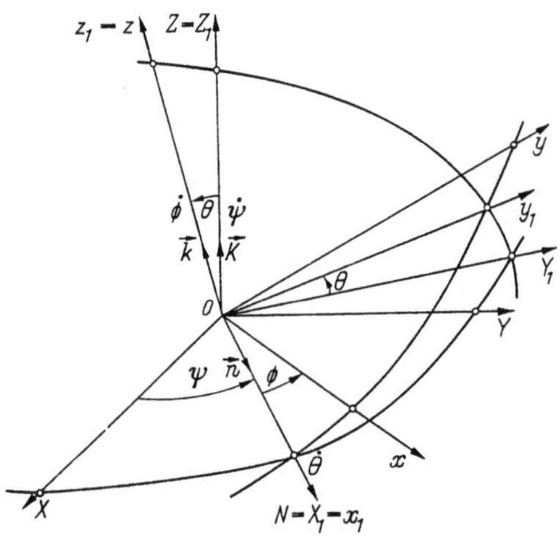

Fig. 1

The transition from coordinates fixed in the rigid body, x, y, z, to coordinates fixed in space, X, Y, Z, can be accomplished by means of an orthogonal matrix A with elements α_{ij} $(i, j = 1, 2, 3)$ connected by six orthogonality conditions

$$\sum_{i=1}^{3} \alpha_{ij}\,\alpha_{ik} = \delta_{jk} \quad \text{or} \quad \sum_{i=1}^{3} \alpha_{ji}\,\alpha_{ki} = \delta_{jk} \qquad (j, k = 1, 2, 3)$$

where δ_{jk} is the Kronecker δ-symbol, defined by

$$\delta_{jk} = 1\,(j = k), \qquad \delta_{jk} = 0\,(j \neq k)$$

The elements of the resulting transformation matrix A can be obtained by writing the matrix as a triple product of the above three rotations, each of which has a relatively simple matrix form:

$$\begin{pmatrix} X \\ Y \\ Z \end{pmatrix} = \begin{pmatrix} \cos\psi & -\sin\psi & 0 \\ \sin\psi & \cos\psi & 0 \\ 0 & 0 & 1 \end{pmatrix} \begin{pmatrix} X_1 \\ Y_1 \\ Z_1 \end{pmatrix} \tag{1.1}$$

$$\begin{pmatrix} X_1 \\ Y_1 \\ Z_1 \end{pmatrix} = \begin{pmatrix} 1 & 0 & 0 \\ 0 & \cos\Theta & -\sin\Theta \\ 0 & \sin\Theta & \cos\Theta \end{pmatrix} \begin{pmatrix} x_1 \\ y_1 \\ z_1 \end{pmatrix} \tag{1.2}$$

$$\begin{pmatrix} x_1 \\ y_1 \\ z_1 \end{pmatrix} = \begin{pmatrix} \cos\Phi & -\sin\Phi & 0 \\ \sin\Phi & \cos\Phi & 0 \\ 0 & 0 & 1 \end{pmatrix} \begin{pmatrix} x \\ y \\ z \end{pmatrix} \tag{1.3}$$

Hence the transformation from body coordinates x, y, z to space coordinates X, Y, Z is given by the formulas

$$\begin{pmatrix} X \\ Y \\ Z \end{pmatrix} = A \begin{pmatrix} x \\ y \\ z \end{pmatrix} \tag{1.4}$$

where

$$A = \begin{pmatrix} \cos\psi\cos\Phi - \cos\Theta\sin\psi\sin\Phi, & -\cos\psi\sin\Phi - \cos\Theta\sin\psi\cos\Phi, & \sin\Theta\sin\psi \\ \sin\psi\cos\Phi + \cos\Theta\cos\psi\sin\Phi, & -\sin\psi\sin\Phi + \cos\Theta\cos\psi\cos\Phi, & -\sin\Theta\cos\psi \\ \sin\Theta\sin\Phi & \sin\Theta\cos\Phi & \cos\Theta \end{pmatrix} \tag{1.5}$$

The inverse transformation from space coordinates X, Y, Z to body coordinates x, y, z is

$$\begin{pmatrix} x \\ y \\ z \end{pmatrix} = A^{-1} \begin{pmatrix} X \\ Y \\ Z \end{pmatrix} \tag{1.6}$$

where the matrix A^{-1} is equal to the transpose A' of A, i.e.

$$A^{-1} = \begin{pmatrix} \cos\psi\cos\Phi - \cos\Theta\sin\psi\sin\Phi, & \sin\psi\cos\Phi + \cos\Theta\cos\psi\sin\Phi, & \sin\Theta\sin\Phi \\ -\cos\psi\sin\Phi - \cos\Theta\sin\psi\cos\Phi, & -\sin\psi\sin\Phi + \cos\Theta\cos\psi\cos\Phi, & \sin\Theta\cos\Phi \\ \sin\Theta\sin\psi & -\sin\Theta\cos\psi & \cos\Theta \end{pmatrix} \tag{1.7}$$

1.3. The components of the angular velocity $\vec{\omega}$ as functions of the Euler angles. Consider two positions of the rigid body, determined by the EULER angles Θ, ψ, Φ and $\Theta + d\Theta$, $\psi + d\psi$, $\Phi + d\Phi$. The increment $d\Theta$ corresponds to an infinitesimal rotation about the line of nodes $O\,N$. Similarly $d\psi$ and $d\Phi$ correspond to infinitesimal rotations about the axes $O\,Z$ and $O\,z$ respectively. Hence the components of the angular velocity $\vec{\omega}$ along the axes $O\,N$, $O\,Z$ and $O\,z$ are $\dot{\Theta}$, $\dot{\psi}$, $\dot{\Phi}$ respectively, and therefore

$$\vec{\omega} = \dot{\Theta}\,\vec{n} + \dot{\psi}\,\vec{K} + \dot{\Phi}\,\vec{k}$$

where $\vec{n}$, $\vec{K}$ and $\vec{k}$ are the unit vectors along the axes $O\,N$, $O\,Z$ and $O\,z$ respectively, and dots denote derivatives with respect to the time t. The table given below is a convenient means of finding the components of $\vec{\omega}$ with respect to the trihedrals $O\,x\,y\,z$ and $O\,X\,Y\,Z$.

		$O\,x$	$O\,y$	$O\,z$	$O\,X$	$O\,Y$	$O\,Z$
	$\vec{\omega}$	p	q	r	P	Q	R
$O\,N$	$\dot{\Theta}$	$\cos\Phi$	$-\sin\Phi$	0	$\cos\psi$	$\sin\psi$	0
$O\,Z$	$\dot{\psi}$	$\alpha_{31} = \alpha$	$\alpha_{32} = \beta$	$\alpha_{33} = \gamma$	0	0	1
$O\,z$	$\dot{\Phi}$	0	0	1	α_{13}	α_{23}	α_{33}

By means of this table we find that

$$\begin{aligned} p &= \dot{\Theta}\cos\Phi + \dot{\psi}\,\alpha \\ q &= -\dot{\Theta}\sin\Phi + \dot{\psi}\,\beta \\ r &= \dot{\psi}\,\gamma + \dot{\Phi} \end{aligned}$$

and

$$\begin{aligned} P &= \dot{\Theta}\cos\psi + \dot{\Phi}\,\alpha_{13} \\ Q &= \dot{\Theta}\sin\psi + \dot{\Phi}\,\alpha_{23} \\ R &= \dot{\Phi}\,\alpha_{33} + \dot{\psi} \end{aligned}$$

Substituting for α, β, γ, and α_{13}, α_{23}, α_{33} their expressions in terms of the Euler angles [see matrix (1.5)], we obtain that

$$\left.\begin{aligned} p &= \dot{\Theta}\cos\Phi + \dot{\psi}\sin\Theta\sin\Phi \\ q &= -\dot{\Theta}\sin\Phi + \dot{\psi}\sin\Theta\cos\Phi \\ r &= \dot{\psi}\cos\Theta + \dot{\Phi} \end{aligned}\right\} \qquad (1.8)$$

and

$$\left.\begin{aligned} P &= \dot{\Theta}\cos\psi + \dot{\Phi}\sin\Theta\sin\psi \\ Q &= \dot{\Theta}\sin\psi - \dot{\Phi}\sin\Theta\cos\psi \\ R &= \dot{\Phi}\cos\Theta + \dot{\psi} \end{aligned}\right\} \qquad (1.9)$$

§ 2. The Euler and Poisson equations of motion

2.1. The dynamical equations of Euler. According to the basic equation of dynamics the derivative with respect to the time of the angular momentum of a rigid body is equal to the moment of the external forces acting on the body, i.e.

$$\frac{d\vec{I}}{dt} = \vec{M}$$

where $\vec{I}$ is the angular momentum of the body with respect to the point O and $\vec{M}$ is the instantaneous moment vector of the external forces with respect to the same point. Here the derivative refers to the fixed space axes since the above equation holds only in an inertial coordinate trihedral. Denoting by $\frac{d\vec{I}}{dt}$ and $\left(\frac{d\vec{I}}{dt}\right)_{\text{rel}}$ the derivatives of $\vec{I}$ with respect to the fixed trihedral $O\,X\,Y\,Z$ and the moving trihedral $O\,x\,y\,z$ respectively, we have that

$$\frac{d\vec{I}}{dt} = \left(\frac{d\vec{I}}{dt}\right)_{\text{rel}} + \vec{\omega} \times \vec{I}$$

and hence

$$\left(\frac{d\vec{I}}{dt}\right)_{\text{rel}} + \vec{\omega} \times \vec{I} = \vec{M} \tag{2.1}$$

Assume that the axes of the trihedral $O\,x\,y\,z$ are directed along the principal axes of the ellipsoid of inertia of the body at O. Then

$$\vec{I} = A\,p\,\vec{i} + B\,q\,\vec{j} + C\,r\,\vec{k} \tag{2.2}$$

where A, B, C denote the moments of inertia of the body about the axes $O\,x$, $O\,y$, $O\,z$ respectively, and $\vec{i}$, $\vec{j}$, $\vec{k}$ are the unit vectors along the same axes. By virtue of formula (2.2) and the fact that

$$\vec{\omega} = p\,\vec{i} + q\,\vec{j} + r\,\vec{k}$$

we can write that

$$\left(\frac{d\vec{I}}{dt}\right)_{\text{rel}} + \vec{\omega} \times \vec{I} = A\frac{dp}{dt}\vec{i} + B\frac{dq}{dt}\vec{j} + C\frac{dr}{dt}\vec{k} + \begin{vmatrix} \vec{i} & \vec{j} & \vec{k} \\ p & q & r \\ A\,p & B\,q & C\,r \end{vmatrix}$$

$$= \left[A\frac{dp}{dt} + (C - B]\,q\,r\right]\vec{i} + \left[B\frac{dq}{dt} + (A - C)\,r\,p\right]\vec{j} +$$

$$+ \left[C\frac{dr}{dt} + (B - A)\,p\,q\right]\vec{k} \tag{2.3}$$

In the case of a heavy rigid body moving about a fixed point the only force applied is gravity $-m\,g\,\vec{K}$, the point of application being the center of mass $G(x_0, y_0, z_0)$. In such a case $\vec{M} = -w\left(\vec{r}_0 \times \vec{K}\right)$, where $w = m\,g$ is the weight of the body, $\vec{K}(\alpha, \beta, \gamma)$ the unit vector along the $O\,Z$-axis and $\vec{r}_0(x_0, y_0, z_0)$ the position vector of the center of mass G. Hence

$$\vec{M} = -m\,g\left(\vec{r}_0 \times \vec{K}\right) = m\,g\left(\vec{K} \times \vec{r}_0\right) = m\,g \begin{vmatrix} \vec{i} & \vec{j} & \vec{k} \\ \alpha & \beta & \gamma \\ x_0 & y_0 & z_0 \end{vmatrix}$$

$$= m\,g\left[(\beta\,z_0 - \gamma\,y_0)\,\vec{i} + (\gamma\,x_0 - \alpha\,z_0)\,\vec{j} + (\alpha\,y_0 - \beta\,x_0)\,\vec{k}\right] \tag{2.4}$$

By virtue of formulas (2.3) and (2.4) the vector equation (2.1) is equivalent to the following system of scalar equations

$$\left.\begin{aligned} A\frac{dp}{dt} + (C - B)\,q\,r &= m\,g(\beta\,z_0 - \gamma\,y_0) \\ B\frac{dq}{dt} + (A - C)\,r\,p &= m\,g(\gamma\,x_0 - \alpha\,z_0) \\ C\frac{dr}{dt} + (B - A)\,p\,q &= m\,g(\alpha\,y_0 - \beta\,x_0) \end{aligned}\right\} \tag{2.5}$$

These equation were given by EULER [2—4]. They constitute the first set of the basic dynamical equations for the motion of a heavy rigid body about a fixed point. Since equations (2.5) are homogeneous with respect to the coefficients A, B, C and $m\,g$, we may introduce new coefficients $A_1 = \frac{A}{m\,g}$, $B_1 = \frac{B}{m\,g}$, $C_1 = \frac{C}{m\,g}$, which are proportional to A, B, C respectively. Therefore in what follows we shall assume that $m\,g = 1$. The same result can also be achieved by a proper choice of units.

L. EULER's own derivation of his equations was rather cumbersome. What L. EULER [2] (1758) obtained only after lengthy calculations, J. L. LAGRANGE [1] (1788) was able to accomplish in a few lines. S. D. POISSON [3, vol. 2] (1811) gave an elementary proof for the existence of EULER's equations without introducing the EULER angles. In 1834 L. POINSOT [1] gave a kinetic interpretation for the EULER equations in terms of fictitious centrifugal forces. For a kinematic interpretation of the EULER equations in terms of relative motions, see P. SAINT-GUILHEM [1], R. B. HAYWARD [1] and J. C. MAXWELL [1]. A simplified derivation of the EULER equations was given by G. SCHMIDT [1], and others. The EULER equations of motion form a system of sixth order. P. V. HARLAMOV [2] introduced a special rectangular coordinate trihedral and reduced the system to two first order ordinary differential equations. A. I. DOKŠEVIČ [1], starting from EULER's equations, has discussed possible ranges for several magnitudes in the body-fixed coordinate trihedral.

V. L. NADOLSCHI [1] showed that the true EULER equations for the motion of a rigid body, namely

$$\begin{aligned} A\,\dot{p} + (C - B)\,q\,r &= M_1 \\ B\,\dot{q} + (A - C)\,r\,p &= M_2 \\ C\,\dot{r} + (B - A)\,p\,q &= M_3 \end{aligned}$$

reduce in the symmetric case $A = B$, to a single linear differential equation of the form

$$2f(t)\,\ddot{p} + \dot{f}(t)\,\dot{p} + 2p = g(t)$$

where $f(t)$ and $g(t)$ are known functions of t if M_1, M_2, M_3 are known functions of t. This last differential equation reduces to the integrable form

$$\frac{d^2p}{d\tau^2} + p = G(\tau)$$

if the new independent variable $\tau = \int [f(t)]^{-1/2}\,dt$ is used.

2.2. The Poisson kinematical equations. The EULER equations contain six functions of the time t, namely the components p, q, r of the angular velocity $\vec{\omega}$ of the rigid body along the axes of the moving trihedral and the direction cosines α, β, γ of the fixed $O\,Z$-axis with respect to the same trihedral. The six constants A, B, C; x_0, y_0, z_0 characterize the mass distribution of the rigid body with respect to the chosen coordinate trihedral. Thus for the complete determination of the motion three additional equations are required. They are obtained by differentiating with respect to the time the vector

$$\vec{K} = \alpha\,\vec{i} + \beta\,\vec{j} + \gamma\,\vec{k}$$

If, as before, $\left(\frac{d\vec{K}}{dt}\right)_{\text{rel}}$ denotes the derivative of $\vec{K}$ with respect to the moving trihedral $O\,x\,y\,z$, then

$$0 = \frac{d\vec{K}}{dt} = \left(\frac{d\vec{K}}{dt}\right)_{\text{rel}} + \vec{\omega} \times \vec{K}$$

or

$$\left(\frac{d\vec{K}}{dt}\right)_{\text{rel}} = \vec{K} \times \vec{\omega} \tag{2.6}$$

or

$$\frac{d\alpha}{dt}\vec{i} + \frac{d\beta}{dt}\vec{j} + \frac{d\gamma}{dt}\vec{k} = \begin{vmatrix} \vec{i} & \vec{j} & \vec{k} \\ \alpha & \beta & \gamma \\ p & q & r \end{vmatrix}$$

The last vector equation is equivalent to the following set of three scalar equations

$$\left.\begin{aligned} \frac{d\alpha}{dt} &= \beta\,r - \gamma\,q \\ \frac{d\beta}{dt} &= \gamma\,p - \alpha\,r \\ \frac{d\gamma}{dt} &= \alpha\,q - \beta\,p \end{aligned}\right\} \tag{2.7}$$

which are known as the POISSON [3] equations. They constitute the second set of the basic equations for the determination of the motion of a heavy rigid body about a fixed point.

F. SBRANA [1] has shown that the EULER and POISSON equations can be replaced by equations which contain only the components p, q, r of the angular velocity $\vec{\omega}$ but do not contain any direction cosines.

O. BOTTEMA and H. J. E. BETH [1, 2] have deduced in n-dimensional space a certain system of equations as the appropriate generalization of the EULER equations and have studied for the special case $n = 4$ the stationary motions of a rigid body under no forces.

2.3. Finding of the first integrals. Scalar multiplication of the EULER vector differential equation

$$\left(\frac{d\vec{I}}{dt}\right)_{\text{rel}} + \vec{\omega} \times \vec{I} + \vec{r}_0 \times \vec{K} = 0$$

by $\vec{\omega}$ and $\vec{K}$ respectively and application of the POISSON equation (2.6) gives the first integrals of motion

$$\vec{r}_0 \cdot \vec{K} = h_1 - T \tag{2.8}$$

and

$$\vec{I} \cdot \vec{K} = c \tag{2.9}$$

Here

$$T = \tfrac{1}{2} \vec{\omega} \cdot \vec{I}$$

is the kinetic energy of the body and h_1 and c are constants of integration. The first integral is the so-called energy integral whereas the second integral expresses the law of conservation of the angular momentum about the vertical.

The third integral

$$K^2 = 1 \tag{2.10}$$

is obtained by scalar multiplication of the POISSON equation (2.6) by $\vec{K}$. It is a trigonometric identity which expresses the fact that the sum of the squares of the direction cosines of $\vec{K}$ is unity.

2.4. On the number of independent integrals. When the integration of the EULER and POISSON equations (2.5) and (2.7) is completed, the EULER angles Φ and Θ are given by the formulas

$$\tan \Phi = \frac{\alpha}{\beta}, \quad \cos \Theta = \gamma \tag{2.11}$$

and the third EULER angle ψ is obtained from the last equation of system (1.8) by a quadrature. Namely

$$\psi = \int_{t_0}^{t} \frac{r - \dot{\Phi}}{\cos \Theta} dt \tag{2.12}$$

Assume that at the initial instant t_0 the EULER angles and the components of the angular velocity $\vec{\omega}$ of the body are given, i.e. for

$t = t_0$ we have

$$\Theta = \Theta_0, \quad \psi = \psi_0, \quad \Phi = \Phi_0$$
$$p = p_0, \quad q = q_0, \quad r = r_0$$

Then the initial values of the unknown functions α, β, γ are

$$\alpha_0 = \sin\Theta_0 \sin\Phi_0$$
$$\beta_0 = \sin\Theta_0 \cos\Phi_0$$
$$\gamma_0 = \cos\Theta_0$$

Given p_0, q_0, r_0; α_0, β_0, γ_0 and determining from the systems of equations (2.5) and (2.7) the functions p, q, r; α, β, γ we will have instead of the required six arbitrary constants only five. This is due to the fact that in one of the known integrals [(2.10)] the arbitrary constant is equal to one. The reason for this inconsistency between the number of initial data and the number of arbitrary constants is due to the fact that the problem of integrating systems (2.5) and (2.7) can be reduced to that of integrating a system of five equations followed by an additional quadrature. This can be seen as follows.

Consider an arbitrary function

$$F(t; p, q, r; \alpha, \beta, \gamma)$$

and calculate its total time derivative

$$\frac{dF}{dt} = \frac{\partial F}{\partial t} + \frac{\partial F}{\partial p}\dot{p} + \frac{\partial F}{\partial q}\dot{q} + \frac{\partial F}{\partial r}\dot{r} + \frac{\partial F}{\partial \alpha}\dot{\alpha} + \frac{\partial F}{\partial \beta}\dot{\beta} + \frac{\partial F}{\partial \gamma}\dot{\gamma}$$

If, after substituting for $\dot{p}$, $\dot{q}$, $\dot{r}$; $\dot{\alpha}$, $\dot{\beta}$, $\dot{\gamma}$ their expressions as given by systems (2.5) and (2.7), we obtain zero on the right-hand side of the last equation, then the equation

$$F(t; p, q, r; \alpha, \beta, \gamma) = \text{const.}$$

is an integral of systems (2.5) and (2.7). The set of six independent integrals

$$F_i(t; p, q, r; \alpha, \beta, \gamma) = C_i \qquad (i = 1, \ldots, 6)$$

gives the general solution of systems (2.5) and (2.7).

Transpose in system (2.5) the second terms from the left to the right and denote the right-hand sides so obtained by AP, BQ, CR respectively. Similarly denote the right-hand sides of equations (2.7) by A, B, Γ respectively. Since the functions P, Q, R; A, B, Γ do not contain the time t explicitly, systems of equations (2.5) and (2.7) can be replaced by the following system of five equations

$$\frac{dp}{P} = \frac{dq}{Q} = \frac{dr}{R} = \frac{d\alpha}{\mathrm{A}} = \frac{d\beta}{\mathrm{B}} = \frac{d\gamma}{\Gamma}\ (= dt) \qquad (2.13)$$

Choosing one of the variables (for example p) as the independent variable, we need to determine only five unknown functions. Hence, in order to integrate system (2.13), it is sufficient to find five independent integrals of this system. The time t is given by the quadrature

$$t = \int \frac{dp}{P(p)} = V(p) \tag{2.14}$$

where the function P must be expressed in terms of p.

Solving (2.14) for p in terms of t and substituting this expression in the previously obtained expressions for $q, r; \alpha, \beta, \gamma$, we shall obtain them as functions of t. In other words, we shall obtain the general solution of system (2.13) or, what is the same, the general solution of the EULER and POISSON equations (2.5) and (2.7). Hence, in order to integrate systems (2.5) and (2.7) it is sufficient to integrate system (2.13) and evaluate the quadrature (2.14).

The expressions denoted by P, Q, R; $\mathrm{A}, \mathrm{B}, \Gamma$ have a common property that P does not contain p explicitly, Q does not contain q explicitly, and so on. Therefore

$$\frac{\partial P}{\partial p} = \frac{\partial Q}{\partial q} = \frac{\partial R}{\partial r} = \frac{\partial \mathrm{A}}{\partial \alpha} = \frac{\partial \mathrm{B}}{\partial \beta} = \frac{\partial \Gamma}{\partial \gamma} = 0$$

and

$$\frac{\partial P}{\partial p} + \frac{\partial Q}{\partial q} + \frac{\partial R}{\partial r} + \frac{\partial \mathrm{A}}{\partial \alpha} + \frac{\partial \mathrm{B}}{\partial \beta} + \frac{\partial \Gamma}{\partial \gamma} = 0 \tag{2.15}$$

If four first integrals of system (2.13) are known, then by virtue of condition (2.15) the last multiplier of JACOBI is equal to one, and therefore the fifth first integral can be found by means of a quadrature. Thus for complete integration of the EULER and POISSON equations (2.5) and (2.7) it is sufficient to find, besides the three classical integrals, a fourth integral of these equations.

This remarkable result is a consequence of JACOBI's theory of the last multiplier for a system of differential equations.

As it will be seen from what follows efforts were now made towards finding a fourth integral of the EULER and POISSON equations. In connection with the JACOBI method see, for example, I. V. STANKEVIČ [1–4], who started with the JACOBI-HAMILTON partial differential equation and used contact transformations for finding a fourth integral of the equations of motion. See also F. DE SIMONI [1].

§ 3. Case of Euler and Poinsot

3.1. The first integrals. According to the theory of the last multiplier due to JACOBI the knowledge of a multiplier (= 1) and four independent first integrals of the EULER and POISSON equations enables us to find the fifth integral of these equations and thus reduce the problem of motion of a heavy rigid body about a fixed point to quadratures. After this has been done, it remains to express the quantities which determine the position of the body, such as the EULER angles, as explicit functions of the time t.

EULER [2, 3] considered the case for which the moment of the external forces is zero. This case is realized if the fixed point O coincides with the center of mass G of the body, i.e. $\vec{r}_0 = 0$, and the resultant of the external forces passes through G. Scalar multiplication of the reduced EULER vector equation

$$\left(\frac{d\vec{I}}{dt}\right)_{\text{rel}} + \vec{\omega} \times \vec{I} = 0 \tag{3.1}$$

by $\vec{I}$ and integration then give the fourth integral $I^2 = \text{const.}$ Equation (3.1) expresses the fact that the direction of the angular momentum vector $\vec{I}$ is constant with respect to the fixed coordinate trihedral. Hence we have the following four algebraic integrals:

$$\vec{I} \cdot \vec{K} = A\,p\,\alpha + B\,q\,\beta + C\,r\,\gamma = c \tag{3.2}$$

$$\vec{I} \cdot \vec{\omega} = A\,p^2 + B\,q^2 + C\,r^2 = 2T = h \quad (h = 2h_1) \tag{3.3}$$

$$\vec{K} \cdot \vec{K} = \alpha^2 + \beta^2 + \gamma^2 = 1 \tag{3.4}$$

$$\vec{I} \cdot \vec{I} = A^2\,p^2 + B^2\,q^2 + C^2\,r^2 = I^2 = \text{const.} \tag{3.5}$$

Any first integral which contains only the components p, q, r is a combination of integrals (3.3) and (3.5). For example, multiplying both sides of equations (3.3) and (3.5) by $-I^2$ and h respectively and adding we obtain the equation

$$A\,(A\,h - I^2)\,p^2 + B\,(B\,h - I^2)\,q^2 + C\,(C\,h - I^2)\,r^2 = 0$$

It is well-known that the instantaneous motion of a rigid body about a fixed point O is a rotation about an axis through O and that there are two cones generated by the instantaneous axes of rotation, a moving cone Σ (fixed in the body) and a fixed cone Σ' in space. During a motion of the body about O the moving cone of the instantaneous axes of rotation rolls on the fixed cone of the instantaneous axes.

The equation of the moving cone Σ, being the locus of points such that $x/p = y/q = z/r$, follows immediately from the above equation, namely

$$A\,(A\,h - I^2)\,x^2 + B\,(B\,h - I^2)\,y^2 + C\,(C\,h - I^2)\,z^2 = 0$$

In what follows we shall use the first integrals (3.3) and (3.5) to reduce the integration of the EULER equations to a single quadrature.

3.2. Symmetric notations for the constants I and h. In order to make our formulas more symmetric we shall introduce instead of the constants h and I two new constants (see A. G. GREENHILL [3, p. 101])

$$\lambda = \frac{h}{I} \quad \text{and} \quad D = \frac{I^2}{h}$$

The constant λ has the dimensions of angular velocity. The constant D has the dimensions of the moment of inertia, because h has those of the kinetic energy and I^2 those of the square of the modulus of the angular momentum.

In terms of the new constants the first integrals (3.3) and (3.5) become

$$\begin{aligned} A\,p^2 + B\,q^2 + C\,r^2 &= D\,\lambda^2 \\ A^2\,p^2 + B^2\,q^2 + C^2\,r^2 &= D^2\,\lambda^2 \end{aligned} \tag{3.6}$$

and the equation of the moving cone becomes

$$A\,(A - D)\,x^2 + B\,(B - D)\,y^2 + C\,(C - D)\,z^2 = 0$$

In order that this cone be real, all the coefficients of the left-hand quadratic form must not have the same sign. Assume that $A > B > C$. Then D must be between A and C, i.e. $A > D > C$. The above second degree cone will degenerate if the determinant of the quadratic form is zero, i.e. if D is equal to A, B or C.

If $D = A$, the cone degenerates into the straight line $y = z = 0$, i.e. we have the x-axis. If $D = C$, the cone degenerates into the z-axis. Finally, if $D = B$, the cone degenerates into a pair of planes through the y-axis:

$$A\,(A - B)\,x^2 - C\,(B - C)\,z^2 = 0$$

3.3. Calculation of the instantaneous rotation. Solving equations (3.6) for p and r in terms of q^2 and substituting these expressions into the second scalar equation of EULER

$$B\frac{dq}{dt} + (A - C)\,r\,p = 0$$

we obtain

$$\begin{aligned} &\sqrt{A\,C}\,B\frac{dq}{dt} \\ &= \sqrt{[D\,(D - C)\,\lambda^2 - B\,(B - C)\,q^2]\,[D\,(A - D)\,\lambda^2 - B\,(A - B)\,q^2]} \end{aligned} \tag{3.7}$$

This equation remains valid even if $A = B = C$. Solving equation (3.7) for dt leads us to an elliptic integral. Hence the equations of motion in the case of EULER and POINSOT can be integrated in terms of elliptic functions. If the ellipsoid of inertia of the rigid body at O reduces to an ellipsoid of revolution ($A = B$ or $B = C$), then the elliptic functions degenerate into the circular functions and the integration can be carried out immediately.

Assuming that $A > B > C$, let us restrict ourselves first to a qualitative discussion of the rotation as determined by equation (3.7).

The roots of the polynomial in q^2 under the square root sign are

$$q_1^2 = f^2 = \lambda^2 \frac{D(D-C)}{B(B-C)}$$

$$q_2^2 = g^2 = \lambda^2 \frac{D(A-D)}{B(A-B)}$$

They are definitely positive quantities according to our assumption concerning the order of magnitudes of the coefficients A, B, C and D. Equation (3.7) then assumes the form

$$\frac{dq}{dt} = \sqrt{\frac{(A-B)(B-C)}{A\,C}(f^2 - q^2)(g^2 - q^2)} \tag{3.8}$$

and equations (3.6) give us the following expressions for p^2 and r^2:

$$\begin{aligned} p^2 &= \frac{B(B-C)}{A(A-C)}(f^2 - q^2) \\ r^2 &= \frac{B(A-B)}{C(A-C)}(g^2 - q^2) \end{aligned} \tag{3.9}$$

The difference

$$g^2 - f^2 = \lambda^2 \frac{D(A-C)(B-D)}{B(A-B)(B-C)}$$

has the sign of $B - D$.

Assume that $B > D$. Then $g^2 > f^2$. In order that p be real the variable q must vary between $-f$ and $+f$. Then equation (3.8) can be rewritten in the form

$$\left(\frac{dq}{dt}\right)^2 = (f^2 - q^2)\,\chi(q) \tag{3.10}$$

where

$$\chi(q) = \frac{(A-B)(B-C)}{A\,C}(g^2 - q^2) \tag{3.11}$$

does not vanish for any value of q between $-f$ and $+f$, the endpoints included.

If at the beginning of the motion q lies between $-f$ and $+f$, then q must remain forever between the same limits. In order to study the motion defined by (3.10) we shall introduce a new auxiliary variable w defined by the equation

$$\left(\frac{dq}{dw}\right)^2 = b^2(f^2 - q^2) \quad (b \text{ const.}) \tag{3.12}$$

Then

$$\left(\frac{dw}{dt}\right)^2 = \frac{1}{b^2}\chi(q) \tag{3.13}$$

Since equation (3.10) has been replaced by two equations (3.12) and (3.13), each of which gives rise to a constant of integration, it is permitted to determine one of them arbitrarily. Let us determine in equation (3.13) this constant in such a way that w vanishes when t vanishes. Then it follows from the same equation that w is always real. Because of the relation

$$\left(\frac{dq}{dw}\right)^2 \left(\frac{dw}{dt}\right)^2 = \left(\frac{dq}{dt}\right)^2$$

the sign of dw/dt is also at our disposal at the beginning of the motion. If we assign to dw at the beginning of the motion the sign of dt, then the signs of dw and dt will remain the same for all t since dw/dt can neither vanish nor become infinite. Hence assume that

$$\frac{dw}{dt} = +\frac{1}{b}\sqrt{\chi(q)} \tag{3.13'}$$

The solution of equation (3.12) is

$$q = -f\cos b\,w \tag{3.14}$$

Hence q is a periodic function of w with period $2\pi/b$. Let us prove that q is also periodic in t. The constant b can be determined in such a way that the period of q with respect to w is equal to the period of q with respect to t. By virtue of (3.13') we have

$$t = \int_0^w \frac{b\,dw}{\sqrt{\chi(q)}}$$

If w increases by $2\pi/b$, then q does not change. If $2\mathcal{T}$ is the corresponding increment of t, then

$$2\mathcal{T} = \int_w^{w+2\pi/b} \frac{b\,dw}{\sqrt{\chi(-f\cos b\,w)}}$$

Since $\chi(q)$ is periodic in w, the period being $2\pi/b$, $2\mathcal{T}$ does not depend on w, and consequently $2\mathcal{T}$ is a constant. Hence q is also periodic in t with the period

$$2\mathcal{T} = \int_0^{2\pi/b} \frac{b\,dw}{\sqrt{\chi(-f\cos b\,w)}} \quad \text{or} \quad \mathcal{T} = \int_0^{\pi/b} \frac{b\,dw}{\sqrt{\chi(-f\cos b\,w)}}$$

By virtue of formula (3.14) we can write that

$$\mathcal{T} = \sqrt{\frac{A\,C}{(A-B)\,(B-C)}} \int\limits_{-f}^{f} \frac{dq}{\sqrt{(f^2-q^2)\,(g^2-q^2)}}$$

$$= 2\sqrt{\frac{A\,C}{(A-B)\,(B-C)}} \int\limits_{0}^{f} \frac{dq}{\sqrt{(f^2-q^2)\,(g^2-q^2)}}$$

Assume that the initial conditions are

$$t=0, \quad q_0=0, \quad \dot{q}_0>0 \tag{3.15}$$

Then neither equation (3.8) nor the initial conditions (3.15) change if one replaces t by $-t$ and q by $-q$. Hence the function $q(t)$ is an odd function of the time t. In addition, it is periodic with period $2\mathcal{T}$.

If q is known, r is given by the second formula of (3.9). Since $g^2>f^2$ and q varies between $-f$ and $+f$, r is never zero and keeps the same sign. The component r readmits the same value each time that q^2 takes the same value. Hence r is an even function of t which admits the period $\mathcal{T}$.

Finally p is given by the first formula of (3.9). It vanishes when q assumes the values $\pm f$. In order to discuss the behavior of p and avoid the difficulty of the double sign, it is preferable to use the second equation of EULER which gives for p the expression

$$p = \frac{-B\,\dfrac{dq}{dt}}{(A-C)\,r}$$

This formula shows that p changes its sign simultaneously with dq/dt, which is an even function of t with period $2\mathcal{T}$. The component r is also an even function of t which admits the period $\mathcal{T}$. Hence p is an even function of t with period $2\mathcal{T}$.

After this qualitative discussion of the rotation let us pass now to the solution of equation (3.8). For this purpose put $q=f\,u$, where u is a new variable. Then equation (3.8) assumes the form

$$\frac{du}{dt} = n\sqrt{(1-u^2)\,(1-k^2u^2)}$$

where

$$k^2 = \frac{f^2}{g^2} = \frac{(A-B)\,(D-C)}{(A-D)\,(B-C)} < 1$$

and

$$n = \sqrt{\frac{(A-B)\,(B-C)}{AC}}\,g = \lambda\sqrt{\frac{D\,(B-C)\,(A-D)}{A\,B\,C}}$$

Hence

$$n\,t = \int\limits_0^u \frac{du}{\sqrt{(1-u^2)(1-k^2u^2)}} \tag{3.16}$$

where at the instant $t = 0$ we have $u = q = 0$.

Thus the calculation of $q = f\,u$ is reduced to the inversion of an elliptic integral.

Having obtained u as a function of the time t, $u = \operatorname{sn} n\,t$, the components p and r can be calculated as follows. From formulas (3.9) we have

$$p^2 = \frac{B(B-C)}{A(A-C)}(f^2 - q^2) = \frac{B(B-C)}{A(A-C)} f^2 (1-u^2) = \frac{B(B-C)}{A(A-C)} f^2 v^2$$

$$r^2 = \frac{B(A-B)}{C(A-C)}(g^2 - f^2 u^2) = \frac{B(A-B)}{C(A-C)} g^2 \left(1 - \frac{f^2}{g^2} u^2\right)$$

$$= \frac{B(A-B)}{C(A-C)} g^2 (1 - k^2 u^2) = \frac{B(A-B)}{C(A-C)} g^2 w^2$$

where

$$u^2 + v^2 = 1$$

$$w^2 + k^2 u^2 = 1$$

Hence

$$\begin{aligned} v &= \operatorname{cn} n\,t \\ w &= \operatorname{dn} n\,t \end{aligned} \tag{3.17}$$

and

$$\begin{aligned} p &= f\sqrt{\frac{B(B-C)}{C(A-C)}}\operatorname{cn} n\,t \\ r &= g\sqrt{\frac{B(A-B)}{C(A-C)}}\operatorname{dn} n\,t \end{aligned} \tag{3.18}$$

Consequently

$$\left.\begin{aligned} p &= a_0 \operatorname{cn} n\,t \\ q &= b_0 \operatorname{sn} n\,t \\ r &= c_0 \operatorname{dn} n\,t \end{aligned}\right\} \tag{3.19}$$

where a_0, b_0 and c_0 are certain constants.

In a similar way we can discuss the case $B < D$. Finally, if $D = B$, then $f^2 = g^2 = \lambda^2$, $k^2 = 1$ and equation (3.16) becomes

$$n\,t = \ln\sqrt{\frac{1+u}{1-u}}$$

Thus in the case of Euler and Poinsot the integration of the Euler dynamical equations reduces to the inversion of the elliptic integral (3.16) and calculation of v and w by the formulas (3.17). Then $q = f\,u$, and p and r are to be calculated from formulas (3.18).

3.4. Calculation of the Euler angles. When the components p, q, r of the instantaneous rotation $\vec{\omega}$ are known as functions of the time t, the EULER angles Θ, ψ, Φ can be obtained by integrating system (1.8) which consists of three first order differential equations. It can be proved that the integration of this system is reducible to the integration of a single RICCATI differential equation (see, for example, G. DARBOUX [4, vol. 1]).

So far we have used only two of the four known first integrals. The remaining two integrals can be interpreted as two particular integrals of the RICCATI equation. It is well-known that in such a case its integration can be reduced to a quadrature.

We can verify this fact also by purely mechanical considerations. Equation (3.1) and integral (3.5) express the fact that the direction and the modulus of the angular momentum $\vec{I}$ are constant. Therefore we can choose for the direction and sense of the $O\,Z$-axis those of the vector $\vec{I}$. The components of $\vec{I}$ along the axes of the moving trihedral $O\,x\,y\,z$ are then $I\,\alpha$, $I\,\beta$, $I\,\gamma$ respectively. On the other hand the components of $\vec{I}$ along the same axes are $A\,p$, $B\,q$, $C\,r$ respectively. Hence we can write that

$$\left.\begin{aligned} I\sin\Theta\sin\Phi &= A\,p \\ I\sin\Theta\cos\Phi &= B\,q \\ I\cos\Theta &= C\,r \end{aligned}\right\} \tag{3.20}$$

From the last of the above formulas it follows that

$$\sin\Theta = \sqrt{1-\frac{C^2 r^2}{I^2}}$$

the angle Θ being an angle between 0 and π. The functions $\sin\Theta$ and $\cos\Theta$ have the same period with respect to t as the function $r(t)$, namely $\mathcal{T}$. The functions $\sin\Phi$ and $\cos\Phi$ are given by the expressions

$$\frac{A\,p}{I\sin\Theta} \quad \text{and} \quad \frac{B\,q}{I\sin\Theta}$$

respectively. The first expression is an even function of t, the second an odd function of t, both admitting the period $2\mathcal{T}$. Hence the two EULER angles Θ and Φ can be calculated without any integration.

It remains to calculate the third angle ψ. Elimination of $\dot{\Theta}$ from the first two equations of system (1.8) gives us the expression

$$\dot{\psi} = \frac{p\sin\Phi + q\cos\Phi}{\sin\Theta} \tag{3.21}$$

On the other hand equations (3.20) give us the following relations

$$\frac{p\sin\Phi + q\cos\Phi}{\sin\Theta} = \frac{A\,p^2 + B\,q^2}{I\sin^2\Theta}$$

$$I^2\sin^2\Theta = A^2\,p^2 + B^2\,q^2$$

Hence

$$\dot{\psi} = I \frac{A p^2 + B q^2}{A^2 p^2 + B^2 q^2} = I \frac{h - C r^2}{I^2 - C^2 r^2}$$

and ψ is obtained by a quadrature. The derivative $\dot{\psi}$ being positive, the angle ψ will continue to increase as t increases. The function $\dot{\psi}$ is an even function of t with period $\mathcal{T}$. We can assume that $\psi = 0$ when $t = 0$. The function $\psi(t)$ is then an odd function of t which increases by a constant when t increases by $\mathcal{T}$. In fact, from

$$\dot{\psi}(t + \mathcal{T}) - \dot{\psi}(t) = 0$$

follows that

$$\psi(t + \mathcal{T}) - \psi(t) = a_\psi$$

If t increases by $2\mathcal{T}$, our discussion shows that the variables p, q, r; $\sin\Theta$, $\cos\Theta$; $\sin\Phi$, $\cos\Phi$ assume their previous values while the value of ψ increases by $2a_\psi$ (which depends on the initial conditions). Hence the rigid body has returned back to its initial position after having rotated about the OZ-axis through an angle $2a_\psi$. In order that the motion be periodic it is necessary and sufficient that a_ψ be commensurable with π.

Explicit expressions for the components p, q, r and the angles α, β, γ in terms of elliptic functions of t were given for the first time by C. G. J. JACOBI [1—3] and O. I. SOMOV [1]. See also A. CAYLEY [1] and A. G. GREENHILL [1, 2, 4—9].

While JACOBI gave the position of the body relative to a particular moving coordinate trihedral in terms of the Theta-functions, W. FRAHM [1] solved a similar problem relative to an arbitrary coordinate trihedral fixed in space.

§ 4. Calculation of the Poinsot motion

4.1. Introductory remarks. The motion of a rigid body about a fixed point O, subject to no external forces (except the force of gravity and an equal force of reaction at O, opposite to the gravitational force), admits the elegant geometric representation of L. POINSOT [1]. Namely, the body moves as if it were rigidly connected to its ellipsoid of inertia, and the latter body were to roll without sliding on a fixed plane perpendicular to the angular momentum vector which is fixed in space. The curve which is traced on the ellipsoid of inertia by the point of contact with the fixed plane is called the polhode. The curve which is traced on the fixed plane by the point of contact with the moving ellipsoid is called the herpolhode. The cone with its vertex at the fixed point O and the elements of which pass through the polhode is the rolling cone. The cone with its vertex at O and the elements of which pass through the herpolhode is the base cone. The body moves as if it were rigidly connected to the rolling cone, and the latter were to roll without sliding on the base cone fixed in space.

For a critical examination of POINSOT's memoir [1] see J. REVEILLE [1].
POINSOT's geometrical representation was generalized by J. J. SYLVESTER [1]. For other geometrical representations of the motion, for example, in terms of the ellipsoid of gyration, see J. MACCULLAGH [1], and for problems arising from the representations of POINSOT and J. MACCULLAGH see A. CLEBSCH [1], E. J. ROUTH [1_2, 2_2] and P. APPELL [1]. The properties of the polhode and herpolhode were studied by W. HESS [1], DE SPARRE [1], J. N. FRANKE [1], A. MANNHEIM [1, 2], A. DE ST. GERMAIN [1], H. RESAL [2], A. G. GREENHILL [5], P. BARBARIN [1], G. DARBOUX [1, vol. 2, Notes XVII and XVIII], E. LACOUR [1], A. PETRUS [1], and others. For an extensive geometric study of the EULER case of motion see also P. V. VORONEC [1].

Notwithstanding the fact that each such geometric representation gives us a clear picture of the motion in space, neither of them gives us a satisfactory insight into the motion with respect to the time t. The well known analytic solution by G. KIRCHHOFF [1, seventh lecture] gives us the components of the instantaneous rotation p, q, r and two EULER angles Θ and Φ by means of the JACOBI elliptic functions $\operatorname{cn} n t$, $\operatorname{sn} n t$ and $\operatorname{dn} n t$ while the third EULER angle ψ is given by an integral which is not suitable for computational purposes. Concerning the KIRCHHOFF solution see N. LINDSKOG [1] and W. DE TANNENBERG [2].

The POINSOT motion is an immediate generalization of the regular precession which is the most general motion of a symmetric rigid body under no external forces. The regular precession as well as the POINSOT motion can be represented by the superposition of a rotation about the figure axis which is fixed in the body, and a rotation about the precessional axis, fixed in space. However, in the case of an unsymmetric rigid body these two rotations are no longer uniform, but both oscillate about certain mean values. Discarding all terms but the first term in the series expansions of the JACOBI elliptic functions $\operatorname{sn} n t$, $\operatorname{cn} n t$ and $\operatorname{dn} n t$ in terms of the Theta-functions $\vartheta_1(n t)$, $\vartheta_2(n t)$, $\vartheta_3(n t)$, $\vartheta_4(n t)$, GRAMMEL [5] gives the mean values of the above three angles of rotation Θ, ψ, Φ as well as the FOURIER series for their oscillations about these values.

4.2. The angular velocity components and the Euler angles. Assuming that the principal moments of inertia satisfy the inequalities $A > B > C$ or $A < B < C$, the solution of the EULER dynamical equations is [see formulas (3.19)]

$$\left.\begin{aligned} p &= a_0 \operatorname{cn} n t \\ q &= b_0 \operatorname{sn} n t \\ r &= c_0 \operatorname{dn} n t \end{aligned}\right\} \tag{4.1}$$

In order to determine the constants a_0, b_0, c_0 and the modulus k of the elliptic functions $\operatorname{sn} n t$, $\operatorname{cn} n t$ and $\operatorname{dn} n t$ assume that at the instant $t = 0$ the axis of the angular momentum vector $\vec{I}$ lies in the x, z-plane and the components of the angular velocity $\vec{\omega}$ are p_0, $q_0 = 0$ and r_0. Then we obtain from (4.1) that

$$a_0 = p_0 \quad \text{and} \quad c_0 = r_0 \tag{4.2}$$

Substituting the solution (4.1) into the EULER equations, the following equations are obtained for the determination of b_0, n and k^2:

$$-A\, p_0\, n = (B - C)\, b_0\, r_0 \tag{4.3}$$

$$B\, b_0\, n = (C - A)\, p_0\, r_0 \tag{4.4}$$

$$-C\, r_0\, n\, k^2 = (A - B)\, b_0\, p_0 \tag{4.5}$$

Multiplication and division of equations (4.3) and (4.4) gives us

$$n^2 = \frac{(B - C)(A - C)}{A\, B}\, r_0^2 \tag{4.6}$$

$$b_0^2 = \frac{A(A - C)}{B(B - C)}\, p_0^2 \tag{4.7}$$

Finally, dividing (4.5) by (4.4) and taking formula (4.7) into account we obtain

$$k^2 = \frac{A(A - B)}{C(B - C)}\, \frac{p_0^2}{r_0^2} \tag{4.8}$$

By virtue of the assumption made about the order of magnitudes of the principal moments of inertia A, B and C the constants n, b_0 and k are all real.

The fixed direction of the angular momentum vector $\vec{I}$ and its absolute value are determined by the components $A\, p_0$ and $C\, r_0$.

Assume that at the instant $t = 0$ the angle $\Phi_0 = \pi/2$ and the angle $\psi_0 = 0$. By virtue of formulas (4.1) and (4.2) it follows from (3.20) that

$$\cos\Theta = \frac{C\, r_0}{I} \operatorname{dn} n t \tag{4.9}$$

$$\cot\Phi = \sqrt{l}\, \frac{\operatorname{sn} n t}{\operatorname{cn} n t} \tag{4.10}$$

where

$$l = \frac{B(A - C)}{A(B - C)} \tag{4.11}$$

Finally from formulas (3.21) and (3.20) we obtain that

$$\dot{\psi} = \frac{I}{A} \sin^2\Phi + \frac{I}{B} \cos^2\Phi \tag{4.12}$$

The trivial case, when $\sin\Theta = 0$ and we have a permanent rotation about the z-axis, must be excluded in formula (3.21).

4.3. Proper rotation of the Poinsot motion. Let us decompose now the "proper rotation" Φ into its mean motion and oscillations about the latter. For this purpose we shall make use of a well known formula which connects the functions $\operatorname{sn} n t$, $\operatorname{cn} n t$ and the Theta-functions $\vartheta_1(n t)$, $\vartheta_2(n t)$, namely,

$$\frac{\operatorname{sn} n t}{\operatorname{cn} n t} = \frac{1}{\sqrt{k'}}\frac{\vartheta_1(n t)}{\vartheta_2(n t)} \qquad \left(k' = \sqrt{1 - k^2}\right) \tag{4.13}$$

where the Theta-functions are given by the following series expansions

$$\begin{aligned}\vartheta_1(n t) &= 2\left(q^{1/4}\sin\frac{\pi n t}{2K} - q^{9/4}\sin 3\frac{\pi n t}{2K} + q^{25/4}\sin 5\frac{\pi n t}{2K} - \cdots\right)\\ \vartheta_2(n t) &= 2\left(q^{1/4}\cos\frac{\pi n t}{2K} + q^{9/4}\cos 3\frac{\pi n t}{2K} + q^{25/4}\cos 5\frac{\pi n t}{2K} + \cdots\right)\end{aligned} \tag{4.14}$$

the parameter q is given by the formula

$$q = e^{-\frac{K'}{K}\pi} \tag{4.15}$$

and K, K' are the complete elliptic normal integrals of the first kind with moduli k and $k' = \sqrt{1 - k^2}$ respectively. Using formulas (4.13) and (4.14) we can rewrite formula (4.10) in the form

$$\cot\Phi = \sqrt{\frac{l}{k'}}\,\frac{\sin s - q^2\sin 3s + q^6\sin 5s - \cdots}{\cos s + q^2\cos 3s + q^6\cos 5s + \cdots} \tag{4.16}$$

where

$$s = \frac{\pi n t}{2K}$$

If we neglect all powers of q beginning with the second power, then we obtain the simple formula

$$\cot\Phi = \sqrt{\frac{l}{k'}}\tan s \tag{4.17}$$

In order to obtain the desired decomposition of Φ into its mean value and oscillations about the latter, differentiate formula (4.17) with respect to t. We then obtain

$$-(1 + \cot^2\Phi)\frac{d\Phi}{dt} = \frac{\pi n}{2K}\sqrt{\frac{l}{k'}}(1 + \tan^2 s)$$

or

$$\frac{d\Phi}{dt} = -\frac{\pi n}{2K}\sqrt{\frac{l}{k'}}\,\frac{1 + \tan^2 s}{1 + \frac{l}{k'}\tan^2 s}$$

or

$$\frac{d\Phi}{dt} = -\frac{\pi n}{2K}\sqrt{\frac{k' l}{k' + l}}\, R \tag{4.18}$$

where

$$R = \frac{1}{1 + \varepsilon \cos 2s}, \qquad \varepsilon = \frac{k' - l}{k' + l} \tag{4.19}$$

Since $|\varepsilon| < 1$, R can be expanded into a converging power series

$$R = 1 - \varepsilon \cos 2s + \varepsilon^2 \cos^2 2s - \varepsilon^3 \cos^3 2s + \cdots$$

which can be integrated term-by-term.

Using the formulas

$$\cos^{2n-1} x = \frac{1}{2^{2n-2}}\left[\cos(2n-1)\,x + \binom{2n-1}{1}\cos(2n-3)\,x + \right.$$
$$\left. + \binom{2n-1}{2}\cos(2n-5)\,x + \cdots + \binom{2n-1}{n-1}\cos x\right]$$

$$\cos^{2n} x = \frac{1}{2^{2n-1}}\left[\cos 2n\,x + \binom{2n}{1}\cos(2n-2)\,x + \right.$$
$$\left. + \binom{2n}{2}\cos(2n-4)\,x + \cdots + \binom{2n}{n-1}\cos 2x + \frac{1}{2}\binom{2n}{n}\right]$$

we obtain

$$\int_0^t R\,dt = A_0 t - \frac{K}{\pi n}\left(A_1 \sin\frac{\pi n t}{K} - A_2 \sin 2\frac{\pi n t}{K} + A_3 \sin 3\frac{\pi n t}{K} - \cdots\right) \tag{4.20}$$

where

$$A_0 = \sum_{\nu=0}^{\infty}\binom{2\nu}{\nu}\left(\frac{\varepsilon}{2}\right)^{2\nu}, \qquad A_1 = 2\sum_{\nu=0}^{\infty}\binom{2\nu+1}{\nu}\left(\frac{\varepsilon}{2}\right)^{2\nu+1}$$
$$A_n = \frac{2}{n}\sum_{\nu=0}^{\infty}\binom{2\nu+n}{\nu}\left(\frac{\varepsilon}{2}\right)^{2\nu+n} \qquad (n > 0) \tag{4.21}$$

Writing out the series for the first three coefficients A_0, A_1 and A_2 we easily obtain the expressions

$$\left.\begin{aligned} A_0 &= \frac{1}{\sqrt{1-\varepsilon^2}} = \frac{k' + l}{2\sqrt{k' l}} \\ A_1 &= \frac{2}{\varepsilon}(A_0 - 1) \\ A_2 &= \frac{A_1}{\varepsilon} - A_0 \end{aligned}\right\} \tag{4.22}$$

For the calculation of the remaining coefficients GRAMMEL has set up a recursion formula.

Integrating (4.18) and taking into account (4.20) and the first formula of system (4.22) for A_0, we obtain Φ in the form

$$\Phi = \frac{\pi}{2} - \omega_r t + \tilde{\Phi}(t) \tag{4.23}$$

where

$$\omega_r = \frac{\pi n}{2K} \tag{4.24}$$

is the mean value of the proper rotation and

$$\tilde{\Phi}(t) = \frac{\sqrt{k' l}}{k' + l}\left(A_1 \sin\frac{\pi n t}{K} - A_2 \sin 2\frac{\pi n t}{K} + A_3 \sin 3\frac{\pi n t}{K} - \cdots\right) \tag{4.25}$$

is the expression for the oscillation of Φ about its mean value $\omega_r t$. The oscillation $\tilde{\Phi}(t)$ is given in the form of a FOURIER series in terms of the period $2K/n$ of the rotation Φ.

In the case of a symmetric rigid body we have $A = B$, and hence $k = 0$, $K = \pi/2$, $\omega_r = n = [(A - C)/A]\, r_0$. These formulas are well known from the theory of regular precession of a symmetric rigid body. In addition, we have $k' = l = 1$, and hence $\varepsilon = 0$, i.e. there is no oscillation as is to be expected.

4.4. Precessional rotation of the Poinsot motion. From formula (4.12) it follows that

$$\frac{d\psi}{dt} = \frac{I}{2AB}\left[(A + B) + (A - B)\frac{1 - \tan^2\Phi}{1 + \tan^2\Phi}\right]$$

or, if we limit ourselves to the approximation (4.17),

$$\frac{d\psi}{dt} = \frac{I}{2AB}\left[(A + B) - (A - B)\frac{k' \cos^2 s - l \sin^2 s}{k' \cos^2 s + l \sin^2 s}\right]$$

or

$$\frac{d\psi}{dt} = \frac{I}{2AB}\left[(A + B) - (A - B)\left(\frac{k' + l}{k' - l} - \frac{4k' l}{k'^2 - l^2} R\right)\right]$$

Integrating and taking into account formula (4.20) and the expression for A_0 from (4.22), we obtain that

$$\psi = \omega_p t + \tilde{\psi}(t) \tag{4.26}$$

where

$$\omega_p = \frac{I}{2AB}\left[(A + B) - (A - B)\frac{\sqrt{k'} - \sqrt{l}}{\sqrt{k'} + \sqrt{l}}\right] \tag{4.27}$$

is the mean value of the velocity of precession and

$$\tilde{\psi}(t) = -\frac{I(A - B)}{AB}\,\frac{2k' l}{k'^2 - l^2}\,\frac{K}{\pi n} \times$$
$$\times\left(A_1 \sin\frac{\pi n t}{K} - A_2 \sin 2\frac{\pi n t}{K} + A_3 \sin 3\frac{\pi n t}{K} - \cdots\right) \tag{4.28}$$

gives the oscillation of ψ about its mean value $\omega_p t$. Also the oscillation $\tilde{\psi}(t)$ is given by a FOURIER series in terms of the period $2K/n$ of the proper rotation Φ.

In the case of a symmetric rigid body we have $A = B$, and the above expression for ω_p reduces to I/A of the regular precession and $\tilde{\psi}(t)$ vanishes identically.

4.5. Nutation of the Poinsot motion. Finally we shall give the expression for the angle Θ between the angular momentum vector $\vec{I}$ and the z-axis. From the theory of elliptic functions it is known that

$$\operatorname{dn} n t = \sqrt{k'}\,\frac{\vartheta_3(n t)}{\vartheta_4(n t)} \tag{4.29}$$

where

$$\begin{aligned}\vartheta_3(n t) &= 1 + 2\left(q \cos\frac{\pi n t}{K} + q^4 \cos 2\frac{\pi n t}{K} + q^9 \cos 3\frac{\pi n t}{K} + \cdots\right)\\ \vartheta_4(n t) &= 1 - 2\left(q \cos\frac{\pi n t}{K} - q^4 \cos 2\frac{\pi n t}{K} + q^9 \cos 3\frac{\pi n t}{K} - \cdots\right)\end{aligned} \tag{4.30}$$

If we retain the first power of q in the last two formulas, then by formula (4.9) we have that

$$\cos\Theta = \sqrt{k'}\cos\Theta_0\,\frac{1 + 2q\cos\frac{\pi n t}{K}}{1 - 2q\cos\frac{\pi n t}{K}} \tag{4.31}$$

where

$$\cos\Theta_0 = \frac{C r_0}{I} \tag{4.32}$$

Here Θ_0 denotes the value of Θ for $t = 0$.

If we expand the right hand side of formula (4.31) into a power series and retain terms up to q^3 we find that

$$\cos\Theta = u_0 + \tilde{u}(t) \tag{4.33}$$

where

$$u_0 = (1 + 4q^2)\sqrt{k'}\cos\Theta_0$$

and

$$\tilde{u}(t) = 4q\sqrt{k'}\cos\Theta_0 \times \\ \times\left[(1 + 3q^2)\cos\frac{\pi n t}{K} + q\cos 2\frac{\pi n t}{K} + q^2\cos 3\frac{\pi n t}{K} + \cdots\right] \tag{4.34}$$

In the case $A = B$ we have $k' = 1$, $K' = \infty$, $q = 0$, and hence $u_0 = \cos\Theta_0$ and $\tilde{u}(t) = 0$.

4.6. Estimation of the validity of the above results. The more the rigid body approaches the symmetric case $A = B$, or the motion of the rigid body approaches the case of a stable permanent rotation about the z-axis, the more rapidly the parameter q, and hence also the error committed in using formula (4.17) and all the following formulas, tends to zero. The error, however, increases the more the motion approaches the particular case $k = 1$, when the axis of rotation coincides exactly or almost with the middle inertia axis. In such a case, however, the stability of the POINSOT motion becomes doubtful, and there is no reason for the discussion of the motion as a precession.

4.7. On some finite relations among the Euler angles. In the EULER and POINSOT case the expressions for the angles Θ, ψ and Φ in terms of the elliptic functions of the time t are too complicated to permit an easy elimination of the time for the purpose of obtaining some finite relations among the EULER angles.

V. V. DOBRONRAVOV [1, 2] has pointed out several relations obtained by indirect elementary considerations. Unfortunately, they are neither new nor do they seem to be of particular interest. For example, two of them are equivalent to the well known theorem (see E. J. ROUTH [2_2, 3_2]) that the fixed angular momentum vector $\vec{I}$ moves, relative to the body, on the invariable cone. Another relation can be derived as follows. Squaring the equalities

$$A p = I \alpha, \qquad B q = I \beta, \qquad C r = I \gamma$$

then multiplying them by $C B$, $A C$, $B A$ respectively, and adding we obtain the equality

$$\frac{\alpha^2}{A} + \frac{\beta^2}{B} + \frac{\gamma^2}{C} = \frac{2T}{I^2} \tag{4.35}$$

In addition, we have the trigonometric identity

$$\alpha^2 + \beta^2 + \gamma^2 = 1 \tag{4.36}$$

Consider the unit vector $\vec{K}(\alpha, \beta, \gamma)$ along the angular momentum vector $\vec{I}$ and assume that $A > B > C$. Then

$$\frac{2A\,T}{I^2} = \frac{A^2 p^2 + A B q^2 + A C r^2}{A^2 p^2 + B^2 q^2 + C^2 r^2} > 1$$

$$\frac{2C\,T}{I^2} = \frac{A C p^2 + B C q^2 + C^2 r^2}{A^2 p^2 + B^2 q^2 + C^2 r^2} < 1$$

Hence the sphere (4.36) intersects the ellipsoid (4.35). Consequently, in the EULER and POINSOT case the heavy rigid body rotates about the fixed point O in such a way that the endpoint of the unit vector $\vec{K}$ moves on a spherical curve lying on the ellipsoid of gyration of the body.

§ 5. Case of Lagrange and Poisson

5.1. The first integrals. LAGRANGE [1, Part 2, Sect. 9] and POISSON [1, 3, vol. 2, p. 207] independently considered the case $A = B$, $x_0 = y_0 = 0$, $z_0 > 0$, i.e. when two of the principal moments of inertia at O are equal and the center of mass G of the body lies on the third axis of inertia of the body.

In this case the EULER differential equations reduce to

$$\left.\begin{aligned} A\frac{dp}{dt} + (C - A)\,q\,r &= \beta\, z_0 \\ A\frac{dq}{dt} + (A - C)\,r\,p &= -\alpha\, z_0 \\ C\frac{dr}{dt} \qquad\qquad &= 0 \end{aligned}\right\} \tag{5.1}$$

Putting

$$\frac{A - C}{A} = m, \qquad \frac{z_0}{A} = a$$

equations (5.1) reduce to the following equations

$$\left.\begin{aligned} \frac{dp}{dt} - m\,q\,r &= \beta\, a \\ \frac{dq}{dt} + m\,p\,r &= -\alpha\, a \\ \frac{dr}{dt} \qquad\quad &= 0 \end{aligned}\right\} \tag{5.2}$$

The last equation gives the fourth integral $r = n = \text{const.}$, which shows that the body rotates about its axis of symmetry Oz with constant angular velocity n.

In the case under consideration the known four first integrals have the form

$$\left.\begin{aligned} p^2 + q^2 + 2a\,\gamma &= h_2 \\ p\,\alpha + q\,\beta + (1 - m)\,r\,\gamma &= k \\ r &= n \\ \alpha^2 + \beta^2 + \gamma^2 &= 1 \end{aligned}\right\} \tag{5.3}$$

5.2. Reduction of the Euler equations of motion. The motion being defined by equations (5.2) and (1.8), let us show that in the case under consideration it is possible to find three additional first integrals, and hence reduce the problem of motion to quadratures.

From the first two equations of system (1.8) and the first equation of system (5.3), it follows that

$$\dot{\Theta}^2 + \dot{\psi}^2 \sin^2\Theta = h_2 - 2a\,\gamma \tag{5.4}$$

Further, substituting into the second equation of system (5.3) for α, β their expressions in terms of the EULER angles, as given by the last column of matrix (1.7), and replacing r by n, the equation

$$\dot{\psi} \sin^2\Theta + (1 - m)\,n\,\gamma = k \tag{5.5}$$

results.

Eliminating $\dot{\psi}$ from equations (5.4) and (5.5) we obtain the equation

$$\dot{\Theta}^2 \sin^2\Theta + [k - (1 - m)\, n\, \gamma]^2 = (h_2 - 2a\,\gamma)\sin^2\Theta$$

or, observing that $\gamma = \cos\Theta$, the equation

$$\dot{\gamma}^2 = (h_2 - 2a\,\gamma)(1 - \gamma^2) - (1 - m)^2 n^2 (\gamma_4 - \gamma)^2 = 2a\, f(\gamma) \tag{5.6}$$

where

$$\gamma_4 = \frac{k}{(1 - m)\, n} \tag{5.7}$$

Hence

$$dt = \frac{d\gamma}{\sqrt{2a}\sqrt{f(\gamma)}} \tag{5.8}$$

the positive sign being taken before the radical if γ increases with t.

Equation (5.5) yields

$$\frac{d\psi}{dt} = (1 - m)\, n\, \frac{\gamma_4 - \gamma}{1 - \gamma^2}$$

or, by virtue of equation (5.8),

$$d\psi = \frac{(1 - m)\, n}{\sqrt{2a}}\, \frac{\gamma_4 - \gamma}{1 - \gamma^2}\, \frac{d\gamma}{\sqrt{f(\gamma)}} \tag{5.9}$$

Finally the third equation of system (1.8) gives us

$$\frac{d\Phi}{dt} = n - \gamma \frac{d\psi}{dt} = n - \gamma (1 - m)\, n\, \frac{\gamma_4 - \gamma}{1 - \gamma^2}$$

which, by means of equation (5.8), assumes the form

$$d\Phi = \frac{n}{\sqrt{2a}} \left[1 - (1 - m)\,\gamma\, \frac{\gamma_4 - \gamma}{1 - \gamma^2}\right] \frac{d\gamma}{\sqrt{f(\gamma)}} \tag{5.10}$$

Hence the problem of integrating the EULER and POISSON equations of motion is reduced to three elliptic quadratures (5.8), (5.9) and (5.10).

The actual integration of the equations of motion in terms of elliptic functions of t was carried out by C. G. J. JACOBI [4, 5], E. LOTTNER [1], O. I. SOMOV [2], C. FRENZEL [1], A. SÖDERBLOM [1] and J. CHRAPAN [1]. The geometrical representation of the motion was given by C. G. J. JACOBI [4], E. LOTTNER [2], J. J. SYLVESTER [1] and G. DARBOUX [1, vol. 2, Notes XIX and XX]. See also N. B. DELONE [1] and W. HESS [2, 4].

Regarding JACOBI's theorem which states that the rotation of a heavy body of revolution about a point on its axis may be replaced by the relative motion of two bodies under the action of no forces (EULER's case), see E. PADOVA [1, 4], G. H. HALPHEN [1], G. DARBOUX [2, 3], W. HESS [5], E. J. ROUTH [1], R. MARCOLONGO [1, 2], A. G. GREENHILL [5, 6], and F. KÖTTER [3].

The polynomial $f(\gamma)$ in equation (5.6) is negative for $\gamma = -\infty, -1$ and $+1$, and positive for $\gamma = 0$ and $\gamma = +\infty$. Since $\gamma = \cos\Theta$ and for real motions Θ is real, there are two real root, γ_3 and γ_2, between $\gamma = -1$ and $\gamma = +1$, and a third root, γ_1, is greater than $+1$.

In terms of the roots the polynomial $f(\gamma)$ can be rewritten in the form

$$f(\gamma) = (\gamma - \gamma_3)(\gamma - \gamma_2)(\gamma - \gamma_1)$$

with

$$-1 < \gamma_3 < \gamma_2 < +1 < \gamma_1$$

Equation (5.6) shows that γ oscillates between the values γ_3 and γ_2.

5.3. The sign of the precession. As γ increases from γ_3 to γ_2 equation (5.9) gives for the half period the expression

$$\frac{\sqrt{2a}}{(1-m)\,n}(\psi - \psi_0) = \int_{\gamma_3}^{\gamma_2} \frac{\gamma_4 - \gamma}{1-\gamma^2}\,\frac{d\gamma}{\sqrt{f(\gamma)}}$$

J. Hadamard [2] has given a proof that the precession for a complete period has the same sign as n (see also A. Métral [1]). His proof uses the theory of residues of functions of a complex variable. Recently J. B. Diaz and F. T. Metcalf [2] have given an elementary proof, not relying upon the theory of residues, to deduce Hadamard's result. This proof will be reproduced here.

If $\gamma_4 \geqq \gamma_2$ or $\gamma_4 < \gamma_3$ (γ_4 can never equal γ_3) the integrand is always non-negative; therefore the integral is non-negative and $\psi - \psi_0$ has the sign of n. But if γ_4 lies between γ_3 and γ_2, the integrand changes sign at $\gamma = \gamma_4$, and it is not immediately evident that the integral is positive. In this case the curve described by the point of intersection of the z-axis with the unit sphere drawn about the fixed point O has loops (see W. D. Macmillan [1], p. 242, Fig. 61).

The problem is to show that the definite integral

$$\frac{\sqrt{2a}}{(1-m)\,n}(\psi - \psi_0) = \int_{\gamma_3}^{\gamma_2} \frac{\gamma_4 - \gamma}{1-\gamma^2}\,\frac{d\gamma}{\sqrt{(\gamma_1-\gamma)(\gamma_2-\gamma)(\gamma-\gamma_3)}}$$

$$-1 < \gamma_3 < \gamma_4 < \gamma_2 < 1 < \gamma_1$$

is positive, the square root being positive as γ increases from γ_3 to γ_2.

Now, since

$$\frac{\gamma_4 - \gamma}{1-\gamma^2} = \frac{1}{2}\left(\frac{1+\gamma_4}{1+\gamma} - \frac{1-\gamma_4}{1-\gamma}\right)$$

it follows that

$$2\int_{\gamma_3}^{\gamma_2} \frac{\gamma_4-\gamma}{1-\gamma^2} \frac{d\gamma}{\sqrt{(\gamma_1-\gamma)(\gamma_2-\gamma)(\gamma-\gamma_3)}}$$

$$= \int_{\gamma_3}^{\gamma_2} \frac{1+\gamma_4}{1+\gamma} \frac{d\gamma}{\sqrt{(\gamma_1-\gamma)(\gamma_2-\gamma)(\gamma-\gamma_3)}} -$$

$$- \int_{\gamma_3}^{\gamma_2} \frac{1-\gamma_4}{1-\gamma} \frac{d\gamma}{\sqrt{(\gamma_1-\gamma)(\gamma_2-\gamma)(\gamma-\gamma_3)}}$$

However,

$$\frac{1}{\sqrt{\gamma_1+1}} < \frac{1}{\sqrt{\gamma_1-\gamma}} < \frac{1}{\sqrt{\gamma_1-1}}$$

$$-1 < \gamma_3 \leqq \gamma \leqq \gamma_2 < 1 < \gamma_1$$

and therefore

$$2\int_{\gamma_3}^{\gamma_2} \frac{\gamma_4-\gamma}{1-\gamma^2} \frac{d\gamma}{\sqrt{(\gamma_1-\gamma)(\gamma_2-\gamma)(\gamma-\gamma_3)}} >$$

$$> \frac{1+\gamma_4}{\sqrt{(1+\gamma_1)(1+\gamma_2)(1+\gamma_3)}} \int_{\gamma_3}^{\gamma_2} \frac{\sqrt{(1+\gamma_2)(1+\gamma_3)}\,d\gamma}{(1+\gamma)\sqrt{(\gamma_2-\gamma)(\gamma-\gamma_3)}} -$$

$$- \frac{1-\gamma_4}{\sqrt{(\gamma_1-1)(1-\gamma_2)(1-\gamma_3)}} \int_{\gamma_3}^{\gamma_2} \frac{\sqrt{(1-\gamma_2)(1-\gamma_3)}\,d\gamma}{(1-\gamma)\sqrt{(\gamma_2-\gamma)(\gamma-\gamma_3)}}$$

By the substitution $\gamma = -u$ the second integral on the right-hand side goes over into the first one, the value of which is π.

Hence

$$\frac{2\sqrt{2a}}{(1-m)n}(\psi-\psi_0) >$$

$$> \pi\left[\frac{1+\gamma_4}{\sqrt{(1+\gamma_1)(1+\gamma_2)(1+\gamma_3)}} - \frac{1-\gamma_4}{\sqrt{(\gamma_1-1)(1-\gamma_2)(1-\gamma_3)}}\right] = 0$$

since (see W. D. Macmillan [1], p. 221)

$$\frac{1-\gamma_4}{1+\gamma_4} = \frac{\sqrt{(\gamma_1-1)(1-\gamma_2)(1-\gamma_3)}}{\sqrt{(1+\gamma_1)(1+\gamma_2)(1+\gamma_3)}}$$

when

$$\gamma_3 < \gamma_4 < \gamma_2$$

5.4. Upper and lower bounds for the apsidal angle. Under certain conditions the constrained motion of a particle on the surface of a sphere, subject only to the force of gravity, consists of an oscillation between two horizontal planes cutting the surface. The position of the particle can be specified by the azimuthal angle ψ and the vertical height Z above a horizontal plane, the azimuth being defined as the angle between the horizontal projection of the position vector and the same projection at some initial instant. As the particle moves from the lowest level to the highest, the corresponding increase in the azimuth is called the apsidal angle and it will be denoted by Ψ.

The problem of motion of a particle on a surface of revolution is in general non-integrable. Therefore the qualitative aspects of such motions are of the greatest interest. The two aspects which have been most frequently investigated are the periodicity and stability of the motion and the establishment of bounds for the apsidal angle.

While G. KOBB [1], O. STAUDE [1] and M. PUGLISI [1, 2] have mainly studied the first aspect of the motion for certain surfaces of revolution, a number of writers have also been concerned with the problem of finding the bounds for the apsidal angle Ψ. The only surface of revolution for which precise bounds have been established is the sphere. These bounds constitute the famous inequalities of V. PUISEUX [1] and G. HALPHEN [2, p. 128] for the spherical pendulum, namely that $\frac{\pi}{2} < \Psi < \pi$. Later A. DE ST. GERMAIN [2, 3] established both inequalities anew by use of contour integration and L. GÉRARD [1] without the use of the latter. Further, A. WEINSTEIN [1] simplified the method of contour integration for obtaining PUISEUX's lower bound and W. KOHN [1], using WEINSTEIN's technique, obtained an extension of PUISEUX's inequality. KOHN also established the corresponding inequalities for the LAGRANGE and POISSON case of the motion of a heavy rigid body about a fixed point. J. L. SYNGE [1] found that bounds for the apsidal angle do not exist for all surfaces of revolution. Quite recently J. B. DIAZ and F. T. METCALF [1, 3] have developed a simple method for obtaining upper and lower bounds of the apsidal angle for a spherical pendulum and a heavy symmetric gyroscope.

For reasons of convenience the fixed point of the body is taken at the point $(0, 0, \frac{1}{2})$ of the fixed coordinate trihedral $O\,X\,Y\,Z$, and the motion of the rigid body is described by considering the motion of the point of intersection of that semi-axis of symmetry of the body with the sphere of unit diameter described about the fixed point which passes through the mass center G.

The energy integral and the angular momentum integral about the vertical assume the form

$$\tfrac{1}{2}A(\dot{\Theta}^2 + \dot{\psi}^2 \sin^2\Theta) + \tfrac{1}{2}C\,n^2 + \tfrac{1}{2} + a\cos\Theta = h_1$$
$$A\,\dot{\psi}\sin^2\Theta + C\,n\cos\Theta = c$$

where a is the distance from the fixed point to the center of mass of the body. Putting

$$Z = \tfrac{1}{2}(1 + \cos\Theta)$$

the above equations become

$$\frac{1}{2} A \left[4Z(1-Z)\left(\frac{d\psi}{dt}\right)^2 + \frac{1}{Z(1-Z)}\left(\frac{dZ}{dt}\right)^2\right] + \frac{1}{2} C n^2 + \\ + 2aZ + \frac{1}{2} - a = h_1 \qquad (5.11)$$

$$4A\, Z(1-Z)\frac{d\psi}{dt} = c - C n(2Z-1) \qquad (5.12)$$

Physically speaking Z represents the height above the horizontal X, Y-plane of the point determined by the intersection of the axis of the gyroscope with the sphere of unit diameter described about the fixed point. Without loss of generality we shall assume in what follows that $n > 0$. If $n = 0$, the problem reduces to that of a spherical pendulum.

Eliminating $d\psi/dt$ from equations (5.11) and (5.12) we obtain

$$\left(\frac{dZ}{dt}\right)^2 = H(Z) \qquad (5.13)$$

where

$$H(Z) = \frac{2}{A} Z(1-Z)\left[h_1 - \frac{1}{2} C n^2 - \left(2aZ + \frac{1}{2} - a\right)\right] - \\ - \left[\frac{c - C n(2Z-1)}{2A}\right]^2 \qquad (5.14)$$

We shall assume that the constants c and n are not zero. By virtue of equation (5.12) this implies that Z cannot assume both values 1 and 0. In other words, in the course of its motion the representative point will never pass through both the north and south poles. Also $d\psi/dt$ cannot be identically zero. Furthermore, we shall assume that dZ/dt is not identically zero, excluding thereby the so-called conical pendulum motion when the representative point rotates on a parallel of latitude.

Since during the motion of the point $H(Z)$ must be positive, it follows that the three roots of $H(Z) = 0$ are all real and satisfy the inequalities

$$0 \leqq Z_1 \leqq Z_2 \leqq 1 \leqq Z_3$$

From expression (5.14) it follows that $Z = 0$ and $Z = 1$ are roots of $H(Z) = 0$ if and only if $c = -C n$ and $c = C n$ respectively.

Putting

$$P = \frac{2}{A}\left(h_1 - \frac{1}{2} C n^2 + a - \frac{1}{2}\right)$$

expanding $H(Z) = \frac{4a}{A}(Z - Z_1)(Z - Z_2)(Z - Z_3)$ in powers of Z, calculating the symmetric functions of the roots, $Z_1 + Z_2 + Z_3$,

$Z_1 Z_2 + Z_1 Z_3 + Z_2 Z_3$, $Z_1 Z_2 Z_3$, and finally eliminating $A/4a$, we obtain the equation

$$Z_1 Z_2 + Z_1 Z_3 + Z_2 Z_3 = \frac{4\varrho}{(1+\varrho)^2} Z_1 Z_2 Z_3 + Z_1 + Z_2 + Z_3 - 1 \quad (5.15)$$

where $\varrho = c/C n$. Calculating Z_3 and $Z_3 - 1$ from the last equation we finally obtain the equation

$$\left(\frac{Z_3 - 1}{Z_3}\right)^{1/2} = \left|\frac{1-\varrho}{1+\varrho}\right| \left[\frac{Z_1 Z_2}{(1-Z_1)(1-Z_2)}\right]^{1/2}$$

which can be rewritten in the form

$$|1-\varrho| (Z_1 Z_2 Z_3)^{1/2} = |1+\varrho| [(1-Z_1)(1-Z_2)(Z_3 - 1)]^{1/2} \quad (5.16)$$

If $\varrho = 1$, then at least one of the roots Z_2 and Z_3 equals one. If $\varrho = -1$, then $Z_1 = 0$.

Solving equation (5.15) for $Z_1 + Z_2$ we obtain

$$Z_1 + Z_2 = 1 - \frac{Z_1 Z_2}{Z_3 - 1} \left[1 - \frac{4\varrho}{(1+\varrho)^2} Z_3\right]$$

Hence for $\varrho \leqq 0$ and $\varrho \neq -1$ it follows that

$$Z_1 + Z_2 < 1 \quad (5.17)$$

When $\varrho = -1$, we have $Z_1 = 0$, and the last inequality still holds.

Consider now the motion of a gyroscope under the assumption that $dZ/dt \geqq 0$, i.e. as the gyroscope rises from its lowest level of motion $Z = Z_1$ to its highest $Z = Z_2$. Elimination of t between equations (5.12) and (5.13) gives for the increment Ψ of the precession ψ the expression

$$\Psi = \int_{Z_1}^{Z_2} \frac{[c - C n(2Z - 1)]\, dZ}{4A\, Z(1-Z) \sqrt{\frac{4a}{A}(Z - Z_1)(Z_2 - Z)(Z_3 - Z)}}$$

which, by virtue of the relation

$$Z_1 Z_2 Z_3 = \frac{A}{4a} \frac{(c + C n)^2}{4A^2}$$

can be rewritten in the form

$$\Psi = \frac{1}{2|1+\varrho|} \int_{Z_1}^{Z_2} \frac{\sqrt{Z_1 Z_2 Z_3}\, h(Z)\, dZ}{Z(1-Z)\sqrt{(Z - Z_1)(Z_2 - Z)}} \quad (5.18)$$

where

$$h(Z) = \frac{1 + \varrho - 2Z}{\sqrt{Z_3 - Z}} \quad (5.19)$$

Using the decomposition of $\frac{1+\varrho-2Z}{Z(1-Z)}$ into partial fractions we obtain

$$\frac{1+\varrho-2Z}{Z(1-Z)} = \frac{1+\varrho}{Z} + \frac{\varrho-1}{1-Z}$$

and formula (5.18) assumes the form

$$\Psi = \operatorname{sgn}(1+\varrho)\frac{1}{2}\int_{Z_1}^{Z_2} \frac{\sqrt{Z_1 Z_2 Z_3}}{Z\sqrt{H_1(Z)}}\,dZ + \frac{\varrho-1}{|1+\varrho|}\,\frac{1}{2}\int_{Z_1}^{Z_2} \frac{\sqrt{Z_1 Z_2 Z_3}}{(1-Z)\sqrt{H_1(z)}}\,dZ \tag{5.20}$$

where

$$H_1(Z) = (Z-Z_1)(Z_2-Z)(Z_3-Z)$$

By virtue of equation (5.16) the last formula can be rewritten as follows

$$\Psi = \operatorname{sgn}(\varrho+1)\frac{1}{2}\int_{Z_1}^{Z_2} \frac{\sqrt{Z_1 Z_2 Z_3}}{Z\sqrt{H_1(Z)}}\,dZ +$$

$$+ \operatorname{sgn}(\varrho-1)\frac{1}{2}\int_{Z_1}^{Z_2} \frac{\sqrt{(1-Z_1)(1-Z_2)(Z_3-1)}}{(1-Z)\sqrt{H_1(Z)}}\,dZ \tag{5.21}$$

Assuming that sgn $0 = 0$, it follows from formula (5.20) that formula (5.21) remains valid also for the case $\varrho = 1$.

If $\varrho \neq -1$, $Z_1 < \frac{1}{2}$ and $Z_2 \geqq \frac{1}{2}$, then

$$\Psi_0 = \frac{1}{2|1+\varrho|}\int_{Z_1}^{1/2} \frac{\sqrt{Z_1 Z_2 Z_3}\,h(Z)\,dZ}{Z(1-Z)\sqrt{(Z-Z_1)(Z_2-Z)}} \tag{5.22}$$

gives the increment in ψ corresponding to the motion from the lowest level $Z = Z_1$ to the equator $Z = \frac{1}{2}$ (if $\varrho \leqq 0$, $\varrho \neq -1$, then from inequality (5.17) it follows that by necessity $Z_1 < \frac{1}{2}$). Formula (5.21) holds also for $\varrho = -1$.

For the actual calculation of the bounds for Ψ and Ψ_0 we would like to refer the reader to the original paper by DIAZ and METCALF. However, as an illustration of the method we shall consider in the next section the motion of a spherical pendulum.

5.41. The spherical pendulum

In the case of a spherical pendulum the equations of motion are obtained from those of the gyroscope by putting $n = 0$, $A = \frac{1}{4}$ and $a = \frac{1}{2}$. Expression (5.14) for $H(Z)$ then reduces to

$$H(Z) = 8Z(1-Z)(h_1-Z) - 4c^2 \tag{5.23}$$

Assuming that $c > 0$, the equation $H(Z) = 0$ has three roots which satisfy the inequalities

$$0 < Z_1 < Z_2 < 1 < Z_3$$

Letting $\varrho \to \infty$ in formula (5.16) we obtain

$$\sqrt{Z_1 Z_2 Z_3} = \sqrt{(1 - Z_1)(1 - Z_2)(Z_3 - 1)} \tag{5.24}$$

Similarly, the inequality

$$Z_1 + Z_2 < 1$$

holds, and hence $Z_1 < \frac{1}{2}$.

Again, letting $\varrho \to \infty$ in formulas (5.18) and (5.19) we obtain

$$\Psi = \frac{1}{2} \int_{Z_1}^{Z_2} \frac{\sqrt{Z_1 Z_2 Z_3}\, dZ}{Z(1 - Z)\sqrt{H_1(Z)}} \tag{5.25}$$

where, as before, $H_1(Z) = (Z - Z_1)(Z_2 - Z)(Z_3 - Z)$.

Also, if $Z_2 \geqq \frac{1}{2}$, then

$$\Psi_0 = \frac{1}{2} \int_{Z_1}^{1/2} \frac{\sqrt{Z_1 Z_2 Z_3}\, dZ}{Z(1 - Z)\sqrt{H_1(Z)}} \tag{5.26}$$

In order to obtain bounds for Ψ, the function $h(Z) = 1/\sqrt{Z_3 - Z}$ must be studied. It can be easily verified that $h'(Z)$ and $h''(Z)$ are both positive for $0 \leqq Z \leqq 1$. Hence, $h(Z)$ is an increasing convex function of Z in this interval (Fig. 2). Thus

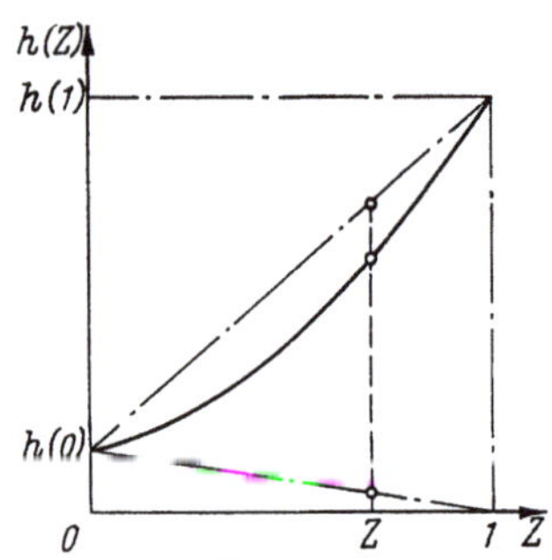

Fig. 2

$$h(0)(1 - Z) \leqq h(Z) \leqq h(0)(1 - Z) + h(1) Z, \quad 0 \leqq Z \leqq 1 \tag{5.27}$$

since $h(0)(1 - Z)$ represents the height above the Z-axis of a point on the straight line connecting the points $[0, h(0)]$ and $[1, 0]$. Similarly, $h(0)(1 - Z) + h(1) Z$ represents the height above the Z-axis of a point on the straight line joining $[0, h(0)]$ and $[1, h(1)]$. Using the actual values for $h(0)$ and $h(1)$, inequality (5.27) can be rewritten in the form

$$\frac{1}{\sqrt{Z_3}}(1 - Z) \leqq \frac{1}{\sqrt{Z_3 - Z}} \leqq \frac{1}{\sqrt{Z_3}}(1 - Z) + \frac{1}{\sqrt{Z_3 - 1}} Z, \quad 0 \leqq Z \leqq 1 \tag{5.28}$$

Substituting the lower bound for $1/\sqrt{Z_3 - Z}$ in formula (5.25) we obtain that

$$\Psi > \frac{1}{2} \int\limits_{Z_1}^{Z_2} \frac{\sqrt{Z_1 Z_2}\, dZ}{Z\sqrt{(Z - Z_1)(Z_2 - Z)}} = I_1 = \frac{\pi}{2}$$

which is PUISEUX's lower bound for Ψ. Similarly,

$$\Psi < I_1 + \frac{\sqrt{Z_1 Z_2 Z_3}}{\sqrt{(1 - Z_1)(1 - Z_2)(Z_3 - 1)}} I_2$$

where

$$I_2 = \frac{1}{2} \int\limits_{Z_1}^{Z_2} \frac{\sqrt{(1 - Z_1)(1 - Z_2)}\, dZ}{(1 - Z)\sqrt{(Z - Z_1)(Z_2 - Z)}} = \frac{\pi}{2}$$

Using formula (5.24) the last inequality assumes the form

$$\Psi < I_1 + I_2 = \pi$$

thereby arriving at HALPHEN's upper bound for Ψ.

If $Z_2 \geqq \frac{1}{2}$, formula (5.26) can be rewritten in the form

$$\Psi_0 = \frac{1}{2} \int\limits_{Z_1}^{1/2} \frac{\sqrt{Z_1(1 - Z_1)}}{Z(1 - Z)\sqrt{(Z - Z_1)(1 - Z_1 - Z)}} \times$$

$$\times \sqrt{\frac{Z_2}{1 - Z_1} \frac{1 - Z_1 - Z}{Z_2 - Z} \frac{Z_3}{Z_3 - Z}}\, dZ$$

Since for $0 \leqq Z \leqq \frac{1}{2}$ the inequalities

$$\frac{1 - Z_1 - Z}{Z_2 - Z} \geqq \frac{1 - Z_1}{Z_2}$$

$$\frac{Z_3}{Z_3 - Z} \geqq 1$$

hold, we have that

$$\Psi_0 > \frac{1}{2} \int\limits_{Z_1}^{1/2} \frac{\sqrt{Z_1(1 - Z_1)}\, dZ}{Z(1 - Z)\sqrt{(Z - Z_1)(1 - Z_1 - Z)}}$$

On the other hand the decomposition of $1/Z(1 - Z)$ into partial fractions gives

$$\frac{1}{Z(1 - Z)} = \frac{1}{Z} + \frac{1}{1 - Z}$$

and therefore

$$\Psi_0 > I_3 + I_4 = \frac{\pi}{2}$$

where

$$I_3 = \arc\tan \sqrt{\frac{1 - Z_1}{Z_1}} = \frac{1}{2} \int\limits_{Z_1}^{1/2} \frac{\sqrt{Z_1(1 - Z_1)}\, dZ}{Z\sqrt{(Z - Z_1)(1 - Z_1 - Z)}}$$

$$I_4 = \frac{1}{2} \int\limits_{Z_1}^{1/2} \frac{\sqrt{Z_1(1 - Z_1)}\, dZ}{(1 - Z)\sqrt{(Z - Z_1)(1 - Z_1 - Z)}} = \frac{\pi}{2} - I_3$$

The above inequality is KOHN's extension of PUISEUX's inequality.

5.5. Stability of a particular solution. In the case of LAGRANGE and POISSON the motion corresponding to the particular solution

$$p = q = 0, \qquad r = n, \qquad \alpha = \beta = 0, \qquad \gamma = 1 \tag{5.29}$$

of the EULER and POISSON equations is of particular interest.

The stability of this solution relative to the functions $p, q, r; \alpha, \beta, \gamma$ has been investigated by ČETAEV [1, 4, p. 430—431] using the method of LYAPUNOV's V-functions. According to this method the particular solution (5.29) is stable provided that there exists a positive definite integral for the equations of perturbations relative to this solution.

The known first integrals (5.3) can be rewritten in the form (introducing again the factor $m g$ which earlier was put equal to one)

$$\begin{aligned} A(p^2 + q^2) + 2 m g z_0 \gamma &= h \\ A(p\alpha + q\beta) + C r \gamma &= c \\ \alpha^2 + \beta^2 + \gamma^2 &= 1 \\ r &= n \end{aligned}$$

Let the perturbed motion be

$$\begin{aligned} p &= x, & q &= y, & r &= n + z \\ \alpha &= \alpha', & \beta &= \beta', & \gamma &= 1 + \delta \end{aligned}$$

Then the equations for the perturbations have the integrals

$$\left.\begin{aligned} A(x^2 + y^2) + 2 m g z_0 \delta &= V_1 \\ A(x\alpha' + y\beta') + C(z\delta + z + n\delta) &= V_2 \\ \alpha'^2 + \beta'^2 + \delta^2 + 2\delta = V_3, \quad z &= V_4 \end{aligned}\right\} \tag{5.30}$$

Further, let us construct a LYAPUNOV V-function in the form of a quadratic bundle of integrals (5.30) with λ as a parameter

$$\left.\begin{aligned} V &= V_1 + 2\lambda V_2 - (m g z_0 + C n \lambda) V_3 + \frac{C^2}{A} V_4^2 - 2 C \lambda V_4 \\ &= A x^2 + 2\lambda A x \alpha' - (m g z_0 + C n \lambda)\alpha'^2 + \\ &\quad + A y^2 + 2\lambda A y \beta' - (m g z_0 + C n \lambda)\beta'^2 + \\ &\quad + \frac{C^2}{A} z^2 + 2\lambda C z \delta - (m g z_0 + C n \lambda)\delta^2 \end{aligned}\right\} \tag{5.31}$$

In order that the first two quadratic forms in x, α' and y, β' be positive definite it is necessary and sufficient to select the parameter λ in such a way that their discriminant be positive, i.e.

$$\begin{vmatrix} A & A\lambda \\ A\lambda & -(m g z_0 + C n \lambda) \end{vmatrix} > 0$$

or

$$A\lambda^2 + C n \lambda + m g z_0 < 0 \tag{5.32}$$

This will be the case if the quadratic polynomial on the left of (5.32) has two different real roots, i.e. if

$$C^2 n^2 - 4A\, m g z_0 > 0 \tag{5.33}$$

The third quadratic form in terms of z, δ which occurs in V will be positive definite if its discriminant is positive, i.e. if

$$\begin{vmatrix} \dfrac{C^2}{A} & C\lambda \\ C\lambda & -(m g z_0 + C n \lambda) \end{vmatrix} > 0$$

or inequality (5.32) holds.

Consequently, if condition (5.33) is satisfied, the parameter λ can be selected in such a way that inequality (5.32) holds. For such a choice of λ all three quadratic forms which constitute V are positive definite, each form with respect to its own variables, and the integral V is positive definite relative to all perturbations $x, y, z, \alpha', \beta', \delta$. According to a wellknown theorem of LYAPUNOV the particular solution (5.29) is stable relative to the variables $p, q, r; \alpha, \beta, \gamma$ provided that

$$C^2 n^2 - 4A\, m g z_0 > 0$$

It is interesting to note that E. J. ROUTH [2_2, 3_2] solved the stability problem for uniform precessions about the vertical without restricting the moments of inertia and the position of the mass center G.

5.6. Cases differing slightly from those of Euler and Poinsot, and Lagrange and Poisson. A. PIGNEDOLI [1] is concerned (i) with the motions of a heavy rigid body about a fixed point O, the distance of which from the center of mass G of the body is infinitesimal, or, in other words, with motions which are close to those of POINSOT (when G coincides with O); and (ii) with the motions of a heavy rigid body about a fixed point when G is different from O and is slightly off the z-axis, and when the structure of the body with respect to this axis differs slightly from a gyroscopic structure. Questions of stability are also discussed.

§ 6. Case of Kovalevskaya

6.1. The first integrals. S. V. KOVALEVSKAYA [1—6] considered the case when $A = B = 2C$, and the center of mass G of the body lies in the plane of equal moments of inertia, i.e. $z_0 = 0$. By rotation of the trihedral $O\,x\,y\,z$ about the z-axis we can make the center of mass of the body lie on the x-axis, say, $x_0 > 0$ and $y_0 = 0$. Introduce the notation

$$c_0 = \frac{x_0}{C} \tag{6.1}$$

Then the Euler equations assume the form

$$\left.\begin{aligned} 2\frac{dp}{dt} &= q\,r \\ 2\frac{dq}{dt} &= -p\,r + c_0\,\gamma \\ \frac{dr}{dt} &= -c_0\,\beta \end{aligned}\right\} \qquad (6.2)$$

and the three classical integrals are

$$\left.\begin{aligned} 2(\alpha\,p + \beta\,q) + \gamma\,r &= 2l & \left(2l = \frac{c}{C}\right) \\ 2(p^2 + q^2) + r^2 &= -2c_0\,\alpha + 6l_1 & \\ \alpha^2 + \beta^2 + \gamma^2 &= 1 & (h_1 = 3C\,l_1) \end{aligned}\right\} \qquad (6.3)$$

Kovalevskaya constructs the fourth integral as follows. Multiply the second equation of (6.2) by $i\left(=\sqrt{-1}\right)$ and add the corresponding sides to those of the first equation of the same system. Analogously multiply the second equation of Poisson by i and add the corresponding sides to those of the first Poisson equation. In this way we obtain the following two equations

$$2\frac{d}{dt}(p + i\,q) = -i\,r(p + i\,q) + c_0\,i\,\gamma \qquad (6.4)$$

$$\frac{d}{dt}(\alpha + i\,\beta) = -i\,r(\alpha + i\,\beta) + i\,\gamma(p + i\,q) \qquad (6.5)$$

In order to eliminate γ multiply equation (6.4) by $p + i\,q$ and equation (6.5) by $-c_0$ and add the corresponding sides. We then obtain that

$$\frac{d}{dt}\ln[(p + i\,q)^2 - c_0(\alpha + i\,\beta)] = -i\,r \qquad (6.6)$$

Replacing in the last equation i by $-i$ we obtain another equation, the addition of which to equation (6.6) gives the equation

$$\frac{d}{dt}\ln[(p + i\,q)^2 - c_0(\alpha + i\,\beta)]\,[(p - i\,q)^2 - c_0(\alpha - i\,\beta)] = 0$$

Integrating the last equation we obtain the fourth algebraic integral

$$(p^2 - q^2 - c_0\,\alpha)^2 + (2p\,q - c_0\,\beta)^2 = k^2 \qquad (6.7)$$

N. B. Delone [1, 2], G. G. Appel'rot [2] and B. K. Mlodzeevskii [2] have studied the geometrical representation of the motion in the particular case where the constant k in integral (6.7) vanishes. The general case of motion with $k \neq 0$ has been studied by G. K. Suslov [1], N. E. Žukovskii [3, 4], F. Kötter [1], G. V. Kolosov [4], and W. de Tannenberg [1, 3].

A particular motion of the Kovalevskaya gyroscope has been studied by Yu. A. Arhangel'skii [9].

After the appearance of the Kovalevskaya memoir [1] it seemed plausible that a fourth algebraic integral could be obtained (by the same method as in the case of a heavy rigid body) also in other cases provided that the potential function of the acting forces is suitably modified. E. Padova [4], however, showed that this is not possible; in other words, we are always led to the Kovalevskaya case.

6.2. Introduction of the new variables s_1 and s_2. The Euler equations are solved by completion of two additional quadratures. However, the problem of reducing the latter to the most suitable form, in order that the quantities characterizing the motion admit the most simple expressions, is by no means a simple problem. Kovalevskaya succeeded in solving this problem and expressed all the necessary quantities except the angle of precession ψ in terms of hyperelliptic functions of the first kind, the two arguments of which depend linearly on the time t.

The success of Kovalevskaya is mainly due to the introduction of two new variables s_1 and s_2, which to some extent replace the variables p and q and are related to the time t by means of relatively simple differential expressions. The latter can be derived from the four algebraic integrals and the first two Euler equations. After the death of Kovalevskaya the German mathematician F. Kötter [2] somewhat simplified her calculations.

Kovalevskaya first introduces the notations

$$\begin{aligned} &x_1 = p + i q, \quad x_2 = p - i q \\ &\xi_1 = (p + i q)^2 - c_0(\alpha + i\beta), \quad \xi_2 = (p - i q)^2 - c_0(\alpha - i\beta) \end{aligned} \tag{6.8}$$

by means of which the first four integrals assume the form

$$\left.\begin{aligned} r^2 &= 6l_1 + \xi_1 + \xi_2 - (x_1 + x_2)^2 \\ c_0 r\gamma &= 2c_0 l + x_2\xi_1 + x_1\xi_2 - x_1 x_2(x_1 + x_2) \\ c_0^2\gamma^2 &= c_0^2 - k^2 + x_2^2\xi_1 + x_1^2\xi_2 - x_1^2 x_2^2 \\ \xi_1\xi_2 &= k^2 \end{aligned}\right\} \tag{6.9}$$

Eliminating r and γ from the first three integrals Kovalevskaya arrives at the equation

$$\xi_1 R(x_2) + \xi_2 R(x_1) + R_1(x_1, x_2) + k^2(x_1 - x_2)^2 = 0 \tag{6.10}$$

where

$$\begin{aligned} R(x) &= -x^4 + 6l_1 x^2 - 4c_0 l x + c_0^2 - k^2 \\ R_1(x_1, x_2) &= -6l_1 x_1^2 x_2^2 - (c_0^2 - k^2)(x_1 + x_2)^2 - \\ &\quad + 4c_0 l(x_1 + x_2) x_1 x_2 + 6l_1(c_0^2 - k^2) - 4c_0^2 l^2 \end{aligned}$$

Next rewrite $R(x)$ and $R_1(x_1, x_2)$ in the form

$$R(x) = [6l_1 - (x_1 + x_2)^2]\, x^2 + 2[-2c_0 l + x_1 x_2 (x_1 + x_2)]\, x + \\ + c_0^2 - k^2 - x_1^2 x_2^2 = A_1 x^2 + 2B_1 x + C_1$$

$$R_1(x_1, x_2) = A_1 C_1 - B_1^2$$

The quantities A_1, B_1 and C_1 are introduced in order to abbreviate writing. Further introduce the function

$$R(x_1, x_2) = -x_1^2 x_2^2 + 6l_1 x_1 x_2 - 2c_0 l (x_1 + x_2) + c_0^2 - k^2 \\ = A_1 x_1 x_2 + B_1 (x_1 + x_2) + C_1$$

Then it is easy to verify that the following identity

$$R(x_1)\, R(x_2) - R_1(x_1, x_2)\,(x_1 - x_2)^2 = R^2(x_1, x_2) \tag{6.11}$$

holds.

From the expressions for $R(x)$ and $R_1(x_1, x_2)$ in terms of A_1, B_1, C_1 it is easily seen that for a simultaneous vanishing of these polynomials the existence of a multiple root of the equation $R(x) = 0$ is necessary.

Eliminating r and γ from system (6.9), the latter reduces to two equations, these equations being (6.10) and $\xi_1 \xi_2 = k^2$. Rewriting them in the form

$$(\sqrt{a_1}\sqrt{\xi_1})^2 + (\sqrt{b_1}\sqrt{\xi_2})^2 = c_1, \quad 2\sqrt{a_1 b_1}\sqrt{\xi_1 \xi_2} = 2\sqrt{a_1 b_1}\, k$$

where

$$a_1 = R(x_2)$$

$$b_1 = R(x_1)$$

$$c_1 = -R_1(x_1, x_2) - k^2 (x_1 - x_2)^2$$

we obtain the equations

$$(\sqrt{a_1}\sqrt{\xi_1} \pm \sqrt{b_1}\sqrt{\xi_2})^2 = c_1 \pm 2\sqrt{a_1 b_1}\, k$$

or

$$\left\{\sqrt{\xi_1}\frac{\sqrt{R(x_2)}}{x_1 - x_2} \pm \sqrt{\xi_2}\frac{\sqrt{R(x_1)}}{x_1 - x_2}\right\}^2 = -\frac{R_1(x_1, x_2)}{(x_1 - x_2)^2} \pm 2k\frac{\sqrt{R(x_1)}\sqrt{R(x_2)}}{(x_1 - x_2)^2} - k^2 \tag{6.12}$$

In order to factor the right-handside of the last equation into two linear factors in k, consider the equation

$$w^2 \pm 2w\frac{\sqrt{R(x_1)}\sqrt{R(x_2)}}{(x_1 - x_2)^2} + \frac{R_1(x_1, x_2)}{(x_1 - x_2)^2} = 0$$

which by virtue of identity (6.11) can be rewritten in the form

$$\left\{w \pm \frac{\sqrt{R(x_1)}\sqrt{R(x_2)}}{(x_1 - x_2)^2}\right\}^2 = \frac{R^2(x_1, x_2)}{(x_1 - x_2)^4}$$

and consequently has the roots

$$\begin{aligned} w_1 &= \frac{R(x_1, x_2) - \sqrt{R(x_1)}\sqrt{R(x_2)}}{(x_1 - x_2)^2} \\ w_2 &= \frac{R(x_1, x_2) + \sqrt{R(x_1)}\sqrt{R(x_2)}}{(x_1 - x_2)^2} \end{aligned} \tag{6.13}$$

By virtue of equations (6.13) and (6.11), equation (6.12) assumes the form

$$\begin{aligned} \left\{\sqrt{\xi_1}\frac{\sqrt{R(x_2)}}{x_1 - x_2} \pm \sqrt{\xi_2}\frac{\sqrt{R(x_1)}}{x_1 - x_2}\right\}^2 &= w_1 w_2 \pm k(w_1 - w_2) - k^2 \\ &= (w_1 \mp k)(w_2 \pm k) \end{aligned} \tag{6.14}$$

Obviously w_1 and w_2 can also be considered as roots of the quadratic equation

$$w^2 - \frac{2R(x_1, x_2)}{(x_1 - x_2)^2} w - \frac{R_1(x_1, x_2)}{(x_1 - x_2)^2} = 0 \tag{6.15}$$

which for complex conjugate values of x_1 and x_2 always has real roots. Finally KOVALEVSKAYA introduced the variables

$$s_1 = w_1 + 3l_1 \quad \text{and} \quad s_2 = w_2 + 3l_1 \tag{6.16}$$

6.3. Transformation of the elliptic differential $dx/\sqrt{R(x)}$. In order to better understand the purpose of introducing the new variables, let us recall the following result from the WEIERSTRASS [1] theory of elliptic functions.

Let $R(x)$ denote an arbitrary integral rational function of degree three or four without a quadratic divisor and let $\sqrt{R(x)}$ denote either of the two values which the square root of $R(x)$ can assume. Put

$$\bar{s} = \frac{\sqrt{R(x_0)}\sqrt{R(x)} + R(x_0) + \frac{1}{2}R'(x_0)(x - x_0)}{2(x - x_0)^2} + \frac{1}{24}R''(x_0) \tag{I}$$

and define $\sqrt{R_1(\bar{s})}$, where $R_1(\bar{s}) = 4\bar{s}^3 - g_2\bar{s} - g_3$, by means of the formula

$$\begin{aligned} \sqrt{R_1(\bar{s})} = \left[\frac{R(x)}{(x - x_0)^3} - \frac{1}{4}\frac{R'(x)}{(x - x_0)^2}\right]\sqrt{R(x_0)} - \\ - \left[\frac{R(x_0)}{(x_0 - x)^3} - \frac{1}{4}\frac{R'(x_0)}{(x_0 - x)^2}\right]\sqrt{R(x)} \end{aligned} \tag{II}$$

Then

$$\frac{dx}{\sqrt{R(x)}} = -\frac{d\bar{s}}{\sqrt{R_1(\bar{s})}} \tag{III}$$

i.e. by means of formulas (I) and (II), which contain an arbitrary constant x_0, the elliptic differential $dx/\sqrt{R(x)}$ can be reduced to the

WEIERSTRASS canonical form $-d\bar{s}/\sqrt{R_1(\bar{s})}$. The coefficients of the third degree integral function $R_1(\bar{s}) = 4\bar{s}^3 - g_2\bar{s} - g_3$ do not depend on x_0. If x_0 is considered as a variable, then the total differential $d\bar{s}$ in (III) is to be replaced by a partial differential of $\bar{s}$.

If

$$R(x) = A\,x^4 + 4B\,x^3 + 6C\,x^2 + 4B'\,x + A'$$

then the two invariants g_2 and g_3 of the binary form of degree four $y^4 R(x/y)$ have the expressions

$$g_2 = A\,A' - 4B\,B' + 3\,C^2$$
$$g_3 = A\,C\,A' + 2B\,C\,B' - A'B^2 - A\,B'^2 - C^3$$

If the above-mentioned form is expressed in terms of new variables X, Y, where x, y are linear functions of X, Y, then g_2 and g_3 do not change but acquire only some power of the transformation determinant as a factor.

For some purposes it is useful to transform formula (I) as follows. Replace $R(x_0)$, $R'(x_0)$, $R''(x_0)$ by their expressions and define a new integral function of x and x_0 by the formula

$$R(x, x_0) = A\,x^2 x_0^2 + 2B\,x\,x_0(x + x_0) + 6C\,x\,x_0 + 2B'(x + x_0) + A'$$

Then formula (I) becomes

$$\bar{s} = \frac{\sqrt{R(x)}\,\sqrt{R(x_0)} + R(x, x_0)}{2(x - x_0)^2} + \frac{1}{2}C \qquad \text{(IV)}$$

If, furthermore, the product $\sqrt{R(x)}\,\sqrt{R(x_0)}$ in formula (I) is replaced by

$$\sqrt{R(x)}\,\sqrt{R(x_0)} = \tfrac{1}{2}\left[\sqrt{R(x)} + \sqrt{R(x_0)}\right]^2 - \tfrac{1}{2}R(x) - \tfrac{1}{2}R(x_0)$$

and $-\frac{1}{2}R(x)$ is expanded in powers of $(x - x_0)$, then formula (IV) reduces to

$$\bar{s} = \left[\frac{\sqrt{R(x)} + \sqrt{R(x_0)}}{2(x - x_0)}\right]^2 - \frac{1}{4}A(x + x_0)^2 - B(x + x_0) - C$$

The roots $\bar{e}_1$, $\bar{e}_2$, $\bar{e}_3$ of $R_1(\bar{s}) = 0$ are rational functions of the roots ε_1, ε_2, ε_3, ε_4 of the fourth degree equation $R(x) = 0$. Namely,

$$\bar{e}_1 = A\left[\frac{D}{A^2} - \frac{1}{4}\left(2\frac{B}{A} + \varepsilon_1 + \varepsilon_2\right)^2\right]$$
$$\bar{e}_2 = A\left[\frac{D}{A^2} - \frac{1}{4}\left(2\frac{B}{A} + \varepsilon_1 + \varepsilon_3\right)^2\right]$$
$$\bar{e}_3 = A\left[\frac{D}{A^2} - \frac{1}{4}\left(2\frac{B}{A} + \varepsilon_1 + \varepsilon_4\right)^2\right]$$

where

$$D = B^2 - A\,C$$

6.4. Differential relations between s and x, and s and t. In the case of the polynomial of KOVALEVSKAYA we have

$$A = -1, \quad B = 0, \quad C = l_1, \quad B' = -c_0\, l, \quad A' = c_0^2 - k^2$$
$$D = -A\,C = l_1, \quad \varepsilon_1 + \varepsilon_2 + \varepsilon_3 + \varepsilon_4 = 0$$

Further, if we put

$$x = x_1, \quad x_0 = x_2, \quad 2\bar{s} + 2C = s = w + 3C = w + 3l_1$$

and observe that w_1 [given by (6.13)] and consequently s_1 have the form of WEIERSTRASS' formula (IV), in which the sign of one of the radicals $\sqrt{R(x_1)}$ or $\sqrt{R(x_2)}$ is changed to the opposite, say, the sign of $\sqrt{R(x_1)}$, and if w_2 and as a consequence s_2 have the form of formula (IV) as it stands, then we obtain that

$$\sqrt{R_1(\bar{s}_1)} = \sqrt{R_1\left(\frac{s_1}{2} - l_1\right)} = \frac{[R(x_1) - \frac{1}{4}(x_1 - x_2)\,R'(x_1)]\sqrt{R(x_2)}}{(x_1 - x_2)^3} - \frac{[R(x_2) + \frac{1}{4}(x_1 - x_2)\,R'(x_2)]\sqrt{R(x_1)}}{(x_1 - x_2)^3}$$

$$\sqrt{R_1(\bar{s}_2)} = \sqrt{R_1\left(\frac{s_2}{2} - l_1\right)} = \frac{[R(x_1) - \frac{1}{4}(x_1 - x_2)\,R'(x_1)]\sqrt{R(x_2)}}{(x_1 - x_2)^3} + \frac{[R(x_2) + \frac{1}{4}(x_1 - x_2)\,R'(x_2)]\sqrt{R(x_1)}}{(x_1 - x_2)^3}$$

In determining the signs of the radicals $\sqrt{R_1(\bar{s}_1)}$ and $\sqrt{R_1(\bar{s}_2)}$ the fact must be born in mind that the sign of one of the radicals $\sqrt{R(x_1)}$ and $\sqrt{R(x_2)}$ is arbitrary. Thus, if we denote by $d_1 = \frac{\partial}{\partial x_1} dx_1$ and $d_2 = \frac{\partial}{\partial x_2} dx_2$ respectively the partial differentials with respect to x_1 and x_2, then according to WEIERSTRASS we can write that

$$\frac{d_1 \bar{s}_1}{\sqrt{R_1(\bar{s}_1)}} = \frac{d_1 s_1}{2\sqrt{R_1\left(\frac{s_1}{2} - l_1\right)}} = \frac{dx_1}{\sqrt{R(x_1)}}$$

$$\frac{d_1 \bar{s}_2}{\sqrt{R_1(\bar{s}_2)}} = \frac{d_1 s_2}{2\sqrt{R_1\left(\frac{s_2}{2} - l_1\right)}} = -\frac{dx_1}{\sqrt{R(x_1)}}$$

If we interchange the letters x_1 and x_2 in the expressions of $\bar{s}$, s, $\sqrt{R_1(\bar{s}_1)}$ and $\sqrt{R_1(\bar{s}_2)}$, then the first three do not change while $\sqrt{R_1(\bar{s}_2)}$

changes its sign. Hence we obtain from the last system that

$$\frac{d_2\bar{s}_1}{\sqrt{R_1(\bar{s}_1)}} = \frac{d_2 s_1}{2\sqrt{R_1\left(\frac{s_1}{2} - l_1\right)}} = \frac{d x_2}{\sqrt{R(x_2)}}$$

$$\frac{d_2\bar{s}_2}{\sqrt{R_1(\bar{s}_2)}} = \frac{d_2 s_2}{2\sqrt{R_1\left(\frac{s_2}{2} - l_1\right)}} = \frac{d x_2}{\sqrt{R(x_2)}}$$

Adding the first and the second equations respectively of the last two systems and denoting $R_1\left(\frac{s}{2} - l_1\right)$ by $\frac{1}{2}\varphi(s)$ we obtain the equations

$$\begin{aligned} \frac{d x_1}{\sqrt{R(x_1)}} + \frac{d x_2}{\sqrt{R(x_2)}} &= \frac{d s_1}{\sqrt{2\varphi(s_1)}} \\ -\frac{d x_1}{\sqrt{R(x_1)}} + \frac{d x_2}{\sqrt{R(x_2)}} &= \frac{d s_2}{\sqrt{2\varphi(s_2)}} \end{aligned} \tag{6.17}$$

where

$$\varphi(s) = s[(s - 3l_1)^2 + c_0^2 - k^2] - 2c_0^2 l^2 \tag{6.18}$$

While the equation of degree three $R_1(\bar{s}) = 0$ is the so-called ARONHOLD resolvent of the fourth degree equation $R(x) = 0$, the equation $\varphi(s) = 0$ is another resolvent known as the EULER resolvent of the same equation. The roots e_1, e_2, e_3 of the equation $\varphi(s) = 0$ are connected with the roots $\varepsilon_1, \varepsilon_2, \varepsilon_3, \varepsilon_4$ of $R(x) = 0$, subject to the condition $\varepsilon_1 + \varepsilon_2 + \varepsilon_3 + \varepsilon_4 = 0$, by the relations

$$e_1 = \tfrac{1}{2}(\varepsilon_1 + \varepsilon_2)^2, \quad e_2 = \tfrac{1}{2}(\varepsilon_1 + \varepsilon_3)^2, \quad e_3 = \tfrac{1}{2}(\varepsilon_1 + \varepsilon_4)^2 \tag{6.19}$$

In order to find the differential relations connecting s and t, rewrite the first two EULER equations in terms of $x_1 = p + i q$ and $x_2 = p - i q$. We then obtain the equations

$$2\frac{d x_1}{dt} = -i(r x_1 - c_0\gamma)$$

$$2\frac{d x_2}{dt} = i(r x_2 - c_0\gamma)$$

which can be rewritten in the form

$$\begin{aligned} 2\frac{d x_1}{\sqrt{R(x_1)}} &= -\frac{i(r x_1 - c_0\gamma)}{\sqrt{R(x_1)}} \\ 2\frac{d x_2}{\sqrt{R(x_2)}} &= \frac{i(r x_2 - c_0\gamma)}{\sqrt{R(x_2)}} \end{aligned} \tag{6.20}$$

Expressions $r x_1 - c_0\gamma$ and $r x_2 - c_0\gamma$ can be calculated from the first three algebraic integrals by virtue of which we have the equations

$$(r x_1 - c_0\gamma)^2 = R(x_1) + \xi_1(x_1 - x_2)^2$$

$$(r x_2 - c_0\gamma)^2 = R(x_2) + \xi_2(x_1 - x_2)^2$$

$$(r x_1 - c_0\gamma)(r x_2 - c_0\gamma) = R(x_1, x_2)$$

The last equation shows us that in the domain of real motions we always have $R(x_1, x_2) \geqq 0$.

From equations (6.14) and (6.16) it follows that

$$2\sqrt{\xi_1}\frac{\sqrt{R(x_2)}}{x_1 - x_2} = \sqrt{(s_1 - e_4)(s_2 - e_5)} + \sqrt{(s_1 - e_5)(s_2 - e_4)} \qquad (6.21)$$

$$2\sqrt{\xi_2}\frac{\sqrt{R(x_1)}}{x_1 - x_2} = \sqrt{(s_1 - e_4)(s_2 - e_5)} - \sqrt{(s_1 - e_5)(s_2 - e_4)} \qquad (6.22)$$

where

$$e_4 = 3l_1 - k, \qquad e_5 = 3l_1 + k$$

Finally equations (6.13) show us that

$$\frac{2R(x_1, x_2)}{(x_1 - x_2)^2} = w_1 + w_2 = s_1 + s_2 - 6l_1$$

$$\frac{2\sqrt{R(x_1)}\sqrt{R(x_2)}}{(x_1 - x_2)^2} = -(w_1 - w_2) = -(s_1 - s_2)$$

Hence

$$\frac{4R(x_1)R(x_2)}{(x_1 - x_2)^4} = (s_1 - s_2)^2 \qquad (6.23)$$

and

$$R(x_1) = \frac{(s_1 - s_2)^2 (x_1 - x_2)^4}{4R(x_2)}$$

Consequently

$$(r x_1 - c_0 \gamma)^2 = \frac{(x_1 - x_2)^4}{4R(x_2)}\left[(s_1 - s_2)^2 + \frac{4\xi_1 R(x_2)}{(x_1 - x_2)^2}\right]$$

or, by virtue of formulas (6.21) and (6.23), we have that

$$(r x_1 - c_0 \gamma)^2 = \frac{R(x_1)}{(s_1 - s_2)^2} \times$$
$$\times \left\{(s_1 - s_2)^2 + \left[\sqrt{(s_1 - e_4)(s_2 - e_5)} + \sqrt{(s_1 - e_5)(s_2 - e_4)}\right]^2\right\}$$

Further, it is easy to verify that the following equality

$$(s_1 - s_2)^2 + \left[\sqrt{(s_1 - e_4)(s_2 - e_5)} + \sqrt{(s_1 - e_5)(s_2 - e_4)}\right]^2$$
$$= \left[\sqrt{(s_1 - e_4)(s_1 - e_5)} + \sqrt{(s_2 - e_4)(s_2 - e_5)}\right]^2$$

holds. Using this equality and taking the square root of both sides of the last equation, we obtain that

$$\frac{r x_1 - c_0 \gamma}{\sqrt{R(x_1)}} = \frac{\sqrt{(s_1 - e_4)(s_1 - e_5)} + \sqrt{(s_2 - e_4)(s_2 - e_5)}}{s_1 - s_2} \qquad (6.24)$$

Similarly

$$\frac{r x_2 - c_0 \gamma}{\sqrt{R(x_2)}} = -\frac{\sqrt{(s_1 - e_4)(s_1 - e_5)} - \sqrt{(s_2 - e_4)(s_2 - e_5)}}{s_1 - s_2} \qquad (6.25)$$

The signs of the radicals on the right-hand sides of the last two formulas are completely determined once the sign of one of the radicals $\sqrt{R(x_1)}$, $\sqrt{R(x_2)}$ is chosen. Furthermore, from the same formulas it follows that s_1 must be larger than s_2 and both must lie between the limits

$$s_1 > e_5 > e_4 > s_2$$

in order that, for $x_1 \neq x_2$, r and γ be real besides p and q. That α and β are also real quantities for real p, q, r and γ follows from the first two algebraic integrals.

From equations (6.20), (6.24), (6.25) and (6.17) we easily deduce that

$$\frac{ds_1}{\sqrt{\varphi(s_1)}} = -i\frac{\sqrt{2(s_1 - e_4)(s_1 - e_5)}}{s_1 - s_2}\,dt$$

$$\frac{ds_2}{\sqrt{\varphi(s_2)}} = i\frac{\sqrt{2(s_2 - e_4)(s_2 - e_5)}}{s_1 - s_2}\,dt$$

whence

$$\frac{ds_1}{dt} = -i\frac{\sqrt{2}\sqrt{\Phi(s_1)}}{s_1 - s_2}$$
$$\frac{ds_2}{dt} = i\frac{\sqrt{2}\sqrt{\Phi(s_2)}}{s_1 - s_2} \tag{6.26}$$

or

$$\frac{ds_1}{\sqrt{\Phi(s_1)}} + \frac{ds_2}{\sqrt{\Phi(s_2)}} = 0$$
$$\frac{s_1\,ds_1}{\sqrt{\Phi(s_1)}} + \frac{s_2\,ds_2}{\sqrt{\Phi(s_2)}} = -i\sqrt{2}\,dt \tag{6.27}$$

where

$$\Phi(s) = (s - e_4)(s - e_5)\,\varphi(s) = (s - e_1)(s - e_2)(s - e_3)(s - e_4)(s - e_5)$$

is a polynomial of degree five with respect to s.

From JACOBI's theory of the last multiplier it follows that, eliminating dt from the equations of motion, the problem of integration is reduced by means of the first four integrals to the integration of a single differential equation in total differentials. Obviously the first equation of system (6.27) is the equation whose existence is guaranteed by JACOBI's theory.

By means of quadratures we obtain from (6.27) two integrals

$$F(s_1) + F(s_2) = C_1$$
$$F_1(s_1) + F_1(s_2) = -i\sqrt{2}(t - t_0) = -i\sqrt{2}t + C_2 \tag{6.28}$$

where C_1 and C_2 are constants of integration and

$$F(s) = \int\frac{ds}{\sqrt{\Phi(s)}}, \quad F_1(s) = \int\frac{s\,ds}{\sqrt{\Phi(s)}}$$

From equations (6.28) the variables s_1 and s_2 can be expressed in terms of t and the arbitrary constants C_1 and C_2, i.e.

$$\begin{aligned} s_1 &= f_1(t, C_1, C_2) \\ s_2 &= f_2(t, C_1, C_2) \end{aligned} \tag{6.28'}$$

Integrals of the type

$$R(x, \sqrt{P(x)})\, dx$$

where $R(x, w)$ is a rational function of x and $w = \sqrt{P(x)}$, $P(x)$ being a polynomial of degree higher than four, are called in general hyper-elliptic integrals. In the particular case when $P(x)$ is of degree five or six the corresponding integrals are also called ultra-elliptic integrals. Thus in the case of KOVALEVSKAYA the problem of motion of a heavy rigid body about a fixed point is reduced to the inversion of a system of two ultra-elliptic integrals determined by equations (6.27).

6.5. Expressions for p and q in terms of s_1 and s_2. KOVALEVSKAYA herself obtained these expressions directly from the differential relations (6.17) by application of various formulas from the theory of elliptic functions. However, following KÖTTER, the same results can be obtained by a proper transformation of the basic algebraic equation (6.15). In fact, considering the left-hand side of this equation as a function $f(w)$ of w, we can reduce it to the form

$$\begin{aligned} f(w) = \frac{1}{(x_1 - x_2)^2} \left[\sqrt{2(w + 3l_1)}\, (x_1 x_2 - w) - \frac{2c_0 l}{\sqrt{2(w + 3l_1)}} (x_1 + x_2) \right]^2 + \\ + \frac{1}{(x_1 - x_2)^2} [2(w + 3l_1)(w^2 + c_0^2 - k^2) - 4c_0^2 l^2] \left[\frac{(x_1 + x_2)^2}{2(w + 3l_1)} - 1 \right] \end{aligned}$$

On the other hand, since $f(w)$ is a quadratic polynomial in w, by application of TAYLOR's formula $f(w)$ can be represented in the form

$$f(w) = f(u) + (w - u) f'(u) + (w - u)^2$$

where u is an arbitrary quantity.

Replacing w by $s - 3l_1$ and u by $z - 3l_1$, z being arbitrary, equation (6.15) finally assumes the form

$$\begin{aligned} \left[\sqrt{2z}\, \frac{x_1 x_2 - z + 3l_1}{x_1 - x_2} - \frac{2c_0 l}{\sqrt{2z}} \frac{x_1 + x_2}{x_1 - x_2} \right]^2 + \frac{2\varphi(z)}{(x_1 - x_2)^2} \left[\frac{(x_1 + x_2)^2}{2z} - 1 \right] - \\ - 2(s - z) \frac{R(x_1, x_2) - (z - 3l_1)(x_1 - x_2)^2}{(x_1 - x_2)^2} + \\ + (s - z)^2 = 0 \end{aligned}$$

Since the roots of the last equation are $s_1 - z$ and $s_2 - z$, then according to a well known property of the roots of a quadratic equation we have for the product $(s_1 - z)(s_2 - z)$ the expression

$$(s_1 - z)(s_2 - z) = \left[\sqrt{2z}\,\frac{x_1 x_2 - z + 3l_1}{x_1 - x_2} - \frac{2c_0 l}{\sqrt{2z}}\,\frac{x_1 + x_2}{x_1 - x_2}\right]^2 + \frac{2\varphi(z)}{(x_1 - x_2)^2}\left[\frac{(x_1 + x_2)^2}{2z} - 1\right]$$

If we assign to z the value of any of the roots e_α $(\alpha = 1, 2, 3)$ of the third degree polynomial equation $\varphi(z) = 0$, then the last equation reduces to

$$\sqrt{2e_\alpha}\,\frac{x_1 x_2 - e_\alpha + 3l_1}{x_1 - x_2} - \frac{2c_0 l}{\sqrt{2e_\alpha}}\,\frac{x_1 + x_2}{x_1 - x_2} = \sqrt{(s_1 - e_\alpha)(s_2 - e_\alpha)} \qquad (\alpha = 1, 2, 3) \qquad (6.29)$$

These formulas always exist except in the case of multiple roots of the equation $\varphi(z) = 0$ when one of the above formulas must be rewritten in another form and the limit process applied.

In order to express p and q in terms of s_1 and s_2, put

$$\left.\begin{aligned} X &= \frac{x_1 x_2 + 3l_1}{x_1 - x_2} = \frac{p^2 + q^2 + 3l_1}{2qi} \\ Y &= \frac{1}{x_1 - x_2} = \frac{1}{2qi}, \qquad Z = \frac{x_1 + x_2}{x_1 - x_2} = \frac{p}{qi} \\ P_\alpha &= \sqrt{(s_1 - e_\alpha)(s_2 - e_\alpha)} \qquad (\alpha = 1, 2, 3) \end{aligned}\right\} \qquad (6.30)$$

and rewrite equation (6.29) in the form

$$X - e_\alpha Y - \frac{c_0 l}{e_\alpha} Z = \frac{1}{\sqrt{2e_\alpha}} P_\alpha \qquad (\alpha = 1, 2, 3) \qquad (6.31)$$

Eliminating X and Z from the last system of equations and observing that equation $\varphi(s) = 0$ gives the expression $e_1 e_2 e_3 = 2c_0^2 l^2$, we obtain the equation

$$-Y(e_1 - e_2)(e_2 - e_3)(e_3 - e_1) = \frac{1}{\sqrt{2}}\left[P_1\sqrt{e_1}(e_3 - e_2) + P_2\sqrt{e_2}(e_1 - e_3) + P_3\sqrt{e_3}(e_2 - e_1)\right]$$

On the other hand, if we calculate the derivative $\varphi'(s)$ of $\varphi(s) = (s - e_1)(s - e_2)(s - e_3)$ and the expressions $\varphi'(e_\alpha) \neq 0$, the last equation assumes the form

$$-Y = \frac{1}{\sqrt{2}} \sum_{\alpha=1}^{3} \frac{\sqrt{e_\alpha}}{\varphi'(e_\alpha)} P_\alpha$$

Substituting this expression of Y into the second equation of system (6.30) and solving it for q, we obtain

$$q = \frac{i}{\sqrt{2} \sum_{\alpha=1}^{3} \frac{\sqrt{e_\alpha}}{\varphi'(e_\alpha)} P_\alpha} \qquad (6.32)$$

In a similar way, eliminating X and Y from system (6.31), we obtain for Z the expression

$$Z = \sum \frac{\sqrt{e_\beta}\sqrt{e_\gamma}}{\varphi'(e_\alpha)} P_\alpha \qquad (\alpha, \beta, \gamma = 1, 2, 3;\ \alpha \neq \beta \neq \gamma)$$

The third equation of system (6.30) finally gives us p in the form

$$p = -\frac{\sum \frac{\sqrt{e_\beta}\sqrt{e_\gamma}}{\varphi'(e_\alpha)} P_\alpha}{\sqrt{2}\sum \frac{\sqrt{e_\alpha}}{\varphi'(e_\alpha)} P_\alpha} \qquad (\alpha, \beta, \gamma = 1, 2, 3;\ \alpha \neq \beta \neq \gamma) \tag{6.33}$$

Hence both components p and q are expressed in terms of

$$P_\alpha = \sqrt{(s_1 - e_\alpha)(s_2 - e_\alpha)} = \sqrt{s_1 s_2 - e_\alpha(s_1 + s_2) + e_\alpha^2} \qquad (\alpha = 1, 2, 3)$$

which are symmetric functions of s_1 and s_2, and consequently can be given in terms of JACOBI's ϑ-functions.

6.6. Expressions for r, α, β and γ in terms of s_1 and s_2. The calculations for r, α, β and γ in terms of s_1 and s_2 are somewhat more complicated than those for p and q. The quantity r can be calculated from the first EULER equation

$$2\frac{dp}{dt} = q\,r \tag{6.34}$$

by finding first dp/dt from equation (6.33). On the other hand, taking the logarithmic derivative of P_α we obtain

$$\frac{dP_\alpha}{dt} = \frac{1}{2} P_\alpha \left(\frac{1}{s_1 - e_\alpha}\frac{ds_1}{dt} + \frac{1}{s_2 - e_\alpha}\frac{ds_2}{dt}\right) \tag{6.35}$$

By virtue of formula (6.35) terms containing products of the form P_α^2 cancel out in the numerator of $-\frac{dp}{dt}$, and there remain only terms containing the products $P_\alpha P_\beta$. Finally the numerator of $-\frac{dp}{dt}$ reduces to the expression

$$\frac{1}{\sqrt{2}}\sum \frac{\sqrt{e_\gamma}}{\varphi'(e_\gamma)} P_\alpha P_\beta \left[\frac{1}{(s_1 - e_\alpha)(s_1 - e_\beta)}\frac{ds_1}{dt} + \frac{1}{(s_2 - e_\alpha)(s_2 - e_\beta)}\frac{ds_2}{dt}\right]$$

Substituting for ds_1/dt and ds_2/dt their expressions as given by formulas (6.26), we obtain the following final expression for the numerator of dp/dt:

$$\frac{1}{\sqrt{2}}\sum \frac{\sqrt{e_\gamma}}{\varphi'(e_\gamma)} \frac{P_\alpha P_\beta\, i\sqrt{2}}{s_1 - s_2}\left[\frac{\sqrt{\Phi(s_1)}}{(s_1 - e_\alpha)(s_1 - e_\beta)} - \frac{\sqrt{\Phi(s_2)}}{(s_2 - e_\alpha)(s_2 - e_\beta)}\right]$$

Introduce the notation

$$P_{\beta\gamma} = \frac{P_\beta P_\gamma}{s_1 - s_2}\left[\frac{\sqrt{\Phi(s_1)}}{(s_1 - e_\beta)(s_1 - e_\gamma)} - \frac{\sqrt{\Phi(s_2)}}{(s_2 - e_\beta)(s_2 - e_\gamma)}\right] \tag{6.36}$$

Obviously, $P_{\beta\gamma} = P_{\gamma\beta}$, and since α, β, γ can assume the values 1, 2, 3 and $\alpha \neq \beta \neq \gamma$, there are altogether three similar expressions $P_{\beta\gamma}$, $P_{\gamma\alpha}$, $P_{\alpha\beta}$.

Finally we obtain

$$2\frac{dp}{dt} = \frac{i\,\Sigma \frac{\sqrt{e_\alpha}}{\varphi'(e_\alpha)} P_{\beta\gamma}}{\left(\Sigma \frac{\sqrt{e_\alpha}}{\varphi'(e_\alpha)} P_\alpha\right)^2}$$

and hence

$$r = \frac{2}{q}\,\frac{dp}{dt} = \sqrt{2}\,\frac{\Sigma \frac{\sqrt{e_\alpha}}{\varphi'(e_\alpha)} P_{\beta\gamma}}{\Sigma \frac{\sqrt{e_\alpha}}{\varphi'(e_\alpha)} P_\alpha} \tag{6.37}$$

The easiest way to calculate γ is to use system (6.31) and calculate X first. Multiplying the equations of system (6.31) by e_1, e_2, e_3 respectively, and eliminating first Z and then Y, we obtain for X the expression

$$X = -\frac{1}{\sqrt{2}}\,\Sigma \frac{\sqrt{e_\alpha}}{\varphi'(e_\alpha)}(e_\beta + e_\gamma)\,P_\alpha$$

Since

$$X = (p^2 + q^2 + 3l_1)\,Y$$

then

$$p^2 + q^2 + 3l_1 = \frac{X}{Y} = \frac{\Sigma \frac{\sqrt{e_\alpha}}{\varphi'(e_\alpha)}(e_\beta + e_\gamma)\,P_\alpha}{\Sigma \frac{\sqrt{e_\alpha}}{\varphi'(e_\alpha)} P_\alpha} \tag{6.38}$$

Multiplying the first and second EULER equations by p and q respectively and adding we obtain that

$$q\,c_0\,\gamma = \frac{d}{dt}(p^2 + q^2)$$

If we use formula (6.38) in the right-hand side of the last equation, terms of the numerator containing the factors P_α^2, P_β^2, P_γ^2 all cancel out, and there remain only terms which contain the factors $P_\alpha P_\beta$, $P_\beta P_\gamma$ and $P_\gamma P_\alpha$. In this way we obtain that

$$q\,c_0\,\gamma = -\frac{i}{\sqrt{2}}\,\frac{\Sigma \frac{\sqrt{e_\alpha}\sqrt{e_\beta}}{\varphi'(e_\gamma)} P_{\alpha\beta}}{\left(\Sigma \frac{\sqrt{e_\alpha}}{\varphi'(e_\alpha)} P_\alpha\right)^2}$$

or, after division of both sides by $q\,c_0$,

$$\gamma = -\frac{1}{c_0}\,\frac{\sum \frac{\sqrt{e_\alpha}\sqrt{e_\beta}}{\varphi'(e_\gamma)} P_{\alpha\beta}}{\sum \frac{\sqrt{e_\alpha}}{\varphi'(e_\alpha)} P_\alpha} \tag{6.39}$$

Finally, when $p^2 + q^2$ and r are known, the function α can be determined from the energy integral

$$2(p^2 + q^2) + r^2 = -2c_0\,\alpha + 6l_1$$

and β from the integral

$$2(\alpha p + \beta q) + \gamma r = 2l$$

or from the third EULER equation

$$\frac{dr}{dt} = -c_0\,\beta$$

KOVALEVSKAYA and KÖTTER have given explicit expressions for α and β in terms of P_α and $P_{\alpha\beta}$.

Since all six functions $p, q, r; \alpha, \beta, \gamma$ can be expressed in terms of the variables s_1 and s_2 and the constants l, l_1 and k, we conclude from formulas (6.28′) that these six quantities are functions of the time t and the five arbitrary constants l, l_1, k, C_1 and C_2.

The EULER angles Φ and Θ can be calculated from formulas (2.11). The angle ψ, given by formula (2.12), requires the completion of a quadrature which involves the sixth arbitrary constant. Thus the EULER angles Φ, ψ, Θ will be expressed in terms of the time t and six arbitrary constants. In other words, the general solution of the equations of motion of a heavy rigid body about a fixed point is found for the case of KOVALEVSKAYA.

6.7. Stability of a particular solution. The EULER and the POISSON equations admit a particular solution

$$p = \omega = \text{const.}, \quad q = r = 0; \quad \alpha = 1, \quad \beta = \gamma = 0 \tag{6.40}$$

corresponding to a uniform rotation of the body about the x-axis. Following V. V. RUMYANCEV [1] let us consider this motion as the unperturbed one and let us investigate its stability relative to the variables $p, q, r; \alpha, \beta, \gamma$.

Let the perturbed motion be

$$\begin{aligned} p &= \omega + \xi_1, & q &= \xi_2, & r &= \xi_3 \\ \alpha &= 1 + \zeta_1, & \beta &= \zeta_2, & \gamma &= \zeta_3 \end{aligned} \tag{6.41}$$

Then the equations for the perturbations are

$$\left.\begin{aligned} 2\frac{d\xi_1}{dt} &= \xi_2\,\xi_3 & \frac{d\zeta_1}{dt} &= \xi_3\,\zeta_2 - \xi_2\,\zeta_3 \\ 2\frac{d\xi_2}{dt} &= -\,\xi_3(\omega+\xi_1) + c_0\,\zeta_3 & \frac{d\zeta_2}{dt} &= (\omega+\xi_1)\,\zeta_3 - \xi_3(1+\zeta_1) \\ \frac{d\xi_3}{dt} &= -\,c_0\,\zeta_2 & \frac{d\zeta_3}{dt} &= \xi_2(1+\zeta_1) - (\omega+\xi_1)\,\zeta_2 \end{aligned}\right\} \quad (6.42)$$

They admit the first integrals

$$\left.\begin{aligned} V_1 &= 2(\xi_1^2 + \xi_2^2 + 2\omega\,\xi_1) + \xi_3^2 + 2c_0\,\zeta_1 \\ V_2 &= 2(\xi_1\,\zeta_1 + \xi_2\,\zeta_2 + \xi_1 + \omega\,\zeta_1) + \xi_3\,\zeta_3 \\ V_3 &= \zeta_1^2 + \zeta_2^2 + \zeta_3^2 + 2\zeta_1 \end{aligned}\right\} \quad (6.43)$$

Following a method due to N. G. ČETAEV [1, 4, p. 430—431] let us construct a LYAPUNOV V-function in the form of a linear bundle of integrals (6.43) of the equations for the perturbations. Consider the function

$$\begin{aligned} V = V_1 - 2\omega\,V_2 + (2\omega^2 - c_0)\,V_3 &= 2\,\xi_1^2 - 4\omega\,\xi_1\,\zeta_1 + (2\omega^2 - c_0)\,\zeta_1^2 + \\ &+ 2\,\xi_2^2 - 4\omega\,\xi_2\,\zeta_2 + (2\omega^2 - c_0)\,\zeta_2^2 + \xi_3^2 - 2\omega\,\xi_3\,\zeta_3 + (2\omega^2 - c_0)\,\zeta_3^2 \end{aligned} \quad (6.44)$$

In order that the first two quadratic forms in ξ_1, ζ_1 and ξ_2, ζ_2 on the right be positive definite, it is necessary and sufficient that the inequality

$$\begin{vmatrix} 2 & -2\omega \\ -2\omega & 2\omega^2 - c_0 \end{vmatrix} = -2c_0 > 0 \quad (6.45)$$

holds. The third quadratic form in terms of ξ_3, ζ_3 in V will be positive definite provided that

$$\begin{vmatrix} 1 & -\omega \\ -\omega & 2\omega^2 - c_0 \end{vmatrix} = \omega^2 - c_0 > 0 \quad (6.46)$$

Inequality (6.46) holds provided that inequality (6.45) is satisfied. Thus condition (6.45) is sufficient for stability of motion (6.40) relative to the variables p, q, r; α, β, γ.

Condition (6.45) is also necessary for the stability of motion (6.40) (see R. GRAMMEL [3]). In fact, the characteristic equation corresponding to the system of equations for the perturbations (6.42) is

$$2\lambda^2(2\lambda^2 - c_0)(\lambda^2 - c_0 + \omega^2) = 0 \quad (6.47)$$

It is obvious that for $c_0 > 0$ equation (6.47) will have at least one positive root. Therefore by a well known theorem of LYAPUNOV the unperturbed motion (6.40) is unstable.

Hence the rotation about the x-axis is stable if and only if G lies on the negative x-axis.

For other stationary motions and their stability in the KOVALEVSKAYA case see T. LEVI-CIVITA [2], and Clelia SILVESTRI [1, 2, 3].

6.8. Concluding remarks concerning the Euler and Poisson equations. The solution of the equations of motion in terms of elliptic functions of the time t consists of two steps: (i) integration of the EULER equations, and (ii) calculation of the nine direction cosines of the body-fixed axes relative to space-fixed axes. F. LINDEMANN [1], assuming that the body is fixed but the whole space is in a state of motion, reduced the rotation problem to a problem of hydrodynamics considered by W. VOIGT and O. VENSKE. While LINDEMANN used the JACOBI elliptic functions, J. HAUG [1] achieved a further simplification by introduction of WEIERSTRASS' $\wp$-function.

Concerning a symmetric way of solving the EULER and POISSON equations of motion, after the components p, q, r of the instantaneous velocity of rotation $\bar{\omega}$ have been found, see A. MAYER [1].

H. BRUNS [1] gave a transformation of the differential equations of motion which clearly showed the difficulties involved in the problem of integrating these equations.

T. LEVI-CIVITA [1] and H. LIEBMANN [1] applied LIE's group theory to the investigation of the differential equations of motion, and, in particular, to the problem of finding new integrable cases.

J. DRACH [1] eliminated the coordinates of the mass center by means of the three classical integrals of the problem and studied the rationality group of the remaining differential equations for p, q, r.

F. KLEIN [1] showed that the rotations about a fixed point are represented by those unimodular linear substitutions

$$\zeta = \frac{\alpha Z + \beta}{\gamma Z + \delta}$$

for which α and δ on the one hand, and β and $-\gamma$ on the other hand, are imaginary conjugate quantities. In the case of a heavy rigid body the coefficients α, β, γ, δ are elliptic functions of the second kind which contain in the numerator as well as in the denominator only a single Theta-function.

E. PADOVA [2, 4], B. PALADINI [1], Miss N. LAGERBORG [1], O. TEDONE [1] and G. URZI [1] have been concerned with particular cases of the motion integrable by quadratures or subject to forces with particular potential functions. F. DE BRUN [2, 3] studied the case where the potential function is a homogeneous polynomial of the second degree and higher in terms of the rectangular coordinates.

G. KOENIGS [1], G. KOBB [3] and E. METTLER [1] have studied the existence of periodic and asymptotic solutions for the equations of motion of a rigid body about a fixed point. METTLER uses not the EULERian equations, in spite of their simple and symmetric form, but the HAMILTONian equations which have the advantage that the cyclic coordinate ψ can be readily eliminated and suitable coordinates can be introduced by means of contact transformations. He arrived at the results resembling those relating to periodic and asymptotic solutions in the problem of three bodies.

A. RAUBER [1] applied POINCARÉ's theorem on the dependence of solutions of a system of differential equations upon the parameters occurring in the equations. This method has the advantage that the series which represent the solutions are convergent for all finite values of the time provided that the parameters are sufficiently small. The coefficients of the various powers of the parameters are functions of the time. In RAUBER's case the parameters are the coordinates x_0, y_0, z_0 of the mass center of the body and the coefficients are determined by a recurrence system of linear differential equations. RAUBER showed that the region of variability for x_0, y_0, z_0 can be gradually extended by analytical continuation.

§ 7. Existence of single-valued solutions

7.1. Introduction. After the cases considered by EULER (1758) and POINSOT (1834), LAGRANGE (1788) and POISSON (1811) there was no progress for 100 years towards the solution of the problem of motion of a heavy rigid body about a fixed point. Obviously, a new approach to the problem was needed.

The success of KOVALEVSKAYA (1888) lies in her new and more general reformulation of the problem in terms of the concepts of the theory of analytic functions of a complex variable. In both cases mentioned above the general solution of the problem is expressible in terms of elliptic functions of the time t. The elliptic functions are single-valued analytic functions for all finite values of t, except for certain points in the complex t-plane at which these functions have poles of the first order. These results suggested to KOVALEVSKAYA the problem of finding all cases when the general solution, containing five arbitrary constants, is single-valued and does not possess any singularities other than poles for all finite values of t, t being a complex variable. KOVALEVSKAYA sought the solutions in the form of power series

$$p = \frac{1}{t^{n_1}} \sum_{n=0}^{\infty} p_n t^n, \quad q = \frac{1}{t^{n_2}} \sum_{n=0}^{\infty} q_n t^n, \quad r = \frac{1}{t^{n_3}} \sum_{n=0}^{\infty} r_n t^n$$

$$\alpha = \frac{1}{t^{m_1}} \sum_{n=0}^{\infty} a_n t^n, \quad \beta = \frac{1}{t^{m_2}} \sum_{n=0}^{\infty} b_n t^n, \quad \gamma = \frac{1}{t^{m_3}} \sum_{n=0}^{\infty} c_n t^n$$

where n_i, m_i $(i = 1, 2, 3)$ are positive integers.

In general, in order that the EULER and POISSON equations be integrable by series of the above form which contain five arbitrary constants, it is necessary that the coefficients of these series satisfy certain conditions. One such condition gave a new case of integrability in addition to those already known. KOVALEVSKAYA took for the constants n_i, m_i the values $n_i = 1$, $m_i = 2$ $(i = 1, 2, 3)$ and left the question open whether or not this system of values is unique. Thus she claimed to have shown that only four cases are possible, namely the case of kinetic symmetry when $A = B = C$, the case of EULER and POINSOT $(x_0 = y_0 = z_0 = 0)$, the case of LAGRANGE and POISSON $(A = B,\ x_0 = y_0 = 0)$, and the case discovered by her $(A = B = 2C,\ z_0 = 0)$. However, A. A. MARKOV, Sr. (see A. M. LYAPUNOV [1]) objected to a certain conclusion of KOVALEVSKAYA which was based on an assumption, the legitimacy of which was somewhat doubtful. By methods completely different from those of KOVALEVSKAYA, LYAPUNOV [1] and G. G. APPEL'ROT [1, 3, 4, 5] were able to circumvent the above mentioned difficulty in KOVALEVSKAYA's paper. They gave rigorous proofs to the effect that the above mentioned cases are the only ones when the general solution is single-valued for any initial values of p, q, r; α, β, γ. In addition, LYAPUNOV extended the result of KOVALEVSKAYA by showing that in all other cases the functions p, q, r; α, β, γ may have singularities other than poles, and that for properly chosen initial values of p, q, r; α, β, γ they certainly will be many-valued functions of the time t. LYAPUNOV assumes that the principal moments of inertia A, B, C, calculated at the fixed point, and the coordinates of the center of mass x_0, y_0, z_0 are real and that A, B, C are all different from zero.

APPEL'ROT showed that if $A \neq B \neq C$, neither the general solution nor particular solutions exist which have for p, q, r poles of order higher than one and for α, β, γ poles of order higher than two. If, however, $A = B \neq C$, $y_0 = 0$, $z_0 \neq 0$, there can exist particular solutions, having for p and q poles of order three. APPEL'ROT and LYAPUNOV also pointed out a particular case, the so-called HESS case (1890) (also called the case of loxodromic pendulum), overlooked by KOVALEVSKAYA, when all six functions p, q, r; α, β, γ have poles of the first order.

The theorems of KOVALEVSKAYA and LYAPUNOV which are quoted in Sec. 7.3, complete the problem of finding all cases for which single-valued solutions exist for arbitrary initial conditions (see P. YA. POLUBARINOVA-KOČINA [1]).

7.2. Existence of algebraic integrals. In all the four cases mentioned above the fourth integral is an algebraic function of p, q, r; α, β, γ. This result suggested another problem of a general nature, namely that of finding all the cases when the problem of motion of a heavy rigid body about a fixed point possesses a fourth integral which is algebraic in $p, q, r; \alpha, \beta, \gamma$. This problem was investigated by H. POINCARÉ [2], É. HUSSON [1, 3], P. BURGATTI [2], F. QUATELA [1], and others. About the same time that KOVALEVSKAYA was working on the problem of motion of a heavy rigid body about a fixed point, H. BRUNS [1] was concerned with the problem of the existence of algebraic integrals in the problem of $n(\geqq 3)$-bodies. He proved that the ten classical integrals are the only independent algebraic integrals of the n-body problem when rectangular coordinates are used. Consequently all remaining integrals should be transcendental. POINCARÉ [1] was concerned with a more general problem than BRUNS. He investigated the problem of the existence of single-valued solutions of the canonical equations of dynamics containing a small parameter μ. POINCARÉ established a theorem on the non-existence of integrals (transcendental and algebraic) which are single-valued (uniform) with respect to the KEPLER variables for all sufficiently small values of the parameter μ which depends on the masses of the bodies. Here the property of uniformity of the integrals refers to a certain neighborhood of the trajectories which have the same osculating ellipse. POINCARÉ's theorem is a local theorem in the sense that it does not exclude the existence of integrals which are uniform in domains of a different nature (see T. LEVI-CIVITA [3]). T. M. CHERRY [1] showed that the theorem of POINCARÉ no longer holds in its original formulation if the restriction that the integral is to be developable in powers of μ is dropped. In the problem of motion of a heavy rigid body about a fixed point the parameter μ is the product of the weight of the body and the distance between the fixed point O and the center of mass G of the body.

POINCARÉ showed that if μ is small and the initial conditions are arbitrary, then, in order that a new single-valued first integral (which consequently is algebraic) exists, it is necessary for the ellipsoid of inertia of the body at the fixed point O to be an ellipsoid of revolution. HUSSON arrived at the same conclusion as POINCARÉ, but for arbitrary values of μ. Further, HUSSON [2, 4] proved that if the ellipsoid of inertia relative to the fixed point is not an ellipsoid of revolution, then for arbitrary initial conditions there cannot exist a new algebraic integral of the problem, except in the case of EULER and POINSOT. In the same paper HUSSON proved that if the ellipsoid of inertia at the fixed point is an ellipsoid of revolution, a fourth algebraic integral exists only in the cases of kinetic symmetry, LAGRANGE and POISSON, and that of KOVALEVSKAYA. BURGATTI gave an elementary proof of this theorem in the paper mentioned above. BURGATTI's proof is based on a theorem of R. LIOUVILLE [1] which states that if in the case $A = B$ there exists a first algebraic integral, then there exists also a first integral which has the form of a homogeneous polynomial in $p, q, r; \alpha, \beta, \gamma$, the weight of p, q, r being one and that of α, β, γ being two. Consequently, in the case under consideration every algebraic first integral is an algebraic combination of homogeneous polynomials.

Hence we have the result that a fourth algebraic first integral of the problem of motion of a heavy rigid body about a fixed point exists only in the cases in which the general solution for $p, q, r; \alpha, \beta, \gamma$ is single-valued over the whole t-plane.

If, however, we are concerned with the problem of finding a particular fourth integral, or with the problem of finding a new integral subject to the condition that the constants entering into the known integrals have prescribed values, then in a number of particular cases these integrals are algebraic.

7.3. Lyapunov's theorem. The results obtained by KOVALEVSKAYA can be formulated in the following theorem.

Theorem of Kovalevskaya. In general, with the exception of the case of kinetic symmetry and those of EULER and POINSOT, LAGRANGE and POISSON, and KOVALEVSKAYA, the EULER and POISSON equations do not possess single-valued solutions depending on five arbitrary constants and having no other singularities than poles in the whole t-plane.

LYAPUNOV proved the following more general theorem.

Theorem of Lyapunov. If the constants A, B, C; x_0, y_0, z_0 are all real and A, B, C are all different from zero, the case of kinetic symmetry and those of EULER and POINSOT, LAGRANGE and POISSON, and KOVALEVSKAYA are the only ones in which the parameters $p, q, r; \alpha, \beta, \gamma$ are single-valued functions of the time t for arbitrary initial values of these parameters.

LYAPUNOV's generalization consists in the following. A function can be single-valued and possess essential singularities besides poles. KOVALEVSKAYA was not concerned with solutions of this type. LYAPUNOV showed that such solutions cannot be single-valued over the whole t-plane (except in the above four cases) since by a proper choice of the initial values of the parameters $p, q, r; \alpha, \beta, \gamma$ they can be made multiple-valued functions of the time t.

According to the nature of LYAPUNOV's method it is necessary to have some particular solution of the EULER and POISSON equations with singularities at least for some of the solution functions p, q, r; α, β, γ. Such solutions, in fact, are known and always exist except in the case where $x_0 = y_0 = z_0 = 0$ and the constants A, B, C do not all differ from each other. The simplest solution of this type is

$$\begin{aligned} p &= \frac{a}{t}, \quad q = \frac{b}{t}, \quad r = \frac{c}{t} \\ \alpha &= \frac{f}{t^2}, \quad \beta = \frac{g}{t^2}, \quad \gamma = \frac{h}{t^2} \end{aligned} \tag{7.1}$$

where a, b, c, f, g, h are constants among which some are different from zero.

Substituting the solution sought (7.1) in the EULER and POISSON equations rewritten in the form

$$\left.\begin{aligned} A\frac{dp}{dt} &= A_1 q r + \beta z_0 - \gamma y_0, & \frac{d\alpha}{dt} &= \beta r - \gamma q \\ B\frac{dq}{dt} &= B_1 r p + \gamma x_0 - \alpha z_0, & \frac{d\beta}{dt} &= \gamma p - \alpha r \\ C\frac{dr}{dt} &= C_1 p q + \alpha y_0 - \beta x_0, & \frac{d\gamma}{dt} &= \alpha q - \beta p \end{aligned}\right\} \tag{7.2}$$

where

$$A_1 = B - C, \quad B_1 = C - A, \quad C_1 = A - B$$

we obtain the following equations for the determination of the coefficients a, b, c, f, g, h:

$$\left.\begin{aligned} -A\,a &= A_1\,b\,c + g\,z_0 - h\,y_0, & -2f &= g\,c - h\,b \\ -B\,b &= B_1\,c\,a + h\,x_0 - f\,z_0, & -2g &= h\,a - f\,c \\ -C\,c &= C_1\,a\,b + f\,y_0 - g\,x_0, & -2h &= f\,b - g\,a \end{aligned}\right\} \tag{7.3}$$

Lyapunov is not concerned with the problem of finding all possible solutions of system (7.3). Instead he indicates only two solutions of this system. The first solution holds when the constants A, B, C are all different, and the second solution exists when among the constants x_0, y_0, z_0 at least one is zero and at least one is different from zero.

As the first solution of system (7.3) Lyapunov takes the solution overlooked by Kovalevskaya, namely,

$$\begin{aligned} a &= -B'C', & b &= -C'A', & c &= -A'B' \\ f &= 0, & g &= 0, & h &= 0 \end{aligned} \tag{7.4}$$

where

$$A' = \sqrt{A/A_1}, \quad B' = \sqrt{B/B_1}, \quad C' = \sqrt{C/C_1}$$

and the signs of the radicals are arbitrary.

Assuming that $y_0 = 0$, Lyapunov takes the second solution of system (7.3) in the form

$$\begin{aligned} &a = 0, \quad b = 2i, \quad c = 0 \\ &f = \frac{2B\,i}{z_0 + i\,x_0}, \quad g = 0, \quad h = \frac{2B}{z_0 + i\,x_0} \end{aligned} \tag{7.5}$$

where x_0 and z_0 are not both equal to zero.

For $x_0 = 0$ or $z_0 = 0$ there exist two similar solutions of system (7.3) which have the form (7.5).

In the case where the constants A, B, C differ from each other we can use solution (7.4). In the case where two of them are equal, one of the constants x_0, y_0, z_0 can always be assumed to be equal to zero (x_0 or y_0 when $A = B$, y_0 or z_0 when $B = C$, z_0 or x_0 when $C = A$). In the latter case we can use solution (7.5) or a similar solution.

7.31. Arbitrary initial values

If t_0 denotes the initial value of t, then by virtue of system (7.1) there correspond to particular solutions (7.4) and (7.5) definite initial values for $p, q, r; \alpha, \beta, \gamma$. Since Lyapunov is concerned with the problem of finding necessary conditions for the existence of single-valued solutions corresponding to arbitrary initial values p_0, q_0, r_0; α_0, β_0, γ_0, he starts with a certain set of such initial values. Then the general solution of system (7.2) contains these initial values as parameters and varies continuously with them provided that the latter vary continuously. If, for example, one of these parameters, say p_0, varies by a small quantity, all the functions p, q, r; α, β, γ vary, and as a consequence of this fact p, q, r; α, β, γ must be replaced by $p + \delta p$, $q + \delta q$, $r + \delta r$; $\alpha + \delta\alpha$, $\beta + \delta\beta$, $\gamma + \delta\gamma$ respectively. Since the latter quantities are solutions of system (7.2), they can be substituted into this system instead of the corresponding quantities p, q, r; α, β, γ. Subtracting from the equations so obtained the corresponding equations (7.2), we obtain the following system of equations to be satisfied by the variations δp, δq, δr; $\delta\alpha$, $\delta\beta$, $\delta\gamma$:

$$\left.\begin{aligned}
A\frac{d\,\delta p}{dt} &= A_1(r\,\delta q + q\,\delta r) + z_0\,\delta\beta - y_0\,\delta\gamma\\
B\frac{d\,\delta q}{dt} &= B_1(p\,\delta r + r\,\delta p) + x_0\,\delta\gamma - z_0\,\delta\alpha\\
C\frac{d\,\delta r}{dt} &= C_1(q\,\delta p + p\,\delta q) + y_0\,\delta\alpha - x_0\,\delta\beta\\
\frac{d\,\delta\alpha}{dt} &= r\,\delta\beta + \beta\,\delta r - q\,\delta\gamma - \gamma\,\delta q\\
\frac{d\,\delta\beta}{dt} &= p\,\delta\gamma + \gamma\,\delta p - r\,\delta\alpha - \alpha\,\delta r\\
\frac{d\,\delta\gamma}{dt} &= q\,\delta\alpha + \alpha\,\delta q - p\,\delta\beta - \beta\,\delta p
\end{aligned}\right\}\qquad(7.6)$$

Substituting in the last system for p, q, r; α, β, γ their expressions (7.1), we obtain the system

$$\left.\begin{aligned}
A\,t\frac{d\,\delta p}{dt} &= A_1(c\,\delta q + b\,\delta r) + z_0\,t\,\delta\beta - y_0\,t\,\delta\gamma\\
B\,t\frac{d\,\delta q}{dt} &= B_1(a\,\delta r + c\,\delta p) + x_0\,t\,\delta\gamma - z_0\,t\,\delta\alpha\\
C\,t\frac{d\,\delta r}{dt} &= C_1(b\,\delta p + a\,\delta q) + y_0\,t\,\delta\alpha - x_0\,t\,\delta\beta\\
t^2\frac{d\,\delta\alpha}{dt} &= g\,\delta r - h\,\delta q + c\,t\,\delta\beta - b\,t\,\delta\gamma\\
t^2\frac{d\,\delta\beta}{dt} &= h\,\delta p - f\,\delta r + a\,t\,\delta\gamma - c\,t\,\delta\alpha\\
t^2\frac{d\,\delta\gamma}{dt} &= f\,\delta q - g\,\delta p + b\,t\,\delta\alpha - a\,t\,\delta\beta
\end{aligned}\right\}\qquad(7.7)$$

This system possesses solutions of the form

$$\delta p = K_1 t^k, \qquad \delta q = K_2 t^k, \qquad \delta r = K_3 t^k,$$
$$\delta\alpha = K_4 t^{k-1}, \qquad \delta\beta = K_5 t^{k-1}, \qquad \delta\gamma = K_6 t^{k-1}$$

where k is a root of the following algebraic equation of degree six

$$\begin{vmatrix} A\,k & -A_1\,c & -A_1\,b & 0 & -z_0 & y_0 \\ -B_1\,c & B\,k & -B_1\,a & z_0 & 0 & -x_0 \\ -C_1\,b & -C_1\,a & C\,k & -y_0 & x_0 & 0 \\ 0 & h & -g & k-1 & -c & b \\ -h & 0 & f & c & k-1 & -a \\ g & -f & 0 & -b & a & k-1 \end{vmatrix} = 0 \qquad (7.8)$$

The discussion of the roots of this equation leads to the conditions to be satisfied in order that the solution $p, q, r; \alpha, \beta, \gamma$ of system (7.2) be single-valued over the whole t-plane for arbitrary initial values of these quantities. For this purpose also the variations $\delta p, \delta q, \delta r; \delta\alpha, \delta\beta, \delta\gamma$ of the above quantities must be single-valued functions of t. The necessary and sufficient conditions to be satisfied, in order that the last requirement be fulfilled, are that (i) all the roots of equation (7.8) are real integers (positive, negative or zero) and (ii) each multiple root of order m nullifies all the minors of the determinant on the left-hand side of equation (7.8) obtained by crossing out up to $m-1$ rows and columns.

Case (*i*). If all the constants A, B, C are different and if we substitute the first particular solution (7.4) of system (7.3) into equation (7.8), the latter reduces to the form

$$k(k-1)^3(k^2-4) = 0$$

Hence the first requirement relative to the nature of the roots of equation (7.8) is always satisfied. However, since the same equation has a triple root $k = 1$, then in order that the solution of system (7.7) does not contain terms of the form $t \ln t$ and $t \ln^2 t$ it is necessary that the minors of order four and five vanish for $k = 1$. Expressing this requirement relative to the minor obtained by crossing out the second and third columns and the third and fourth rows, we obtain the equation

$$\begin{vmatrix} A & 0 & -z_0 & y_0 \\ A'B'B_1 & z_0 & 0 & -x_0 \\ 0 & -A'B' & 0 & B'C' \\ 0 & C'A' & -B'C' & 0 \end{vmatrix} = 0$$

or

$$x_0 \sqrt{A(B-C)} + y_0 \sqrt{B(C-A)} + z_0 \sqrt{C(A-B)} = 0$$

Assume that $A > B > C$. Then the middle radical is imaginary, and for real A, B, C; x_0, y_0, z_0 the last equation is possible only if

$$y_0 = 0, \quad x_0 \sqrt{A(B-C)} + z_0 \sqrt{C(A-B)} = 0 \tag{7.9}$$

This case, overlooked by KOVALEVSKAYA, was obtained by APPEL'ROT [1] from the same equation (7.8). He tried to satisfy system (7.2) by series solutions which have poles of the first order and the coefficients of which contain five arbitrary constants. Originally this case was first discovered by HESS [7].

This rediscovery of HESS' case by APPEL'ROT gave a new impetus for further studies which were carried out by P. A. NEKRASOV [1, 2, 3, 4], N. E. ŽUKOVSKII [1, 2, 5], B. K. MLODZEEVSKII [2], and MLODZEEVSKII and NEKRASOV [1]. NEKRASOV showed that if x_0 and z_0 are both different from zero, i.e. the case under consideration does not reduce to that of EULER, then for properly chosen initial values of the functions $p, q, r; \alpha, \beta, \gamma$, the latter will be many-valued functions of the time t. LYAPUNOV shows the same fact in the following way.

For $y_0 = 0$ and a, b, c, f, g, h given by the second particular solution (7.5) of system (7.3), equation (7.8) reduces to the form

$$D_1 D_2 = 0 \tag{7.10}$$

where

$$D_1 = \begin{vmatrix} Bk & z_0 & -x_0 \\ \dfrac{2B}{z_0 + i x_0} & k-1 & 2i \\ -\dfrac{2Bi}{z_0 + i x_0} & -2i & k-1 \end{vmatrix} = B(k-1)(k+2)(k-3) \tag{7.11}$$

and

$$D_2 = \begin{vmatrix} Ak & 2(C-B)i & -z_0 \\ 2(B-A)i & Ck & x_0 \\ -\dfrac{2B}{z_0 + i x_0} & \dfrac{2Bi}{z_0 + i x_0} & k-1 \end{vmatrix}$$

$$= (k-2)\Big[ACk(k+1) + 2(B-A)(C-2B) - - 2B(A-C)\frac{x_0(x_0 + i z_0)}{x_0^2 + z_0^2}\Big] \tag{7.12}$$

If $A \neq C$ and x_0, z_0 are both different from zero, the quadratic polynomial with respect to k in squared brackets does not have real roots.

If $x_0 = z_0 = 0$ (according to equation (7.9) they both vanish simultaneously), we have the case of EULER and POINSOT. Hence if the constants A, B, C are all different, single-valued solutions cannot exist except in the case of EULER and POINSOT.

Case (*ii*). Consider now the case where the constants A, B, C are not all different. In order to be more specific assume that $A = B$. Then by rotation of the coordinate trihedral about the z-axis through a certain angle we can make the center of mass of the body lie on the x-axis, i.e. $y_0 = 0$. Substituting $y_0 = 0$ and expressions (7.5) into equations (7.8) we obtain equation (7.10), where D_1 and D_2 are given by formulas (7.11) and (7.12).

Excluding the case $k = 2$, equation $D_2 = 0$ can have real roots only if $A = C$, i.e. when $A = B = C$ (case of kinetic symmetry) or, if $x_0 = 0$, i.e. when $A = B$, $x_0 = y_0 = 0$ (case of LAGRANGE and POISSON), or, finally, if $z_0 = 0$.

In the last case equation $D_2/(k-2) = 0$ reduces for $A = B$ to

$$k(k+1) = 2(n-1)$$

where $n = A/C$. Hence n must be a positive integer.

For $z_0 = 0$ system (7.3) has a solution

$$a = 0, \quad b = 0, \quad c = 2i$$

$$f = \frac{2C}{x_0 + i y_0}, \quad g = \frac{2Ci}{x_0 + i y_0}, \quad h = 0$$

which can be obtained from solution (7.5) by circular permutation of the letters a, b, c; f, g, h; x_0, y_0, z_0; A, B, C. Consequently, the corresponding equation in k is

$$BA\,k(k+1) + 2(C-B)(A-2C) - 2C(B-A)\frac{y_0(y_0 + i x_0)}{x_0^2 + y_0^2} = 0$$

For $A = B$ this equation reduces to

$$\left(k - \frac{2}{n} + 2\right)\left(k + \frac{2}{n} - 1\right) = 0$$

and the requirement that the roots be integers implies that $2/n$ must be an integer. Hence n and $2/n$ must both be positive integers. This is only possible if $n = 1$ and $n = 2$. The first case leads to the case of LAGRANGE and POISSON, and the second to that of KOVALEVSKAYA.

The investigations of the cases of EULER and POINSOT, LAGRANGE and POISSON, and KOVALEVSKAYA have shown that the corresponding solutions are, in fact, single-valued. Consequently the necessary conditions for the single-valuedness of the solutions are also sufficient. Thus the theorem of KOVALEVSKAYA and its generalization by LYAPUNOV are proved.

7.32. *Real initial values*

By virtue of the above theorems we conclude that, for real values of A, B, C; x_0, y_0, z_0 with A, B, C all different from zero, new cases of single-valued functions p, q, r; α, β, γ arc possible only under the assumption that the initial values of these functions are subject to certain conditions. In particular we can require the initial values to be real. So far the initial values p_0, q_0, r_0; α_0, β_0, γ_0 could assume any arbitrary values. From our preceding analysis it does not follow that there exist only the above mentioned four cases for which p, q, r; α, β, γ are single-valued functions of the time t if we assume the initial values to be real.

Excluding the above four cases, we can say on the basis of the preceding analysis that if for $t = t_0$ we assign to p, q, r; α, β, γ some arbitrary values p_1, q_1, r_1; α_1, β_1, γ_1 which differ little from those given by formulas (7.1), namely

$$\begin{aligned} p_0 &= \frac{a}{t_0}, & q_0 &= \frac{b}{t_0}, & r_0 &= \frac{c}{t_0} \\ \alpha_0 &= \frac{f}{t_0^2}, & \beta_0 &= \frac{g}{t_0^2}, & \gamma_0 &= \frac{h}{t_0^2} \end{aligned} \tag{7.13}$$

where a, b, c, f, g, h satisfy equations (7.3), then among such sets of values p_1, q_1, r_1; α_1, β_1, γ_1 there will always be some for which the functions p, q, r; α, β, γ will not be single-valued. On the other hand for arbitrary t_0 and all possible solutions of equations (7.3) the numbers given by formulas (7.13) cannot all be real. Therefore let us start from some real values for p_0, q_0, r_0; α_0, β_0, γ_0 and let us take instead of particular solution (7.1) of system (7.2) some other solution with the above real initial values for p, q, r; α, β, γ.

LYAPUNOV considers two particular solutions. For the first particular solution let $\alpha = \beta = \gamma = 0$. Then system (7.2) reduces to the system

$$\begin{aligned} A\frac{dp}{dt} &= (B - C)\, q\, r \\ B\frac{dq}{dt} &= (C - A)\, r\, p \\ C\frac{dr}{dt} &= (A - B)\, p\, q \end{aligned}$$

These are the equations of motion in the case of EULER and POINSOT, i.e. when no external forces are acting on the body. For $t = t_0$ let the initial values p_0, q_0, r_0 be real. In order that among the solutions of the above equations there are solutions with singularities, the constants A, B, C must be all different. Then, except for some particular cases, the functions p, q, r have poles of the first order, and in a neighborhood of a pole $t = \tau$ their expansions in terms of $t - \tau$ are of the form

$$\left.\begin{aligned} p &= \frac{a}{t-\tau} + a_1 + a_2(t-\tau) + \cdots \\ q &= \frac{b}{t-\tau} + b_1 + b_2(t-\tau) + \cdots \\ r &= \frac{c}{t-\tau} + c_1 + c_2(t-\tau) + \cdots \end{aligned}\right\} \tag{7.14}$$

Substituting these series expansions for p, q, r into the above system we obtain for a, b, c the values given by formulas (7.4).

If among the numbers x_0, y_0, z_0 one is equal to zero, say, $y_0 = 0$, LYAPUNOV takes for the second particular solution

$$p = r = \beta = 0$$

$$B\frac{dq}{dt} = \gamma x_0 - \alpha z_0$$

$$\frac{d\alpha}{dt} = -q\gamma, \qquad \frac{d\gamma}{dt} = q\alpha$$

a solution which characterizes the motion of a compound pendulum and corresponds to real initial values q_0, α_0, γ_0. In order to have solutions with singularities at least one of the numbers x_0 and z_0 must be different from zero. Then, except for some particular cases, the functions q, α, γ have poles and their expansions in terms of $t - \tau$ in a neighborhood of the pole $t = \tau$ are of the form

$$\left.\begin{aligned} q &= \frac{b}{t-\tau} + b_1 + b_2(t-\tau) + \cdots \\ \alpha &= \frac{f}{(t-\tau)^2} + \frac{f_1}{t-\tau} + f_2 + \cdots \\ \gamma &= \frac{h}{(t-\tau)^2} + \frac{h_1}{t-\tau} + h_2 + \cdots \end{aligned}\right\} \tag{7.15}$$

where b, f, h are given by formulas (7.5).

In what follows solutions of the form (7.14) and (7.15) will replace solutions (7.1) of system (7.2). Substituting

$$\left.\begin{aligned} (t-\tau)\, p &= a + a_1(t-\tau) + \cdots \\ (t-\tau)\, q &= b + b_1(t-\tau) + \cdots \\ (t-\tau)\, r &= c + c_1(t-\tau) + \cdots \\ (t-\tau)^2 \alpha &= f + f_1(t-\tau) + \cdots \\ (t-\tau)^2 \beta &= g + g_1(t-\tau) + \cdots \\ (t-\tau)^2 \gamma &= h + h_1(t-\tau) + \cdots \end{aligned}\right\} \tag{7.16}$$

which is a solution of form (7.14) and (7.15), into the variational equations (7.6), the latter reduce to a system of linear differential equations which for $t = \tau$ have regular solutions.

In fact, if we replace $t - \tau$ by t and introduce the notations $\delta p = x_1$, $\delta q = x_2$, $\delta r = x_3$, $t\,\delta\alpha = x_4$, $t\,\delta\beta = x_5$, $t\,\delta\gamma = x_6$, then system (7.6) assumes the form

$$t\,\frac{dx_s}{dt} = P_{s1}(t)\,x_1 + P_{s2}(t)\,x_2 + \cdots + P_{s6}(t)\,x_6 \qquad (s = 1, 2, \ldots, 6) \tag{7.17}$$

where $P_{sj}(t)$ are functions of t which for sufficiently small $|t|$ can be expanded in terms of positive integral powers of t. For linear differential equations of type (7.17) necessary and sufficient conditions can be easily found in order that all solutions in a neighborhood of the point $t = 0$ be single-valued functions of the parameter t (see L. SAUVAGE [1]).

To obtain necessary conditions in order that the functions x_s be single-valued in a neighborhood of the point $t = 0$ it is sufficient to consider the system of linear differential equations

$$t\,\frac{dx_s}{dt} = P_{s1}(0)\,x_1 + P_{s2}(0)\,x_2 + \cdots + P_{s6}(0)\,x_6 \qquad (s = 1, 2, \ldots, 6) \tag{7.18}$$

System (7.18), however, is the same as system (7.7), the integration of which led LYAPUNOV to the conclusion that there are only four cases in which the solutions of system (7.2) are single-valued functions of the time t. Consequently no new cases are obtained if we assume that the initial values $p_0, q_0, r_0; \alpha_0, \beta_0, \gamma_0$ are real.

7.33. *Real initial values with* $\alpha_0^2 + \beta_0^2 + \gamma_0^2 = 1$

In the last section, assuming that the initial values $p_0, q_0, r_0; \alpha_0, \beta_0, \gamma_0$ are real, we did not subject them to any other conditions.

In mechanics, however, α, β, γ are direction cosines of a certain direction with respect to a rectangular coordinate trihedral. Therefore α, β, γ as well as their initial values $\alpha_0, \beta_0, \gamma_0$ are subject to the condition that the sum of their squares is equal to one. Hence the question arises whether there do not exist new cases for which the functions $p, q, r; \alpha, \beta, \gamma$ are single-valued for any real values $p_0, q_0, r_0; \alpha_0, \beta_0, \gamma_0$, satisfying the condition $\alpha_0^2 + \beta_0^2 + \gamma_0^2 = 1$. The answer to this question, however, is in the negative. Namely, LYAPUNOV showed that if for real A, B, C; x_0, y_0, z_0 and A, B, C all different from zero, we are not faced with one of the four known cases, then the initial values $p_0, q_0, r_0; \alpha_0, \beta_0, \gamma_0$ can be always chosen to be real and to satisfy the condition

$$\alpha_0^2 + \beta_0^2 + \gamma_0^2 = 1$$

in such a way that at least some of the functions $p, q, r; \alpha, \beta, \gamma$ will not be single-valued with respect to t.

In fact, from what has been said before it follows that in any case which is not reducible to one of the four known cases the initial values $p_0, q_0, r_0; \alpha_0, \beta_0, \gamma_0$ can be chosen to be real in such a way that not all of the functions $p, q, r; \alpha, \beta, \gamma$ will be single-valued in t. Let

$$p = f_1(t), \qquad q = f_2(t), \qquad r = f_3(t)$$
$$\alpha = \varphi_1(t), \qquad \beta = \varphi_2(t), \qquad \gamma = \varphi_3(t)$$

be such a many-valued solution corresponding to real initial values $p_0, q_0, r_0; \alpha_0, \beta_0, \gamma_0$. Then, as in the case of any solution of system (7.2), the functions α, β, γ will satisfy the relation

$$\alpha^2 + \beta^2 + \gamma^2 = \sigma^2$$

where σ is a certain positive constant different from zero. If σ were equal to zero, then by virtue of the assumption that $\alpha_0, \beta_0, \gamma_0$ are real it would follow that $\alpha_0 = \beta_0 = \gamma_0 = 0$ and hence $\alpha = \beta = \gamma = 0$. Hence we would have obtained the case of EULER and POINSOT in which case the solution is single-valued contrary to our assumption.

For $\sigma > 0$ the equations

$$p = \frac{1}{\sqrt{\sigma}} f_1\left(\frac{t - t_0}{\sqrt{\sigma}} + t_0\right), \qquad \alpha = \frac{1}{\sqrt{\sigma}} \varphi_1\left(\frac{t - t_0}{\sqrt{\sigma}} + t_0\right)$$
$$q = \frac{1}{\sqrt{\sigma}} f_2\left(\frac{t - t_0}{\sqrt{\sigma}} + t_0\right), \qquad \beta = \frac{1}{\sqrt{\sigma}} \varphi_2\left(\frac{t - t_0}{\sqrt{\sigma}} + t_0\right)$$
$$r = \frac{1}{\sqrt{\sigma}} f_3\left(\frac{t - t_0}{\sqrt{\sigma}} + t_0\right), \qquad \gamma = \frac{1}{\sqrt{\sigma}} \varphi_3\left(\frac{t - t_0}{\sqrt{\sigma}} + t_0\right)$$

determine also a solution of system (7.2) corresponding to real initial values $p_0, q_0, r_0; \alpha_0, \beta_0, \gamma_0$, satisfying the condition

$$\alpha_0^2 + \beta_0^2 + \gamma_0^2 = 1$$

and among the functions $p, q, r; \alpha, \beta, \gamma$ at least some will be many-valued in t.

This completes the proof for the existence of single-valued solutions for arbitrary initial conditions.

B. Particular solutions of the Euler and Poisson equations

§ 8. Particular cases of integrability

8.1. Introduction. From the theorems of KOVALEVSKAYA and LYAPUNOV formulated in Section 7.3 it follows that besides the four cases mentioned in these theorems there cannot exist new cases where the general solution is single-valued for arbitrary initial values. If, however, restrictions are imposed on the energy constant h, or the constant c in the angular momentum integral about the vertical, or on the values of A, B, C; x_0, y_0, z_0, then, in particular cases, a solution of the problem may exist. The general form of such restrictions, however, is not known.

In an unfinished manuscript S. A. ČAPLYGIN [5] made an attempt towards finding all the integrable cases by a unique method. Although much has been written on the motion of a rigid body about a fixed point, it is mainly a literature of special cases. It has been almost the tradition of writers in this field to give one or more new cases of integrability or to point out in some respect the limiting character of their investigations.

Since KOVALEVSKAYA's investigations a number of basic cases of integrability and particular solutions of the EULER and POISSON equations have been found and studied. All these cases have been discovered by a skillful application of the EULER and POISSON equations or by virtue of certain particular properties of the fourth algebraic first integral. In what follows we shall list and discuss these cases.

8.2. Case of a loxodromic pendulum. This case was found by W. HESS [7] (1890), rediscovered by G. G. APPEL'ROT [1] (1892), and P. A. NEKRASOV [1] (1892), and further studied by NEKRASOV [2, 3, 4, 5], B. K. MLODZEEVSKII and P. A. NEKRASOV [1], N. E. ŽUKOVSKII [1, 2, 5], S. A. ČAPLYGIN [1], and R. LIOUVILLE [1]. See also E. WAELSCH [1] and L. N. SRETENSKII [3].

HESS assumes for the mass distribution two conditions, namely

$$y_0 = 0 \quad \text{and} \quad A(B-C)\,x_0^2 - C(A-B)\,z_0^2 = 0$$

ŽUKOVSKII showed that these two conditions imply that the mass center G lies on the perpendicular through the fixed point O to a circular cross section of the reciprocal momental ellipsoid of inertia at O,

$$\frac{x^2}{A} + \frac{y^2}{B} + \frac{z^2}{C} = 1$$

In addition, HESS assumes that at the initial instant the impulse vector $\vec{I}$ lies in the said circular cross section. Under these assumptions a fourth particular integral exists,

$$A\,p\,x_0 + C\,r\,z_0 = 0$$

This integral shows that the impulse vector remains in the said plane throughout the whole course of the motion.

T. MANACORDA [3], seeking to characterize structurally a rigid body which, acted upon by a system of forces the moment of which is perpendicular to $\overrightarrow{OG}$, has the angular momentum also perpendicular to $\overrightarrow{OG}$, arrived at the HESS and APPEL'ROT conditions for a heavy rigid body.

8.3. Permanent rotations. In 1894 O. STAUDE [2] and B. K. MLODZEEVSKII [1], independent of each other, showed that the EULER and POISSON equations admit an ∞^1 of solutions if we make no assumptions concerning the moments of inertia and the position of the mass center G of the body. These solutions correspond to rotations of the body about axes fixed in the body and in space. MLODZEEVSKII showed that such permanent axes of rotation can be either the principal axes of inertia when put in a horizontal position (case of a compound pendulum), or a set of vertical axes. In the latter case the magnitude of the instantaneous angular velocity of the body is constant. The rotations of the body about the vertical axes were investigated in detail by STAUDE. The axes of rotation form a cone of the second order, concentric with the ellipsoid of inertia of the body at the fixed point O but not coaxial with the latter and depending upon the position of the mass center G with respect to the principal axes of inertia of the body at O.

W. VAN DER WOUDE [1] showed that each axis of permanent rotation is a principal axis of inertia of the body at one of its points which is different from O, the x-, y-, and z-axes being principal axes of inertia at O. A simple geometric proof of this fact was given by M. MANARINI [1, 2]. For some refinements of the above classical results see M. MANARINI [3] and E. WAELSCH [1]. For an extension of the concept of permanent rotations to five dimensional space see O. BOTTEMA [2].

L. ALFIERI [1] solved a problem which includes that of O. STAUDE, namely, to determine all possible uniform rotations of a rigid body about a fixed point O different from its mass center G, when the system of acting forces is equivalent to a constant force applied to a point distinct, in general, from both O and G.

8..31 The mass center cone and the mass center curve

Given the EULER and POISSON equations of motion

$$\dot{\vec{I}} + \vec{\omega} \times \vec{I} + \vec{r}_0 \times \vec{K} = 0 \tag{8.1}$$

$$\dot{\vec{K}} + \vec{\omega} \times \vec{K} = 0 \tag{8.2}$$

and their three classical integrals

$$T = -\vec{r}_0 \cdot \vec{K} + h \qquad \left(T = \tfrac{1}{2}\,\vec{\omega} \cdot \vec{I}\right) \tag{8.3}$$

$$\vec{I} \cdot \vec{K} = c \tag{8.4}$$

$$K^2 = 1 \tag{8.5}$$

let us try to find out under what conditions the above equations can be satisfied by a constant unit vector $\vec{K}$.

In such a case equation (8.2) reduces to

$$\vec{\omega} \times \vec{K} = 0 \tag{8.6}$$

or, by virtue of the fact that $\vec{K}$ is constant and $\omega^2 = p^2 + q^2 + r^2$, to

$$\vec{\omega} = \varepsilon\,\omega\,\vec{K} \tag{8.7}$$

where ε is equal to $+1$ or -1. Hence $\vec{\omega}$ must be also a constant vector. Consequently $\vec{I} = \varepsilon\,\omega\,\vec{I}_1$, and equation (8.1) reduces to

$$\omega^2 \left(\vec{K} \times \vec{I}_1\right) + \vec{r}_0 \times \vec{K} = 0 \tag{8.8}$$

where $\vec{I}_1 = \vec{I}_1(A\alpha, B\beta, C\gamma)$. Equation (8.8) is equivalent to

$$\left.\begin{aligned} (B-C)\beta\gamma\omega^2 &= y_0\gamma - z_0\beta \\ (C-A)\gamma\alpha\omega^2 &= z_0 x - x_0\gamma \\ (A-B)\alpha\beta\omega^2 &= x_0\beta - y_0\alpha \end{aligned}\right\} \tag{8.8'}$$

Hence if the constant unit vector $\vec{K}$ and the constant ω satisfy equations (8.5) and (8.8), then $\vec{K} = \text{const.}$ and (8.7) is a solution of equations (8.1) and (8.2). Consequently, if under the above conditions a rigid body with a fixed point O is set in motion by assigning to it a velocity of rotation $\vec{\omega}$ about a directed fixed axis in the body determined by the constant direction cosines α, β, γ and if this axis is put in a vertical position with its unit vector $\vec{K}$ directed upward, then the body remains in a state of permanent rotation about this axis.

Scalar multiplication of equation (8.8) by $\vec{I}_1$ gives [by virtue of (8.7)] the equation of the above cone in the form

$$\vec{\omega} \cdot (\vec{r}_0 \times \vec{I}_1) = 0 \tag{8.9}$$

or in the scalar form

$$(B-C)\beta\gamma x_0 + (C-A)\gamma\alpha y_0 + (A-B)\alpha\beta z_0 = 0 \tag{8.9'}$$

By virtue of this equation the direction cosines α, β, γ of the permanent axis of rotation are subject to condition (8.9) or (8.9'). In other words, the vertical axis of permanent rotation must lie on a cone of the second order defined by the equation

$$(B-C)x_0 y z + (C-A)y_0 z x + (A-B)z_0 x y = 0 \tag{8.10}$$

We shall refer to this cone as "the mass center cone" of the body with respect to the fixed point O.

This cone is not coaxial with the central ellipsoid of inertia

$$A x^2 + B y^2 + C z^2 = 1 \tag{8.11}$$

but depends, on the one hand, upon the magnitudes of the moments of inertia, and, on the other hand, upon the position of the mass center $G(x_0, y_0, z_0)$ with respect to the principal axes of inertia.

If we select an element of the cone (8.10), assign a direction to this element (by means of an arrow), and calculate its direction cosines α, β, γ, then the corresponding ω is determined by equation (8.8). However, by virtue of the particular form of this equation only one of the two oppositely directed elements (α, β, γ) and $(-\alpha, -\beta, -\gamma)$ of the cone will give to ω^2 a positive value. Therefore each element of the cone is to be discussed twice according to its two possible directions.

Equation (8.10) of the cone can be rewritten in the form

$$\begin{vmatrix} x_0 & y_0 & z_0 \\ x & y & z \\ A\,x & B\,y & C\,z \end{vmatrix} = 0 \tag{8.12}$$

If (x, y, z) is a point on the ellipsoid of inertia (8.11), then the ratios of the direction cosines of the perpendicular from O to the tangent plane of this ellipsoid at the point (x, y, z) are $A\,x : B\,y : C\,z$. Hence to any axis through O with direction cosines $x : y : z$ there corresponds a secondary axis with direction cosines $A\,x : B\,y : C\,z$. Therefore, if the first axis satisfies equation (8.12), then the plane through this axis and the secondary axis contains the mass center of the body. Hence we have the following

Theorem 1

The mass center cone is the locus of the axes such that the plane through any of these axes and the corresponding secondary axis contains the mass center of the body.

It has been known since the time of Poisson that in the case of an arbitrary motion of a rigid body about a fixed point O the secondary axis corresponding to the axis of the instantaneous rotation is the axis of the angular momentum. Hence we have the following

Theorem 2

If in the motion of a rigid body about a fixed point at an instant t the instantaneous axis of rotation lies on the mass center cone, then the mass center of the body lies in the plane through this axis and the axis of the angular momentum.

Introduce besides the mass center $G(x_0, y_0, z_0)$ also the point $G'(x_0/A, y_0/B, z_0/C)$. Equation (8.12) shows that the mass center cone passes through five exceptional axes, namely the three coordinate axes $O\,x$, $O\,y$, $O\,z$ and both axes $\overrightarrow{O\,G}$ and $\overrightarrow{O\,G'}$. Since under the assumption

$$B \neq C, \quad C \neq A, \quad A \neq B; \quad x_0 \neq 0, \quad y_0 \neq 0, \quad z_0 \neq 0 \tag{8.13}$$

neither the axis $\overrightarrow{O\,G}$ nor the axis $\overrightarrow{O\,G'}$ lies in one of the coordinate planes, and the plane through the axes $\overrightarrow{O\,G}$ and $\overrightarrow{O\,G'}$ does not contain any of the coordinate axes, we can say that under the assumptions (8.13) the mass center cone is a proper cone of the second order.

In what follows we shall assume without loss of generality that

$$A > B > C; \quad x_0 > 0, \quad y_0 > 0, \quad z_0 > 0 \tag{8.14}$$

and we shall determine on the mass center cone the positions of the five exceptional axes mentioned above.

For this purpose we shall write down the equations of the tangent planes which touch the cone along the coordinate axes $O\,x, O\,y, O\,z$. They are

$$\left.\begin{aligned} T_1(x, y, z) &= (C - A)\, y_0\, z + (A - B)\, z_0\, y = 0 \\ T_2(x, y, z) &= (A - B)\, z_0\, x + (B - C)\, x_0\, z = 0 \\ T_3(x, y, z) &= (B - C)\, x_0\, y + (C - A)\, y_0\, x = 0 \end{aligned}\right\} \tag{8.15}$$

and hence

$$T_1(0, -1, 0) < 0, \quad T_1(0, 0, 1) < 0, \quad T_1(x_0, y_0, z_0) < 0,$$
$$T_1(x_0/A, y_0/B, z_0/C) < 0, \quad T_1(0, 1, 0) > 0$$
$$T_3(1, 0, 0) < 0, \quad T_3(0, -1, 0) < 0, \quad T_3(x_0, y_0, z_0) < 0,$$
$$T_3(x_0/A, y_0/B, z_0/C) < 0, \quad T_3(0, 1, 0) > 0$$

Since the two sheets of the cone are lying on opposite sides of any tangent plane to this cone, we obtain that the five semi-axes x, $-y$,

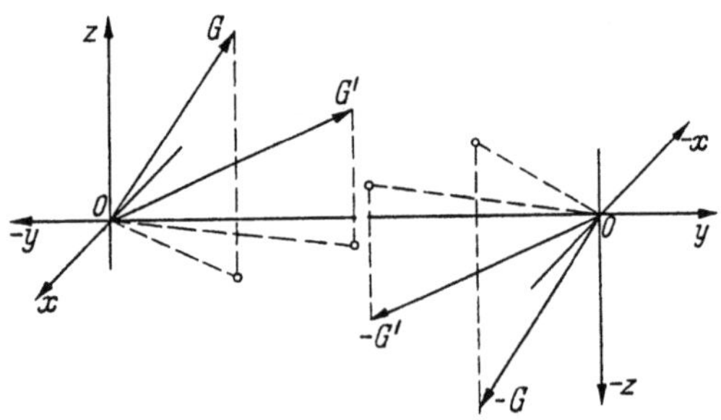

Fig. 3

$z, \overrightarrow{O\,G}, \overrightarrow{O\,G'}$ lie on one sheet and the other five oppositely directed semi-axes $-x$, y, $-z$, $-\overrightarrow{O\,G}$, $-\overrightarrow{O\,G'}$ on the second sheet of the mass center cone (Fig. 3).

In order to determine the order of the five axes mentioned above on the first sheet of the mass center cone, introduce the three planes through G and one of the coordinate axes:

$$\left.\begin{aligned} P_1(x, y, z) &= y_0\, z - z_0\, y = 0 \\ P_2(x, y, z) &= z_0\, x - x_0\, z = 0 \\ P_3(x, y, z) &= x_0\, y - y_0\, x = 0 \end{aligned}\right\} \tag{8.16}$$

Then

$$P_3(0, -1, 0) < 0, \quad P_3(x_0, y_0, 0) = 0, \quad P_3(1, 0, 0) < 0,$$
$$P_3(x_0/A, y_0/B, z_0/C) > 0$$

Therefore, if we pass the planes through the z-semi-axis and one of the other four semi-axes on the first sheet of the mass center cone, then these planes appear in the order $(z, -y)$, (z, x), (z, G), (z, G'). Hence we obtain that on the first sheet of the mass center cone, which contains the mass center G, the five exceptional semi-elements appear in the order

$$-y, x, \overrightarrow{OG}, \overrightarrow{OG'}, z \tag{8.17}$$

The curve of intersection of the mass center cone with the unit sphere described about the fixed point O we shall call the "mass center curve". Therefore, if we think of α, β, γ as the coordinates of a point on the unit sphere, then equation (8.9) or (8.9') gives us the equation of the mass center curve. To every point α, β, γ of this curve there

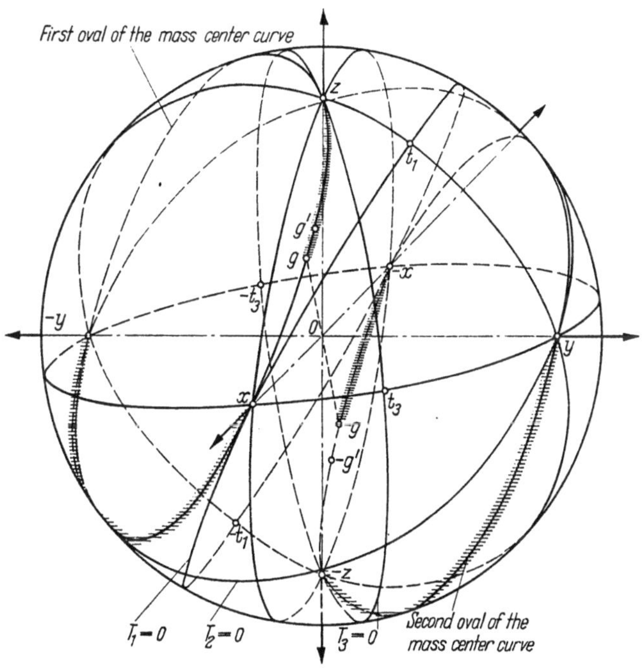

Fig. 4. $A > B > C$; $x_0 > 0$, $y_0 > 0$, $z_0 > 0$

corresponds an element of the mass center cone with vertex at O, and conversely. This element starts at O and is directed toward the point (α, β, γ).

In Fig. 4 are represented the two ovals of the mass center curve as well as the three great circles of intersection of the three planes $T_1 = 0$, $T_2 = 0$, $T_3 = 0$ with the unit sphere. The points of intersection of the coordinate axes with the sphere are x, y, z, $-x$, $-y$, $-z$. In addition, there are represented the points

$$g\left(\frac{x_0}{\varrho_0}, \frac{y_0}{\varrho_0}, \frac{z_0}{\varrho_0}\right), \qquad \varrho_0 = \sqrt{x_0^2 + y_0^2 + z_0^2}$$

$$g'\left(\frac{x_0}{A\,\sigma_0}, \frac{y_0}{B\,\sigma_0}, \frac{z_0}{C\,\sigma_0}\right), \qquad \sigma_0 = \sqrt{\frac{x_0^2}{A^2} + \frac{y_0^2}{B^2} + \frac{z_0^2}{C^2}}$$

which are the points of intersection of the semi-axes $\overrightarrow{OG}$ and $\overrightarrow{OG'}$ with the unit sphere, and their diametrically opposite points $-g$ and $-g'$. The points $x, g, g', z, -y$ are lying on the first oval in the above order while the points $-x, -g, -g', -z, y$ lie on the second oval in the indicated order.

To every semi-axis, drawn from O to a point (α, β, γ) on the mass center curve, equation (8.8) gives the value for ω^2. Accordingly as this value is positive or negative this axis is a permissible or a nonpermissible axis of rotation when put in a vertical position. Similarly the point (α, β, γ) will be said to be an admissible or a nonadmissible point of the mass center curve. Since diametrically opposite points by equation (8.8) are, in general, of a different character, it is sufficient to study the nature of the points along one oval of the mass center curve.

Consider the first oval of the mass center curve. By virtue of (8.9) and (8.17) this oval is circumscribed about the spherical triangle $(x, z, -y)$ in the sense that it cuts the three adjacent octants of the sphere in such a way that the arcs $(-y, z)$, (z, x), $(x, -y)$ of this oval lie in the octants $(\alpha, \beta, \gamma) = (-, -, +), (+, +, +), (+, -, -)$ respectively. This remark together with (8.14) permits us to determine the sign of the left-hand sides of equations (8.8′) for a point (α, β, γ) on the oval. The right-hand sides of these equations are $P_1(\alpha, \beta, \gamma)$, $P_2(\alpha, \beta, \gamma)$ and $P_3(\alpha, \beta, \gamma)$ respectively. These quantities change their signs along the oval at the points x, $-y$, z and g, at which their values are respectively $P_1 = 0, +, +, 0$; $P_2 = +, 0, -, 0$; $P_3 = -, -, 0, 0$. Hence the signs of the left- and right-hand sides of equations (8.8′) along the arcs of the first oval indicated below are as follows:

First oval arcs	First equation		Second equation		Third equation	
	l. h. s.	r. h. s.	l. h. s.	r. h. s.	l. h. s.	r. h. s.
$-y, z$	$-$	$+$	$+$	$-$	$+$	$-$
z, g', g	$+$	$+$	$-$	$-$	$+$	$+$
g, x	$+$	$-$	$-$	$+$	$+$	$-$
$x, -y$	$+$	$+$	$+$	$+$	$-$	$-$

Consequently the signs of both sides of each equation (8.8′) are the same for the second and the fourth arcs and opposite for the first and third arcs of the first oval. The diametrically opposite arcs of the second oval are of opposite character.

If we imagine that the permissible arcs of both ovals are shaded, then we obtain the following

Theorem 3

Along each oval of the mass center curve two shaded arcs alternate with two unshaded arcs. The shaded arcs along the first oval are $(x, -y)$ and (z, g', g), and along the second oval are $(y, -z)$ and $(-g, -x)$.

Concerning the permissible axes of the mass center cone we can now state the following

Theorem 4

Those and only those elements of the mass center cone which connect the fixed point O and a point on a shaded arc, when put in the vertical position, directed upward and given a definite angular velocity ω, are the permanent axes of rotation of a body about a fixed point O.

The value of ω is to be determined for each point (α, β, γ) of a shaded arc from one of the equations (8.8'). However, under the assumptions (8.13) and (8.14) we have the following particular cases. For the points $\pm x$ or $(\alpha, \beta, \gamma) = (\pm 1, 0, 0)$ the first equation of system (8.8') is identically satisfied and the other two equations give $\omega = \infty$. We have similar results for the points $\pm y$ and $\pm z$. Hence we have the following

Theorem 5

Each principal axis of inertia, when put in the vertical position and assigned with an arrow directed upward or downward, is an axis of permanent rotation, the angular velocity being infinite.

For the points g and $-g$ we have $\alpha : \beta : \gamma = x_0 : y_0 : z_0$, and hence from equations (8.8') we obtain that $\omega = 0$.

Hence we have the following

Theorem 6

The axis through the mass center G, assigned with an arrow and put in the vertical position, directed upward or downward, can always be a permanent axis of rotation, the angular velocity ω being zero.

The last two theorems are concerned with axes (α, β, γ) corresponding to the endpoints of the shaded arcs. Therefore we can say that for all other axes to which arrows are assigned, when put in the vertical position directed upward, the angular velocity ω has a value different from zero and ∞.

Instead of the assumptions made in (8.14) in what follows we shall keep only the double inequality

$$C > A > B$$

obtained by circular permutation of the letters A, B, C in (8.14), but we shall drop partially or totally the assumptions concerning x_0, y_0, z_0, i.e. concerning the position of the mass center G. In other words, the mass center G can lie in one of the principal planes or along one of the principal axes of the ellipsoid of inertia at O or it can coincide with the center of the ellipsoid.

Case 1. The mass center G lies in one of the principal planes of inertia. If one of the coordinates x_0, y_0, z_0 is zero, then from equation (8.10) of the mass center cone results the following

Theorem 7

If the mass center G lies in one of the principal planes of inertia at the point O, then the mass center cone degenerates into a pair of

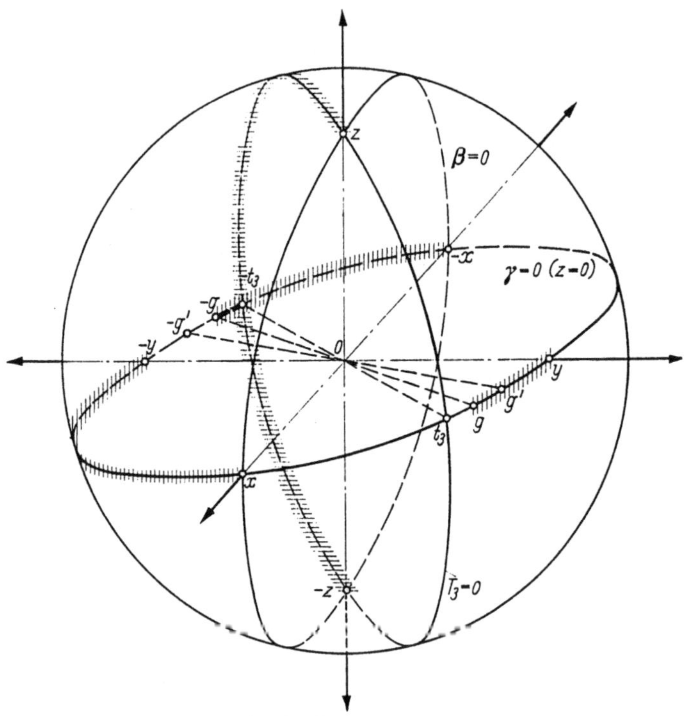

Fig. 5. $C > A > B$; $x_0 > 0$, $y_0 > 0$, $z_0 = 0$

planes, one of which is the above principal plane of inertia and the other plane passes through the third principal axis of inertia of the body.

In order to be more specific assume that $z_0 = 0$ and that as before $y_0 > 0$, $x_0 > 0$. Then the mass center cone (8.10) degenerates into the planes $z = 0$ and $T_3 = 0$ and the mass center curve into the corresponding great circles of the unit sphere. The two mutually orthogonal circles intersect at the points t_3, $-t_3$ which are in the quadrants (y, x) and $(-y, -x)$ respectively.

Once the positions of the two great circles into which the mass center curve degenerates are determined, the shaded parts of these circles can be determined by a passage to the limit or directly by application of formulas (8.8').

By a passage to the limit the two ovals reduce to the two spherical lunes $(z, -t_3, -g, -g', -y, x, t_3, z)$ and $(-z, t_3, g, g', y, -x, -t_3, -z)$ respectively, and the shaded parts are determined by application of Theorem 3 (Fig. 5).

However, instead of Theorem 3 the following theorem can also be used to determine the shaded parts of both circles of the mass center curve (Fig. 5).

Theorem 8

The shaded parts are the arcs $(x, -y)$, $(-g, -x)$, (y, g) of the circle $z = 0$ and the half-circle $(z, -t_3, -z)$ of the circle $T_3 = 0$. The unshaded arcs are $(-y, -g)$, $(-x, y)$, (g, x) of the first circle and the half-circle $(-z, t_3, z)$ of the second circle.

The analytical proof of this theorem follows from formulas (8.8') if we substitute in those equations $z_0 = 0$.

Case 2. The mass center G lies on one of the principal axes of inertia at O. If two of the coordinates x_0, y_0, z_0 are zero, equation (8.10) immediately gives the following

Theorem 9

If the mass center G lies on one of the principal axes of inertia of the body at O, then the mass center cone degenerates into two principal planes of inertia passing through the axis under consideration.

Concerning the shaded and unshaded parts we have now the following

Theorem 10

One half of each of the great circles into which the mass center curve degenerates is shaded, the other half is unshaded.

In fact, putting $y_0 = z_0 = 0$ in formulas (8.8') we obtain the system

$$\left.\begin{aligned} (B - C)\,\beta\,\gamma\,\omega^2 &= 0 \\ (C - A)\,\gamma\,\alpha\,\omega^2 &= -x_0\,\gamma \\ (A - B)\,\alpha\,\beta\,\omega^2 &= x_0\,\beta \end{aligned}\right\} \qquad (8.18)$$

subject to the assumption $x_0 > 0$. The first equation is satisfied for all points (α, β, γ) of the mass center curve which consists of the circles $\beta = 0$ and $\gamma = 0$. For the circle $\beta = 0$ the third equation is identically satisfied and the second equation gives $\alpha < 0$. For the circle $\gamma = 0$ the second equation is identically satisfied while the third equation shows that $\alpha > 0$. Hence the half-circles $(-y, x, y)$ and $(z, -x, -z)$ must be shaded (Fig. 6).

About the value of ω for each point (α, β, γ) of the mass center curve the following is to be said. While the second equation gives a

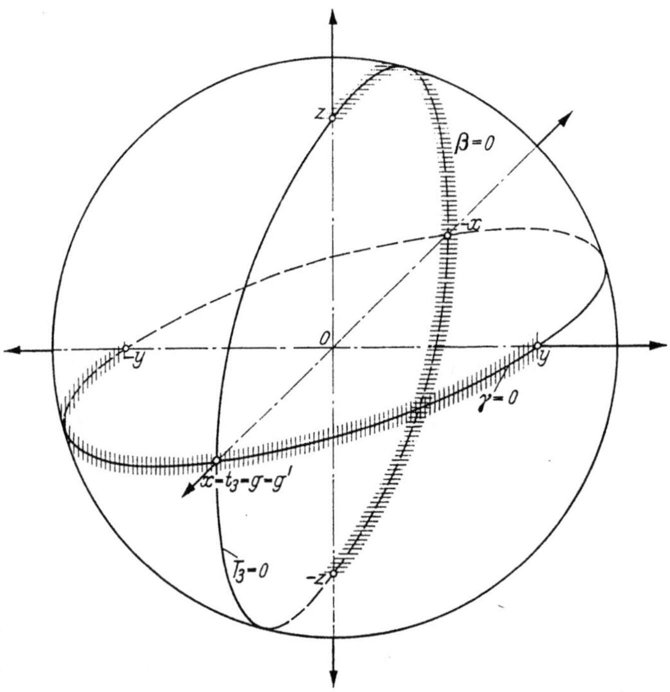

Fig. 6. $C > A > B$; $x_0 > 0$, $y_0 = 0$, $z_0 = 0$

definite value for ω for all points of the circle $\beta = 0$ and the third equation for all points of the circle $\gamma = 0$, and in particular the value $\omega = \infty$ for the points $\pm z$ and $\pm y$, the value of ω remains undetermined for the points $\pm x$. Hence instead of the previous theorems we have the following

Theorem 11

One principal axis of inertia, directed upward or downward, can always be the vertical axis of permanent rotation, namely that one which passes through the mass center. The value of the angular velocity about this axis is arbitrary; for the remaining two axes the angular velocity is infinite.

Case 3. The mass center coincides with the center of the principal ellipsoid of inertia.

With the assumption $x_0 = y_0 = z_0 = 0$ the mass center curve (8.10) ceases to exist. As a consequence not ∞^1 but ∞^2 axes through O can be considered as the axes of permanent rotations. The angular velocity ω by virtue of equations (8.8′) is zero except in the cases when two of the quantities α, β, γ vanish and consequently ω becomes undetermined. This is what happens with the three principal axes of inertia which are the only axes of permanent rotations for any position of the principal axes of inertia.

8.32. Special cases $B = C$ and $A = B$

Finally let us consider the case of an ellipsoid of inertia with $A > B = C$ and the case of an oblate ellipsoid with $A = B > C$, the axes of rotation being the x- and the z-axis respectively. In each of these cases three subcases are to be considered.

Subcase 1. The mass center has an arbitrary position, i.e. it lies neither in the plane of the equator nor on the axis of rotation. Without loss of generality we can assume that in the case $B = C$ we have $z_0 = 0$ and as in (8.14) $x_0 > 0$, $y_0 > 0$, and in the case $A = B$ the coordinate $x_0 = 0$ and $y_0 > 0$, $z_0 > 0$. In both cases equation (8.10) reduces to $y_0 z x = 0$, i.e. we have

Theorem 12

The mass center cone consists of the plane of the equator and the meridian plane passing through the mass center.

In the case $A > B = C$, $z_0 = 0$, formulas (8.8′) reduce to

$$\left.\begin{aligned} 0 \cdot \beta\gamma\omega^2 &= y_0\gamma \\ (A - B)\gamma\alpha\omega^2 &= x_0\gamma \\ (A - B)\alpha\beta\omega^2 &= x_0\beta - y_0\alpha \end{aligned}\right\} \tag{8.19}$$

For the meridian circle $\gamma = 0$ the first two equations are satisfied identically while the third equation requires shading the arcs $(-y, x)$, (g, y), $(-x, -g)$ of this circle. At the same time this equation assigns to each point $(\alpha, \beta, 0)$ of this circle a definite value of ω different from zero and ∞, except for the points g and $-g$ at which $\omega = 0$ and the four points $\pm x$, $\pm y$, where $\omega = \infty$. For the points $(0, \beta, \gamma)$ of the equator $\alpha = 0$ no condition is imposed by equations (8.19). However, for these points the above equations are satisfied only for $\omega = \infty$.

Similar results are obtained for the case $A = B$, $x_0 = 0$. Hence we have

Theorem 13

Of the two circles into which the mass center curve degenerates the equatorial circle must be completely shaded while along the circle of the meridian three shaded and three unshaded arcs alternate.

Subcase 2. The mass center lies in the plane of the equator. By a discussion similar to that given above we conclude that in this case the mass center curve also degenerates into two circles of which the entire equatorial circle but only one half the meridian circle must be shaded. One of the principal axes of inertia, directed upward or downward, can always be used as a permanent axis of rotation when put in the vertical position. That one which passes through the mass center can rotate with an arbitrary angular velocity, all the others with infinite angular velocity.

Subcase 3. The mass center lies on the axis of rotation. In both cases $B = C$, $y_0 = z_0 = 0$ and $A = B$, $x_0 = y_0 = 0$ the mass center cone ceases to exist. We have ∞^2 vertical axes of permanent rotations. In the first case the system of equations (8.8′) reduces to

$$(A - B)\,\alpha\,\gamma\,\omega^2 = x_0\,\gamma$$

$$(A - B)\,\alpha\,\beta\,\omega^2 = x_0\,\beta$$

These two equations give no values of ω for the points $\pm x$; for any other point (α, β, γ) of the sphere they give $\alpha > 0$ and a finite value for ω which for $\alpha = 0$ becomes ∞. Instead of shading the mass center curve, we must now shade the whole surface of that hemisphere which is separated from the mass center by the equatorial plane of the extended ellipsoid of revolution. In the case of an oblate ellipsoid of inertia, that surface of the hemisphere must be shaded which is on the same side of the equatorial plane as the mass center.

8.33. Stability of permanent rotations

The stability of permanent rotations has been studied by J. Hadamard [1], R. Grammel [3, 4_1], P. Stoewa [1], O. Bottema [1], and others. Hadamard tried to obtain explicit conditions of stability for the given mass distribution. Such a problem, however, turned out to be hopeless. Grammel considered the point O, the axis of rotation, and the principal moments of inertia at O as given, and determined regions in which the mass center must lie in order that a uniform rotation be stable. Stoewa's discussion is based on three classical stability criteria, such as the criterion of Lagrange and Lejeune-Dirichlet, the criterion of Lord Kelvin and Routh, and a criterion based on first order variations from a known permanent rotation. He objected to some conclusions of R. Grammel. Bottema approached the stability problem directly by asking what axes of permanent rotation are stable if a rigid body with a fixed point O is given. He was able to solve the problem under the assumption that the mass center lies on one of the principal axes of inertia at O.

The most complete results concerning the stability of permanent rotation are due to V. V. RUMYANCEV [2]. In what follows we shall discuss his method which is based on the construction of LYAPUNOV's V-functions.

Case 1. Arbitrary mass distribution of the body. Under the assumptions (8.14) consider an arbitrary admissible half-element (α, β, γ) of the cone (8.9′) (different from the principal inertia axes of the body) which when put in the vertical position and directed upward is a permanent axis of rotation. The velocity of rotation is then determined by any one of the equations of system (8.8′) and the components of $\vec{\omega}$ are

$$p_0 = \alpha\,\omega, \qquad q_0 = \beta\,\omega, \qquad r_0 = \gamma\,\omega \tag{8.20}$$

Let the perturbed motion be

$$\begin{aligned} p &= p_0 + \xi_1, & q &= q_0 + \xi_2, & r &= r_0 + \xi_3 \\ \gamma_1 &= \alpha + \eta_1, & \gamma_2 &= \beta + \eta_2, & \gamma_3 &= \gamma + \eta_3 \end{aligned} \tag{8.21}$$

Substituting (8.21) into the EULER and POISSON equations of motion (in terms of p, q, r; $\gamma_1, \gamma_2, \gamma_3$) and observing relations (8.20) and equations (8.8′), we obtain the equations for the perturbations in the form

$$\left.\begin{aligned} A\frac{d\xi_1}{dt} + (C-B)(r_0\,\xi_2 + q_0\,\xi_3 + \xi_2\,\xi_3) &= z_0\,\eta_2 - y_0\,\eta_3 \\ B\frac{d\xi_2}{dt} + (A-C)(p_0\,\xi_3 + r_0\,\xi_1 + \xi_3\,\xi_1) &= x_0\,\eta_3 - z_0\,\eta_1 \\ C\frac{d\xi_3}{dt} + (B-A)(q_0\,\xi_1 + p_0\,\xi_2 + \xi_1\,\xi_2) &= y_0\,\eta_1 - x_0\,\eta_2 \end{aligned}\right\} \tag{8.22}$$

$$\left.\begin{aligned} \frac{d\eta_1}{dt} &= r_0\,\eta_2 + \beta\,\xi_3 + \xi_3\,\eta_2 - (q_0\,\eta_3 + \gamma\,\xi_2 + \xi_2\,\eta_3) \\ \frac{d\eta_2}{dt} &= p_0\,\eta_3 + \gamma\,\xi_1 + \xi_1\,\eta_3 - (r_0\,\eta_1 + \alpha\,\xi_3 + \xi_3\,\eta_1) \\ \frac{d\eta_3}{dt} &= q_0\,\eta_1 + \alpha\,\xi_2 + \xi_2\,\eta_1 - (p_0\,\eta_2 + \beta\,\xi_1 + \xi_1\,\eta_2) \end{aligned}\right\} \tag{8.23}$$

Systems (8.22) and (8.23) possess the first integrals

$$\left.\begin{aligned} V_1 &= A(\xi_1^2 + 2p_0\,\xi_1) + B(\xi_2^2 + 2q_0\,\xi_2) + C(\xi_3^2 + 2r_0\,\xi_3) + \\ &\quad + 2(x_0\,\eta_1 + y_0\,\eta_2 + z_0\,\eta_3) = \text{const.} \\ V_2 &= A(p_0\,\eta_1 + \alpha\,\xi_1 + \xi_1\,\eta_1) + B(q_0\,\eta_2 + \beta\,\xi_2 + \xi_2\,\eta_2) + \\ &\quad + C(r_0\,\eta_3 + \gamma\,\xi_3 + \xi_3\,\eta_3) = \text{const.} \\ V_3 &= \eta_1^2 + \eta_2^2 + \eta_3^2 + 2(\alpha\,\eta_1 + \beta\,\eta_2 + \gamma\,\eta_3) = 0 \end{aligned}\right\} \tag{8.24}$$

Following N. G. ČETAEV [1] let us construct a LYAPUNOV V-function in the form

$$\begin{aligned} V &= V_1 - 2\omega V_2 + \lambda V_3 + \tfrac{1}{4}\mu V_3^2 \\ &= A\,\xi_1^2 + B\,\xi_2^2 + C\,\xi_3^2 - 2\omega(A\,\xi_1\eta_1 + B\,\xi_2\eta_2 + C\,\xi_3\eta_3) + \\ &\quad + 2\mu(\alpha\beta\,\eta_1\eta_2 + \alpha\gamma\,\eta_1\eta_3 + \beta\gamma\,\eta_2\eta_3) + (\lambda + \mu\alpha^2)\,\eta_1^2 + \\ &\quad + (\lambda + \mu\beta^2)\,\eta_2^2 + (\lambda + \mu\gamma^2)\,\eta_3^2 + \tfrac{1}{4}\mu f(\eta_1, \eta_2, \eta_3) \end{aligned} \tag{8.25}$$

where

$$\lambda = A\,\omega^2 - \frac{x_0}{\alpha} = B\,\omega^2 - \frac{y_0}{\beta} = C\,\omega^2 - \frac{z_0}{\gamma} \tag{8.26}$$

holds by virtue of equations (8.8′),

$$f(\eta_1, \eta_2, \eta_3) = (\eta_1^2 + \eta_2^2 + \eta_3^2)\,[\eta_1^2 + \eta_2^2 + \eta_3^2 + 4(\alpha\,\eta_1 + \beta\,\eta_2 + \gamma\,\eta_3)],$$

and μ is an arbitrary constant.

The function V will be positive definite with respect to the variables $\xi_1, \xi_2, \xi_3, \eta_1, \eta_2, \eta_3$, if its quadratic part, i.e. $V - \frac{\mu}{4} f(\eta_1, \eta_2, \eta_3)$ is positive definite.

By SYLVESTER's criterion necessary and sufficient conditions for $V - \frac{\mu}{4} f$ to be positive definite are

$$\begin{aligned} &\lambda - A\,\omega^2 + \mu\,\alpha^2 > 0 \\ &(\lambda - A\,\omega^2)(\lambda - B\,\omega^2) + \mu[\alpha^2(\lambda - B\,\omega^2) + \beta^2(\lambda - A\,\omega^2)] > 0 \\ &\lambda^3 - (A + B + C)\,\omega^2\lambda^2 + (B\,C + A\,C + A\,B)\,\omega^4\lambda - \\ &\quad - A\,B\,C\,\omega^6 + \mu[\lambda^2 - A\,\omega^2(\beta^2 + \gamma^2)\,\lambda - B\,\omega^2(\alpha^2 + \gamma^2)\,\lambda - \\ &\quad - C\,\omega^2(\alpha^2 + \beta^2)\,\lambda + (B\,C\,\alpha^2 + A\,C\,\beta^2 + A\,B\,\gamma^2)\,\omega^4] > 0 \end{aligned}$$

If we replace λ by one of its expressions (8.26) and ω^2 by its value calculated from one of equations (8.8′), the above three conditions reduce to

$$\left.\begin{aligned} \mu\,\alpha^2 - \frac{x_0}{\alpha} > 0 \\ \frac{x_0 y_0}{\alpha\beta} - \mu\left(\frac{x_0}{\alpha}\beta^2 + \frac{y_0}{\beta}\alpha^2\right) > 0 \\ \mu\left(\frac{y_0 z_0}{\beta\gamma}\alpha^2 + \frac{x_0 z_0}{\alpha\gamma}\beta^2 + \frac{x_0 y_0}{\alpha\beta}\gamma^2\right) - \frac{x_0 y_0 z_0}{\alpha\beta\gamma} > 0 \end{aligned}\right\} \tag{8.27}$$

If these conditions are satisfied, then V is a positive definite integral of the perturbed equations of motion of a heavy rigid body, and the unperturbed motion (8.20) of the body will be stable with respect to the variables $p, q, r;\ \gamma_1, \gamma_2, \gamma_3$.

For $\mu = 0$ the above conditions reduce to

$$\frac{x_0}{\alpha} > 0, \qquad -\frac{y_0}{\beta} > 0, \qquad -\frac{z_0}{\gamma} > 0 \tag{8.28}$$

Under the assumption (8.14) conditions (8.28) are satisfied for all points of the arc $(-g, -x)$ of the mass center curve (except the point $-x$ itself), or, what is the same, for all half-elements of the cone (8.9′) which lie between $-\overrightarrow{OG}$ and $-\overrightarrow{Ox}$ (Fig. 4).

Let us note that ω^2 is equal to zero for the half-element $-\overrightarrow{OG}$ which corresponds to the stable equilibrium of the body when the mass center lies on the vertical below the point of support O.

Assume now that $\mu \neq 0$. Since on all admissible arcs except $(-g, -x)$ we have $\alpha > 0$, then it is obvious that the first condition of (8.27) can be satisfied only by a proper choice of $\mu > 0$.

The second condition of (8.27) can be satisfied for a positive μ and α, β such that

$$\frac{x_0}{\alpha}\beta^2 + \frac{y_0}{\beta}\alpha^2 < 0 \tag{8.29}$$

This condition holds along a certain part of the admissible arc $(x, -y)$ adjoining the point x, in a small neighbourhood of which the values of α are close to $+1$, while those of β are negative and close to zero.

The third condition of (8.27) is also satisfied by a proper choice of a positive value for μ provided that the points (α, β, γ) of the arc $(x, -y)$ satisfy the condition

$$\left(\frac{x_0}{\alpha}\beta^2 + \frac{y_0}{\beta}\alpha^2\right)\frac{z_0}{\gamma} + \frac{x_0 y_0}{\alpha\beta}\gamma^2 > 0 \tag{8.30}$$

This condition is satisfied along a certain part of the arc $(x, -y)$ adjoining the point x. Since $\alpha > 0$, $\beta < 0$, $\gamma < 0$ along such an arc, then condition (8.29) is *a fortiori* satisfied provided that condition (8.30) holds.

In order that all three conditions of (8.27) be satisfied for $\alpha > 0$, $\beta < 0$, $\gamma < 0$ subject to condition (8.30), certain positive values must be chosen for the arbitrary constant μ. Therefore, if the largest of all the μ-values specified above is selected, then conditions (8.27) will be satisfied simultaneously.

Consider another Lyapunov V-function,

$$\begin{aligned} V &= V_1 - 2\omega V_2 + \lambda V_3 + A\frac{\omega}{\alpha}\xi_1 V_3 \\ &= A\,\xi_1^2 + B\,\xi_2^2 + C\,\xi_3^2 - 2\omega(B\,\xi_2\eta_2 + C\,\xi_3\eta_3) + \\ &\qquad + 2A\frac{\omega}{\alpha}\xi_1(\beta\eta_2 + \gamma\eta_3) + \lambda(\eta_1^2 + \eta_2^2 + \eta_3^2) + \\ &\qquad + A\frac{\omega}{\alpha}\xi_1(\eta_1^2 + \eta_2^2 + \eta_3^2) \end{aligned} \tag{8.31}$$

the time derivative of which by virtue of the equations for the perturbations (8.22) and (8.23) is equal to zero. The constant λ as before is determined by formulas (8.26).

In order that V given by (8.31) be a positive definite quadratic form, the following inequalities must be satisfied

$$\lambda > 0, \quad \lambda - B\,\omega^2 > 0, \quad \lambda - C\,\omega^2 > 0$$

$$(\lambda - B\,\omega^2)\,(\lambda - C\,\omega^2) - A\,\frac{\omega^2}{\alpha^2}\,[(\lambda - B\,\omega^2)\,\gamma^2 + (\lambda - C\,\omega^2)\,\beta^2] > 0$$

After replacing λ by one of its expressions given by (8.26) and ω^2 by its value calculated from one of equations (8.8′), the sufficient conditions for the stability of the unperturbed motion reduce to

$$\begin{gathered} B\,\frac{x_0}{\alpha} - A\,\frac{y_0}{\beta} > 0, \quad -\frac{y_0}{\beta} > 0, \quad -\frac{z_0}{\gamma} > 0 \\ \frac{1}{\beta\,\gamma}\left[y_0\,z_0\,\alpha^2 + \frac{A}{B - C}\,(y_0\,\gamma - z_0\,\beta)\left(\frac{y_0}{\beta}\,\gamma^2 + \frac{z_0}{\gamma}\,\beta^2\right)\right] > 0 \end{gathered} \tag{8.32}$$

The first three inequalities of (8.32) are satisfied by all points of the admissible arc $(x, -y)$ while the fourth is only satisfied for a certain part of this arc adjoining the point x. If we consider the plane

$$P_1(\beta, \gamma) = y_0\,\gamma - z_0\,\beta = 0 \tag{8.16′}$$

which passes through the x-axis and the mass center of the body, then the fourth condition of (8.32) is satisfied undoubtedly for all those points of the arc $(x, -y)$ which are on that side of the plane (8.16′) for which $P_1(\beta, \gamma) \leqq 0$.

Thus we have established that the axes of stable permanent rotations correspond to the half-elements of the cone (8.9′) which pass through an arbitrary point of the admissible arc $(-g, -x)$ of the mass center curve and through the points (α, β, γ) of the admissible arc $(x, -y)$ for which condition (8.30) or the fourth condition of (8.32) is satisfied.

In certain cases it is advisable to make use of the sufficiency conditions for the instability of permanent rotations, derived from the first approximation to the equations for the perturbations (8.22) and (8.23). The corresponding determinental equation has the form (see R. Grammel [4₁])

$$\sigma^2\,(g_0\,\sigma^4 + g_1\,\sigma^2 + g_2) = 0 \tag{8.33}$$

where

$$\begin{aligned} g_0 &= A\,B\,C, \quad g_1 = (A\,B\,C - \textstyle\sum A\,B_1\,C_1\,\alpha^2)\,\omega^2 - \sum A\,(B + C)\,x_0\,\alpha \\ g_2 &= -\omega^4 \textstyle\sum A\,B_1\,C_1\,\alpha^2 + \omega^2 \sum [A\,B_1\,\alpha\,(x_0\,\alpha^2 + x_0\,\gamma^2 + y_0\,\alpha\,\beta) - \\ &\quad - A\,C_1\,\alpha\,(x_0\,\alpha^2 + x_0\,\beta^2 + z_0\,\alpha\,\gamma) + 2A_1 B_1\,\gamma^2\,(x_0\,\alpha + y_0\,\beta)] + \\ &\quad + (A\,x_0\,\alpha + B\,y_0\,\beta + C\,z_0\,\gamma)\,(x_0\,\alpha + y_0\,\beta + z_0\,\gamma) \end{aligned} \tag{8.34}$$

The symbol Σ stands for the sum of three terms obtained from the expression under this sign by circular permutation of the letters $A, B, C; A_1, B_1, C_1; \alpha, \beta, \gamma; x_0, y_0, z_0$, and

$$A_1 = B - C, \quad B_1 = C - A, \quad C_1 = A - B$$

If one of the inequalities

$$g_1 < 0, \quad g_2 < 0, \quad g_1^2 - 4 g_0 g_2 < 0 \tag{8.35}$$

is satisfied, then among the roots of the determinental equation there are roots with positive real parts. Hence according to a theorem of LYAPUNOV the unperturbed motion (8.20) is unstable.

For $\overrightarrow{OG}$ which is defined by the equations

$$\frac{\alpha}{x_0} = \frac{\beta}{y_0} = \frac{\gamma}{z_0}$$

and which can be a permanent axis of rotation only for $\omega = 0$ the first condition of (8.35) is satisfied for $\alpha > 0, \beta > 0, \gamma > 0$. By continuity it can be assumed that this condition will hold also in a certain neighbourhood of the element $\overrightarrow{OG}$ of the cone (8.9′), i.e. on the admissible arc (g, z) of the mass center curve there will exist a certain domain of instability.

Case 2. The mass center of the body lies in one of the principal inertia planes at the fixed point O. Let $x_0 > 0, y_0 > 0, z_0 = 0$.

In such a case the cone (8.9′) degenerates into the coordinate plane $z = 0$ and the plane

$$T_3(\alpha, \beta) = (B - C) \beta x_0 + (C - A) \alpha y_0 = 0 \tag{8.9''}$$

Let us investigate first the stability of permanent axes of rotation which are different from the principal axes of inertia but are lying in the plane $T_3(\alpha, \beta) = 0$.

The components of the angular velocity are given by formulas (8.20), where α, β satisfy equation (8.9″), $-1 \leqq \gamma \leqq 1$, and ω is determined by any of the equations (8.8′) with $z_0 = 0$.

Consider again the LYAPUNOV V-function (8.25). The first two conditions of (8.27), which must be satisfied in order that this function be positive definite, retain their form also for $z_0 = 0$, while the third condition assumes the form $\mu \frac{x_0 y_0}{\alpha \beta} \gamma^2 > 0$.

These conditions are satisfied for an arbitrary value of $\mu > 0$ provided that $\alpha < 0, \beta < 0, \gamma \neq 0$. Under the assumption $C > A > B$ the latter conditions hold for any admissible permanent axis of rotation lying in the plane $T_3(\alpha, \beta) = 0$ except the axis for which $\gamma = 0$. Consequently rotations about these axes are stable (Fig. 5).

Let us indicate also unstable axes of permanent rotation. Consider, for example, the second instability condition of (8.35). Using the third equation of system (8.8′) we find that for $z_0 = 0$ this condition assumes the form

$$g_2 \equiv [-A_1 B_1 C\,\omega^2 + 3A_1 B_1 (x_0\,\alpha + y_0\,\beta)]\,\omega^2\,\gamma^2 + \\ + (A_1\,\beta\,x_0 + B_1\,\alpha\,y_0)\left[3\,(x_0\,\beta - y_0\,\alpha) - \frac{1}{C_1}\left(A\,y_0\frac{\alpha}{\beta^2} + B\,x_0\frac{\beta}{\alpha^2}\right) + \\ + \frac{1}{C_1}\left(A\frac{x_0}{\beta} + B\frac{y_0}{\alpha}\right)\gamma^2\right] < 0 \tag{8.36}$$

For all axes lying in the plane

$$T_3(\alpha, \beta) \equiv A_1\,\beta\,x_0 + B_1\,\alpha\,y_0 = 0$$

except the axis for which $\gamma = 0$, inequality (8.36) reduces to

$$A_1 B_1\left[3\,(x_0\,\alpha + y_0\,\beta) - \frac{C}{C_1}\,\frac{x_0\,\beta - y_0\,\alpha}{\alpha\,\beta}\right] < 0$$

For $A > B > C$ the latter condition reduces to the inequality

$$3\,(x_0\,\alpha + y_0\,\beta) > \frac{C}{A - B}\,\frac{x_0\,\beta - y_0\,\alpha}{\alpha\,\beta} \tag{8.37}$$

and for $B > C > A$ to the opposite inequality

$$3\,(x_0\,\alpha + y_0\,\beta) < \frac{C}{A - B}\,\frac{x_0\,\beta - y_0\,\alpha}{\alpha\,\beta} \tag{8.38}$$

Thus we have established that permanent rotations about admissible axes lying in the plane $T_3(\alpha, \beta) = 0$ (except the principal axis of inertia $O\,z$ and the axis for which $\gamma = 0$) are stable if the mass center of the body lies in the plane determined by the largest and middle axes of the ellipsoid of inertia at O, and that they are unstable if the mass center of the body lies in the plane passing through the smallest and middle (or largest) axes of the ellipsoid of inertia and condition (8.37) [or (8.38)] holds.

Let us note that in the case $B > C > A$ condition (8.38) holds provided that

$$A_1\,x_0^2 \geq B_1\,y_0^2$$

This condition is satisfied in the case of a loxodromic pendulum for which the mass distribution satisfies the conditions

$$B(C - A)\,y_0^2 - A\,(B - C)\,x_0^2 = 0, \qquad z_0 = 0 \tag{8.39}$$

and in the case of G. Grioli [1] for which

$$(B - C)\,x_0^2 - (C - A)\,y_0^2 = 0, \qquad z_0 = 0 \tag{8.40}$$

Consequently in these cases permanent rotations about axes lying in the plane $T_3(\alpha, \beta) = 0$ are unstable.

Let us pass now to the consideration of stability of permanent rotations about axes lying in the plane $z = 0$ and different from the principal axes of inertia. The components of the angular velocity along the axes are given by (8.20), where α, β satisfy the condition $\alpha^2 + \beta^2 = 1$,

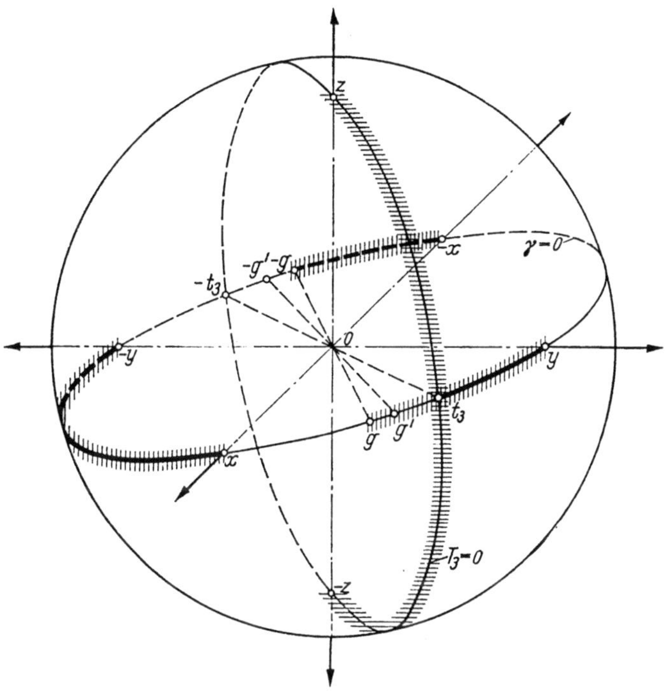

Fig. 7. $A > B > C$; $x_0 > 0$, $y_0 > 0$, $z_0 = 0$

$\gamma = 0$ and ω is determined by the third equation of system (8.8′). Consider again the LYAPUNOV V-function (8.25) and substitute into the latter $\gamma = 0$ and $\lambda = A\,\omega^2 - x_0/\alpha = B\,\omega^2 - y_0/\beta$.

In such a case the sufficient conditions for positive definiteness of V assume the form

$$\frac{1}{C_1\,\alpha\,\beta}(A_1\,\beta\,x_0 + B_1\,\alpha\,y_0) > 0, \qquad \mu\,\alpha^2 - \frac{x_0}{\alpha} > 0$$
$$\frac{x_0\,y_0}{\alpha\,\beta} - \mu\left(x_0\frac{\beta^2}{\alpha} + y_0\frac{\alpha^2}{\beta}\right) > 0 \tag{8.41}$$

The first condition holds on the admissible arcs

$(x, -y)$, $(-g, -x)$, (y, t_3)	if $A > B > C$	(Fig. 7)
$(-y, -g)$, $(-t_3, y)$	if $B > C > A$	(Fig. 8)
$(-g, -t_3)$	if $C > A > B$	(Fig. 5)

while the second and third conditions are satisfied for arbitrary non-negative values of μ provided that $\alpha < 0$, $\beta < 0$, and also for some positive value of μ if (8.29) holds.

Consequently permanent rotations are stable about axes passing through an arbitrary point [except the points $-x$, $-y$] of the ad-

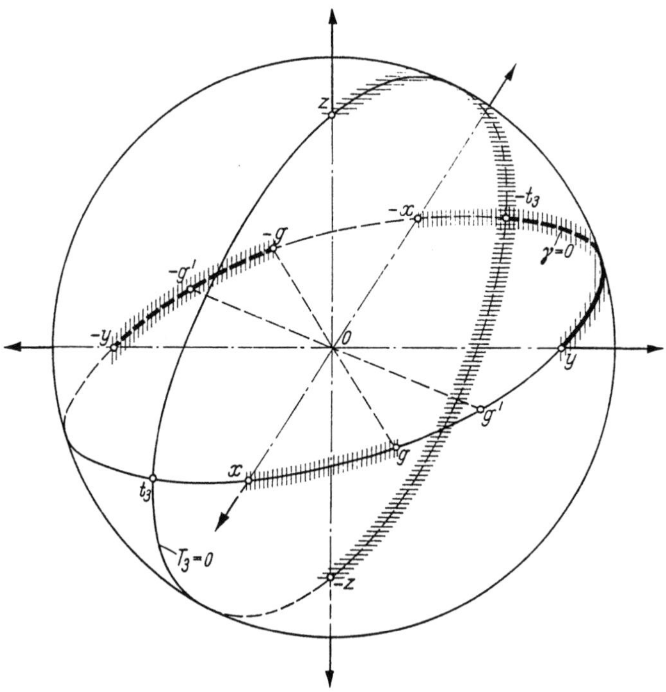

Fig. 8. $B > C > A$; $x_0 > 0$, $y_0 > 0$, $z_0 = 0$

missible arcs $(-g, -x)$, $(-y, -g)$, $(-g, -t_3)$ in the cases $A > B > C$, $B > C > A$ and $C > A > B$ respectively, as well as through those points of the admissible arcs $(x, -y)$ and $(y, -t_3)$ for $A > B > C$ and $B > C > A$ respectively for which the coordinates α, β satisfy condition (8.29).

Let us note that this condition for points of the arc $(x, -y)$ in the case $A > B > C$ can be rewritten in the form of the inequality

$$\frac{x_0}{y_0} < -\frac{\alpha^3}{\beta^3} \tag{8.42}$$

while for points of the arc $(y, -t_3)$ under the assumption $B > C > A$ in the form of an inequality opposite to (8.42).

Consider the straight line

$$S_1(\alpha, \beta) = x_0 \alpha + y_0 \beta = 0$$

orthogonal to the straight line through the mass center G and denote by s_1 the point of intersection of the straight line $S_1(\alpha, \beta) = 0$ with the arc $(x, -y)$ of the circle $\alpha^2 + \beta^2 = 1$. Then condition (8.29) holds for all points of the arc (x, s_1) in the case $A > B > C$ if $x_0 < y_0$, and for all points of the arc $(y, -s_1)$ in the case $B > C > A$ if $x_0 > y_0$ (Figs 9 and 10).

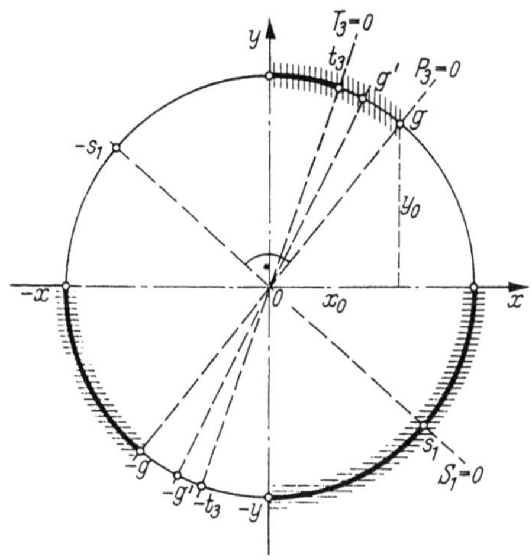

Fig. 9. $A > B > C$; $x_0 < y_0$

If $B > C > A$, then for a rigid body whose mass distribution satisfies conditions (8.40), the points t_3 and s_1 coincide. If, in addition,

$$C - A > B - C \quad (x_0 > y_0)$$

then permanent rotations about axes passing through an arbitrary point of the arc $(y, -t_3)$, except the point y, are stable. For a rigid body whose mass distribution satisfies the conditions (8.39), the above mentioned permanent axes of rotation will be stable provided that $A B_1^2 > A_1^2 B$.

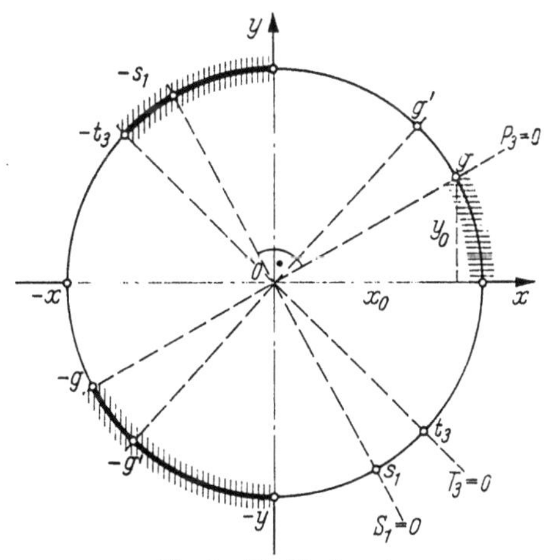

Fig. 10. $B > C > A$; $x_0 > y_0$

In the case of an ellipsoid of revolution, say $B = C \gtrless A$, it can be established by means of the sufficient conditions for stability (8.41) that permanent rotations are stable about axes passing through arbitrary points (except the points $\pm x, -y$) of the admissible arcs $(-x, -g)$ and $(-g, -y)$ for $A > B$ and $B > A$ respectively, as well as through those points of the arc $(x, -y)$ for which inequality (8.42) holds (in the case $A > B$).

Case 3. The mass center of the body lies on a principal axis of inertia at the fixed point O. Let

$$x_0 > 0, \quad y_0 = z_0 = 0 \tag{8.43}$$

In such a case the mass center cone (8.10) degenerates into two principal planes of inertia $y = z = 0$ which intersect along the principal inertia axis $O\,x$, passing through the mass center G.

Let us investigate the stability of permanent rotations about axes lying in the plane $y = 0\,(\beta = 0)$ and different from the principal inertia axes.

The components of the angular velocity $\vec{\omega}$ along the coordinate axes are determined by formulas (8.20) in which β is put equal to zero, while ω is determined by the second equation of system (8.8′).

Consider again the LYAPUNOV V-function (8.25) with $\beta = 0$, $\lambda = A\,\omega^2 - x_0/\alpha = C\,\omega^2$.

The conditions for the positive definiteness of this function can be reduced to the inequalities (for $\mu > 0$)

$$(C - A)\,\omega^2\,\gamma^2 > 0, \quad C > B$$

Consequently, if $C > A$ and $C > B$, then the permanent rotations about axes lying in the plane $y = 0$ are stable.

Also unstable rotations can be indicated. In fact, in case (8.43) the second instability condition of (8.35) reduces to the inequality

$$(B - C)\,[3\,(C - A)\,\alpha + C/\alpha] < 0 \tag{8.44}$$

It is obvious that this condition is satisfied in the following cases:

(i) for all admissible axes ($\alpha < 0$) lying in the plane $y = 0$ if $B > C > A$;

(ii) for all admissible axes ($\alpha > 0$) whose direction cosines α with respect to the x-axis satisfy the inequality

$$\alpha^2 < C/[3\,(A - C)] = \alpha_1^2 \tag{8.45}$$

if $A > C > B$;

(iii) for all admissible axes ($\alpha > 0$) whose direction cosines α with respect to the x-axis satisfy the inequality

$$\alpha^2 > \alpha_1^2 \tag{8.46}$$

if $A > C$ and $B > C$.

Thus it is proved that in the case (8.43) permanent rotations about axes lying in the plane $y = 0$ are: (i) stable if the x-axis, passing through the mass center G of the body, is the largest or middle axis while the z-axis is the smallest axis of the ellipsoid of inertia; (ii) unstable if the x-axis and z-axis are the largest and middle axes respectively of the inertia ellipsoid, or if the x-axis and z-axis are the smallest and middle axes respectively of the inertia ellipsoid and the direction cosine α of the permanent axis satisfies inequality (8.45), or if the x-axis and z-axis are correspondingly the smallest (or middle) and largest axes of the inertia ellipsoid and the direction cosine α of the permanent axis satisfies inequality (8.46).

The results concerning stability and instability of permanent rotations about admissible axes lying in the plane $z=0$ are analogous to the above formulated, and can be obtained from them by interchanging the axes z and y and the moments of inertia C and B.

Next let us consider the stability of permanent rotations in the case where the ellipsoid of inertia at the fixed point O is an ellipsoid of revolution, say,

$$B = C, \quad x_0 = z_0 = 0, \quad y_0 > 0 \tag{8.47}$$

The cone of permanent axes degenerates into two principal planes of inertia $x=0$, $z=0$, whereas the angular velocity of the axes lying in the plane $x=0$ becomes infinite.

For investigation of stability of permanent rotations about axes lying in the plane $z=0$ let us make use of the LYAPUNOV V-function (8.25) with $\gamma=0$ and $\lambda = A\,\omega^2 = B\,\omega^2 - y_0/\beta$.

The conditions for positive definiteness of this function are

$$\mu\,\alpha^2 (A - B)\,\omega^2 > 0$$

$$\mu\,\alpha^2 > 0$$

Obviously these conditions are satisfied for an arbitrary $\mu > 0$ if $A > B = C$; β, the cosine of the angle made by the permanent axis with the y-axis, will be negative.

Consider now the second instability condition of (8.35) which under assumptions (8.47) and $B > A$, where $\beta > 0$, reduces to

$$\beta^2 > A/[3\,(B - A)] = \beta_1^2 \tag{8.48}$$

Consequently permanent rotations about admissible axes lying in the plane $z=0$ subject to assumptions (8.47) are:

(i) stable if $A > B$;

(ii) unstable if $B > A$ and inequality (8.48) holds.

Finally let us investigate the stability of rotation about the principal inertia axis $O\,x$ which passes through the mass center G of the body and which can be a permanent vertical axis of rotation for an arbitrary value of the angular velocity ω. Assume that

$$x_0 \gtrless 0, \quad y_0 = z_0 = 0$$

In order to study the stability of the unperturbed motion

$$p_0 = \omega, \quad q_0 = r_0 = 0; \quad \alpha = 1, \quad \beta = \gamma = 0$$

consider again the LYAPUNOV V-function (8.25), taking into account the above assumptions and putting $\lambda = A\,\omega^2 - x_0$.

Conditions for positive definiteness of this function are

$$\mu > x_0, \quad (A - B)\,\omega^2 > x_0, \quad (A - C)\,\omega^2 > x_0 \qquad (8.49)$$

The first condition can be satisfied by a proper choice of μ. In discussing the last two conditions we shall distinguish the following subcases.

(i) *Short-axis gyroscope:* $A > B > C$. If $x_0 < 0$, then conditions (8.49) are satisfied for an arbitrary ω. Consequently a suspended short-axis gyroscope is stable for any value of the angular velocity ω.

Obviously this conclusion holds also for the case of an ellipsoid of revolution when $A = B \geqq C$ or $A \geqq B = C$.

If $x_0 > 0$, then conditions (8.49) are satisfied for

$$\omega^2 > x_0/(A - B) = \omega_1^2 \qquad (8.50)$$

Consequently if the angular velocity of rotation $\omega > \omega_1$, then the short-axis gyroscope is stable. This statement holds also in the case of LAGRANGE and POISSON, when $B = C$.

(ii) *Long-axis gyroscope:* $B > C > A$. For $x_0 < 0$ conditions (8.49) are satisfied only if $\omega^2 < \omega_1^2$. Thus a long-axis suspended gyroscope is stable if its angular velocity $\omega < \omega_1$. This conclusion holds also for an ellipsoid of revolution when $B = C > A$ or $B > C = A$.

(iii) *Middle-axis gyroscope:* $C > A \geqq B$. Conditions (8.49) are satisfied only in the case $x_0 < 0$ provided that

$$\omega^2 < x_0/(A - C) = \omega_2^2 \qquad (8.51)$$

Since in this case the necessary condition for stability (see R. GRAMMEL [4₁]) has also the form (8.51), then a suspended middle-axis gyroscope is stable only in the case where its angular velocity satisfies condition (8.51).

Let us note that analogous sufficient conditions for the stability of rotation of a heavy rigid body about the principal axes of inertia were obtained also by V. N. SKIMEL' [1] by means of a LYAPUNOV V-function of the form (8.31).

8.34. *Applications to particular cases of motion*

Let us apply the results obtained in Case 3 concerning the stability of permanent rotations to rigid bodies whose mass distribution is subject to conditions corresponding to some known cases of integrability.

(i) *Case of S. Kovalevskaya*: $B = C = 2A$, $x_0 = z_0 = 0$, $y_0 > 0$. Permanent rotations about admissible axes ($\beta > 0$) lying in the plane $z = 0$ are unstable if β, the cosine of the angle between the axis of permanent rotation and the y-axis, satisfies the inequality

$$\beta^2 > \beta_1^2 = \tfrac{1}{3}$$

It is known that the rotation of the KOVALEVSKAYA gyroscope about the principal axis of inertia through the mass center G is stable only in the case where the mass center lies on the vertical below the point of suspension ($\beta = -1$) (see V. V. RUMYANCEV [1]).

(ii) *Case of Steklov and Bobylev* (see Sec. 8.4): $A = 2B$, $y_0 = z_0 = 0$, $x_0 > 0$. Consider the following two subcases:

(a) $C > A = 2B$. Rotations about admissible axes ($\alpha < 0$) lying in the plane $y = 0$ are stable, while rotations about axes ($\alpha > 0$) lying in the plane $z = 0$ are unstable if α, the cosine of the angle between the permanent axis of rotation and the x-axis, satisfies the inequality

$$\alpha^2 > \alpha_2^2 = \tfrac{1}{3}, \qquad \alpha_2^2 = B/[3(A-B)]$$

Rotations about the negative x-axis ($\alpha = -1$) are stable only for $\omega < \tilde{\omega}_2$, where $\tilde{\omega}_2^2 = x_0/(C-A)$; rotations about the positive x-axis ($\alpha = 1$) are unstable (see R. GRAMMEL [4_1]).

(b) $A > C > B$. Rotations about the permanent axes ($\alpha > 0$) are unstable if the direction cosines α for the axes lying in the plane $y = 0$ satisfy the condition $\alpha^2 < \alpha_1^2$, while for those lying in the plane $z = 0$ the condition $\alpha^2 > \alpha_2^2 = \frac{1}{3}$. Rotations are stable about the negative x-axis ($\alpha = -1$) for arbitrary values of the angular velocity and about the positive x-axis ($\alpha = 1$) for $\omega > \tilde{\omega}_2$.

(iii) *Second case of Steklov* (see Sec. 8.62): $B > A > 2C$, $x_0 > 0$, $y_0 = z_0 = 0$. Permanent rotations about axes lying in the plane $z = 0$ are stable, while rotations about those in the plane $y = 0$ are unstable if $\alpha^2 > \alpha_1^2$. Rotations about the negative x-axis ($\alpha = -1$) are stable only for $\omega < \tilde{\omega}_1$, where $\tilde{\omega}_1^2 = x_0/(B-A)$. Rotations about the positive x-axis ($\alpha = 1$) are unstable (see R. GRAMMEL [4_1]).

P. A. KUZ'MIN [1] noticed that there exists a solution analogous to that of V. A. STEKLOV for certain initial values and a different mass distribution; namely, $2B > A > 2C$, $A > B$. In such a case permanent rotations are unstable if the corresponding axes, lying in the plane $y = 0$, satisfy the condition $\alpha^2 > \alpha_1^2$, while those lying in the plane $z = 0$ satisfy the condition $\alpha^2 < \alpha_2^2$.

Rotations about the negative x-axis ($\alpha = -1$) are stable for arbitrary values of the angular velocity, while rotations about the positive x-axis ($\alpha = 1$) are stable for $\omega > \tilde{\omega}_1$.

(iv) *Second case of Goryačev* (see Sec. 8.61): $AC = 8(A-2B)(B-C)$, $x_0 > 0$, $y_0 = z_0 = 0$. Consider two possible subcases:

(a) $C > A > B$, $2B > A$. Permanent rotations about admissible axes lying in the plane $y = 0$ are stable, while those about axes in the plane $z = 0$ are unstable if $\alpha^2 > \alpha_2^2$. Rotations about the negative x-axis ($\alpha = -1$) are stable only for $\omega < \tilde{\omega}_2$.

(b) $C > B > A$. Permanent rotations about axes lying in the plane $y = 0$ are stable, while those about axes in the plane $z = 0$ are unstable. Rotations about the negative x-axis ($\alpha = -1$) are stable if $\omega < \tilde{\omega}_2$.

(v) *Case of N. Kowalewski* (see Sec. 8.9): $AC = 9(B-C)(A-2B)$, $x_0 > 0$, $y_0 = z_0 = 0$. In the case $2B > A > C > B$ the rotations are unstable if the corresponding axes, lying in the plane $y = 0$, satisfy the condition $\alpha^2 < \alpha_1^2$, while those lying in the plane $z = 0$ satisfy the condition $\alpha^2 > \alpha_2^2$. Rotations about the negative x-axis ($\alpha = -1$) are stable for arbitrary values of the angular velocity, while those about the positive x-axis ($\alpha = 1$) are stable for $\omega > \tilde{\omega}_2$.

(vi) *Second case of Čaplygin* (see Sec. 8.63): $x_0 > 0$, $y_0 = z_0 = 0$, $0.5965 < C/A < 0.6$, $1.5 < B/A < 1.5965$ ($B > A > C$).

Permanent rotations about admissible axes lying in the plane $z = 0$ are stable, while those lying in the plane $y = 0$ are unstable if $\alpha^2 > \alpha_1^2$.

Permanent rotations about the negative x-axis ($\alpha = -1$) are stable only for $\omega < \tilde{\omega}_1$; rotations about the positive x-axis ($\alpha = 1$) are unstable (see R. GRAMMEL [4_1]).

(vii) *Case of Goryačev and Čaplygin* (see Sec. 8.5): $B = C = 4A$, $y_0 > 0$, $x_0 = z_0 = 0$.

Permanent rotations about the admissible axes lying in the plane $z = 0$ are unstable if β, the cosine of the angle between the axis under consideration and the y-axis, satisfies the condition $\beta^2 > \beta_1^2 = \frac{1}{9}$.

Rotations about the negative y-axis ($\beta = -1$) are stable for arbitrary values of the angular velocity.

Remark. So far only restrictions concerning the mass distribution of the body have been taken into account disregarding the initial conditions. The latter in certain cases exclude the possibility of permanent rotations about the vertical (for example, in the case of GORYAČEV and ČAPLYGIN).

8.4. Case of Steklov and Bobylev. This case was discovered simultaneously by V. A. STEKLOV [1] (1896) and D. BOBYLEV [1] (1896) who showed that when $A = 2B$, and the mass center of the body is on the x-axis ($x_0 > 0$) of the ellipsoid of inertia at O, i.e. when $y_0 = z_0 = 0$, the EULER and POISSON equations can be integrated, and the solution depends upon three arbitrary constants.

8.5. Case of Goryačev and Čaplygin. D. N. GORYAČEV [2] (1900) showed that when $A = B = 4C$, the mass center of the body is in the plane of equal moments of inertia at O, i.e. $z_0 = 0$ (by a rotation of the trihedral $Oxyz$ about the z-axis we can make the center of mass of the body lie on the x-axis, i.e. $y_0 = 0$ also), and the projection of the angular momentum vector of the body on the vertical through O is zero, i.e. $c = 0$, then a fourth particular algebraic integral exists and the solution depends upon three arbitrary constants.

S. A. ČAPLYGIN [3] (1901) improved GORYAČEV's result and obtained a solution containing four arbitrary constants. See also L. N. SRETENSKII [3].

L. N. SRETENSKII [2] and YU. A. ARHANGEL'SKII [1] have investigated the motion of the GORYAČEV and ČAPLYGIN gyroscope under the assumption that the body is rotating with a very large angular velocity about the principal axis of inertia through the mass center of the body. The resulting motions are analogous to the quasi-regular precessional motions of the LAGRANGE gyroscope. See also YU. A. ARHANGEL'SKII [7, 10, 11] and P. G. KROTOVA [1].

The EULER and POISSON equations in the case under consideration assume the form

$$\begin{aligned} &4\frac{dp}{dt} = 3qr, && 4\frac{dq}{dt} = -3rp + a\gamma, && \frac{dr}{dt} = -\beta a \\ &\frac{d\alpha}{dt} = r\beta - q\gamma, && \frac{d\beta}{dt} = p\gamma - r\alpha, && \frac{d\gamma}{dt} = q\alpha - p\beta \end{aligned} \tag{8.52}$$

where

$$a = \frac{x_0}{C}$$

and the three classical first integrals are

$$\left.\begin{aligned} 4(p^2 + q^2) + r^2 + 2a\alpha &= k \qquad \left(k = \frac{2h_1}{C}\right) \\ 4(p\alpha + q\beta) + r\gamma &= h \qquad \left(h = \frac{c}{C}\right) \\ \alpha^2 + \beta^2 + \gamma^2 &= 1 \end{aligned}\right\} \quad (8.53)$$

Let

$$s = (p^2 + q^2)\, r - a\, p\, \gamma$$

Then by virtue of the above equations we obtain that

$$\frac{ds}{dt} = -\frac{a\,q}{4}\,[4(p\,\alpha + q\,\beta) + r\,\gamma]$$

If $h = 0$ (i.e. $c = 0$), then $ds/dt = 0$, and $s = \text{const}$. Hence

$$(p^2 + q^2)\, r - a\, p\, \gamma = C_4 \qquad (C_4 \text{ const.}) \quad (8.54)$$

is the fourth integral, the left-hand side of which is of degree three with respect to the components of the angular velocity $\vec{\omega}$ of the body.

The reduction of the problem to quadratures can now be accomplished as follows. Putting $r = u - v$, $4(p^2 + q^2) = u\,v$, introducing the notations

$$\left.\begin{aligned} U &= u^3 - k\,u - 4C_4, & V &= v^3 - k\,v + 4C_4 \\ U_1^2 &= U - 2a\,u, & V_1^2 &= V - 2a\,v \\ -U_2^2 &= U + 2a\,u, & -V_2^2 &= V + 2a\,v \end{aligned}\right\} \quad (8.55)$$

and using integrals (8.53) and (8.54) we obtain the following expressions for $p, q, r; \alpha, \beta, \gamma$ in terms of u and v:

$$\begin{aligned} &8a\,p = U_1 V_2 - U_2 V_1, \quad 8a\,q = U_1 V_1 + U_2 V_2, \quad r = u - v \\ &2a\,\alpha = -\frac{U+V}{u+v}, \quad 2a\beta = -\frac{U_1 U_2 - V_1 V_2}{u+v}, \quad 2a\gamma = -\frac{U_1 V_2 + U_2 V_1}{u+v} \end{aligned} \quad (8.56)$$

The new variables u and v as functions of t satisfy the differential equations

$$2(u + v)\frac{du}{dt} = U_1\, U_2, \qquad 2(u + v)\frac{dv}{dt} = V_1 V_2$$

Hence

$$\frac{du}{U_1\, U_2} - \frac{dv}{V_1 V_2} = 0, \qquad \frac{2u\,du}{U_1\, U_2} + \frac{2v\,dv}{V_1 V_2} = dt \qquad (8.57)$$

and the problem is solved in terms of hyperelliptic integrals. Recently Miss V. A. Toropova [1] has applied this method to the general case of rotation of a heavy rigid body about a fixed point.

If the polynomials U_1^2 and U_2^2 have a common root, the nby necessity $C_4 = 0$, and equations (8.57) assume the form

$$\frac{du}{u\sqrt{(2a+k-u^2)(2a-k+u^2)}} - \frac{dv}{v\sqrt{(2a+k-v^2)(2a-k+v^2)}} = 0$$
$$\frac{du}{\sqrt{(2a+k-u^2)(2a-k+u^2)}} + \frac{dv}{\sqrt{(2a+k-v^2)(2a-k+v^2)}} = \frac{dt}{2} \qquad (8.58)$$

Integration of the first equation gives the algebraic integral

$$\sqrt{(2a+k-u^2)(2a-k+v^2)} - \sqrt{(2a+k-v^2)(2a-k+u^2)} = C_5\, u\, v$$

where C_5 is a new constant of integration.

In terms of the old variables the above integral assumes the form

$$2a\,p = C_5 (p^2+q^2)^{3/2}$$

which is one of GORYAČEV's first integrals.

R. MARCOLONGO [3] pointed out that the six quantities $p, q, r; \alpha, \beta, \gamma$ can be expressed in terms of Theta-functions of two arguments, both of which are linear functions of the time t. See also O. OLSSON [1].

The stability of rotation of a rigid body in the case of GORYAČEV and ČAPLYGIN has been investigated by M. P. GULYAEV and M. OŠIBAEV [1], and E. J. ROUTH.

A particular solution of system (8.52) is

$$p = \omega = \text{const.}, \quad q = r = 0; \quad \alpha = 1, \quad \beta = \gamma = 0 \qquad (8.59)$$

being a permanent rotation of the body about the x-axis coinciding with the vertical through the fixed point. If this motion is considered as the unperturbed motion of the body, then the problem is to find necessary and sufficient conditions for the stability of this solution with respect to the variables $p, q, r; \alpha, \beta, \gamma$.

Let the perturbed motion be

$$p = \omega + \xi_1, \quad q = \xi_2, \quad r = \xi_3; \quad \alpha = 1 + \eta_1, \quad \beta = \eta_2, \quad \gamma = \eta_3 \qquad (8.60)$$

Then the differential equations for the perturbations are

$$\left.\begin{aligned}
4\frac{d\xi_1}{dt} &= 3\,\xi_2\,\xi_3\\
4\frac{d\xi_2}{dt} &= -3(\omega+\xi_1)\,\xi_3 + a\,\eta_3\\
\frac{d\xi_3}{dt} &= -a\,\eta_2\\
\frac{d\eta_1}{dt} &= \xi_3\,\eta_2 - \xi_2\,\eta_3\\
\frac{d\eta_2}{dt} &= (\omega+\xi_1)\,\eta_3 - (1+\eta_1)\,\xi_3\\
\frac{d\eta_3}{dt} &= (1+\eta_1)\,\xi_2 - (\omega+\xi_1)\,\eta_2
\end{aligned}\right\} \qquad (8.61)$$

The above system possesses the following three first integrals:

$$\left.\begin{aligned} V_1 &= 4(2\omega\,\xi_1 + \xi_1^2 + \xi_2^2) + \xi_3^2 + 2a\,\eta_1 \\ V_2 &= 4(\xi_1 + \xi_1\,\eta_1 + \xi_2\,\eta_2 + \omega\,\eta_1) + \xi_3\,\eta_3 \\ V_3 &= \eta_1^2 + \eta_2^2 + \eta_3^2 + 2\eta_1 \end{aligned}\right\} \tag{8.62}$$

According to LYAPUNOV's direct method the particular solution (8.59) will be stable provided that there exists an integral with definite sign.

For this purpose construct a LYAPUNOV V-function as the following linear combination of integrals (8.62)

$$\begin{aligned} V &= V_1 - 2\omega\,V_2 + (4\omega^2 - a)\,V_3 \\ &= \{4\,\xi_1^2 - 8\omega\,\xi_1\,\eta_1 + (4\omega^2 - a)\,\eta_1^2\} + \\ &\quad + \{4\,\xi_2^2 - 8\omega\,\xi_2\,\eta_2 + (4\omega^2 - a)\,\eta_2^2\} + \\ &\quad + \{\xi_3^2 - 2\omega\,\xi_3\,\eta_3 + (4\omega^2 - a)\,\eta_3^2\} \\ &= (\mathrm{I}) + (\mathrm{II}) + (\mathrm{III}) \end{aligned}$$

In order that the quadratic forms (I) and (II) be positive definite it is necessary and sufficient that the inequality

$$\begin{vmatrix} 4 & -4\omega \\ -4\omega & 4\omega^2 - a \end{vmatrix} = -4a > 0 \tag{8.63}$$

holds. Similarly, in order that the quadratic form (III) be positive definite it is necessary and sufficient that

$$\begin{vmatrix} 1 & -\omega \\ -\omega & 4\omega^2 - a \end{vmatrix} = 3\omega^2 - a > 0 \tag{8.64}$$

If condition (8.63) is satisfied, i.e. $a < 0$, then condition (8.64) also holds. Condition (8.63) is sufficient for the stability of the particular solution (8.59) since by virtue of system (8.61) $d\,V/dt$ is identically zero.

It is easy to insure that condition (8.63) is also necessary for the stability of solution (8.59). In fact, the characteristic equation corresponding to the first approximation of system (8.61) is

$$\lambda^2[4\lambda^4 + (4\omega^2 - 5a)\,\lambda^2 - 3a\,\omega^2 + a^2] = 0$$

For $a > 0$ this equation has at least one positive root, and consequently the particular solution (8.59), by virtue of a well known theorem by LYAPUNOV, is unstable.

It is interesting to note that ROUTH solved the stability problem for uniform precessions about the vertical without any restrictions on the moments of inertia and the position of the mass center of the body.

8.6. Other cases of Goryačev, Steklov and Čaplygin. At the turn of the century several authors, such as D. N. GORYAČEV, V. A. STEKLOV and S. A. ČAPLYGIN, set out to find new integrable cases for which the time-independent first integrals are algebraic and do not contain arbitrary constants. Conditions under which such integrals exist impose restrictions upon the principal moments of inertia of the body, either in the form of an equality or in the form of inequalities restricting the values of the ratios of two principal moments of inertia. The most general results obtained by these authors are those by S. A. ČAPLYGIN.

Assuming that $y_0 = z_0 = 0$, he sought conditions in order that integrals of the form

$$-x_0\beta = \delta p q + \lambda p^n q, \qquad -x_0\gamma = \varepsilon p r + \mu p^n r$$

exist for the EULER and POISSON equations. Here δ, ε, λ and μ are constants to be determined. ČAPLYGIN showed that only the following three cases are possible.

8.61. Second case of Goryačev [1] (1898): $\lambda = 0$, $n = 3$

Under the assumption

$$A C = 8(A - 2B)(B - C)$$

particular integrals

$$-2x_0\beta = (4B - 3A) p q, \qquad -x_0\gamma = (\varepsilon + \mu p^2) p r$$

exist, where

$$\varepsilon = \frac{(3A - 4B)(2B - C)(2B - 3C)}{BC}, \qquad \mu = \frac{\varepsilon A(3A - 4B)(4B - 3C)(5C - 4B)}{32x_0 BC(B - C)}$$

The solution contains a single arbitrary parameter.

A. A. BOGOYAVLENSKII [1] showed that in this case we have

$$c = 0, \qquad h = \frac{4x_0(A - 2B)(9A^2 - 56AB + 64B^2)}{(3A - 4B)(15A^2 - 64AB + 64B^2)}$$

8.62. Second case of Steklov (1899): $\lambda = \mu = 0$

This is the case where $B > A > 2C$, and $y_0 = z_0 = 0$. V. A. STEKLOV [2] showed that in such a case p, q, r are proportional to $\mathrm{cn}\,\varkappa t$, $\mathrm{sn}\,\varkappa t$, $\mathrm{dn}\,\varkappa t$ respectively, where $\varkappa$ is determined by A, B, C and the distance between the points O and G, and that

$$-\beta = \frac{(C - A)(B - A)}{x_0(2C - A)} p q, \qquad -\gamma = \frac{(C - A)(B - A)}{x_0(2B - A)} p r$$

P. A. KUZ'MIN [1] carried over STEKLOV's results to the subcase $2B > A > 2C$, $A > B$. In both subcases the solution depends upon a single arbitrary constant t_0.

A. A. BOGOYAVLENSKII [1] showed that in the first case we have

$$c = 0, \qquad h = \frac{x_0(A^2 - 2AB - 2AC + 2BC)}{(A - B)(C - A)}$$

8.63. Second case of Čaplygin [4] (1904)

Under the assumptions

$$n = -\frac{1}{3}\left(0.5965 < \frac{C}{A} < 0.6, \quad 1.5 < \frac{B}{A} < 1.5965\right)$$

particular algebraic integrals

$$-x_0 \beta = (\delta + \lambda p^{-4/3})\, p\, q, \qquad -x_0 \gamma = (\varepsilon + \mu p^{-4/3})\, p\, r$$

exist, where

$$\delta = \frac{(B-A)(C-A)}{2C-A}, \qquad \varepsilon = \frac{(B-A)(C-A)}{2B-A}$$

$$\lambda = \frac{C(3A-2B)}{2C-A}\, s, \qquad \mu = \frac{B(3A-2C)}{2B-A}\, s$$

and s is determined by the equation

$$A^3(2B+2C-3A)\, s^3 = \frac{4(2B-A)^2(2C-A)^2}{9(3A-2B)(3A-2C)}\, x_0^2$$

The moments of inertia are subject to the relation

$$9(2B-A)(2C-A) = 4BC$$

The remaining integrals of the problem are

$$(2B-A)\frac{B-C}{A} q^2 = (C-A)\, p^2 - 3(3A-2C)\, s\, p^{2/3}$$

$$(2C-A)\frac{B-C}{A} r^2 = (A-B)\, p^2 + 3(3A-2B)\, s\, p^{2/3}$$

$$-(2B-A)(2C-A)\, x_0 \alpha = A(B-A)(C-A)\, p^2 + \\ + \tfrac{3}{2} A(3A^2 - 4BC)\, s\, p^{2/3}$$

$$P\, dt = \frac{dz}{\sqrt{(1-b^4 z^4)(1-c^4 z^4)}}$$

where

$$b^4 = \frac{B-A}{(3A-2B)\, s}, \qquad c^4 = \frac{A-C}{(2C-3A)\, s}$$

$$P = \sqrt{\frac{(2B-3A)(3A-2C)}{(2B-A)(2C-A)}}\, s, \qquad p = z^3$$

By means of the HERMITE transformation

$$1 + b^2 c^2 z^4 = z^2 u$$

the time is finally given by the sum of two elliptic integrals, namely

$$4P(t+t_0) = \int \frac{du}{\sqrt{(u+2bc)[u^2-(b^2+c^2)^2]}} - \\ - \int \frac{du}{\sqrt{(u-2bc)[u^2-(b^2+c^2)^2]}}$$

Expressed in terms of p, the variable u is given by the formula

$$u = b^2 c^2 p^{2/3} + p^{-2/3}$$

For a geometrical interpretation of the motion see P. V. Harlamov [3].

8.7. Case of Mercalov. Under the same first two assumptions as those of Goryačev and Čaplygin, i.e. (i) $A = B = 4C$, and (ii) the center of mass of the body is in the equatorial plane of the ellipsoid of inertia at the fixed point O, N. I. Mercalov [1] obtained a fourth integral of the equations of motion when $c \neq 0$. Discarding the case $q = q_0 = \text{const.}$ (case of Steklov and Bobylev) and assuming that the center of mass of the body lies on the y-axis of the ellipsoid of inertia at O, i.e. $x_0 = z_0 = 0$, he introduced a new variable ζ, where $d\zeta/dt = p$, and his fourth integral reads

$$(p^2 + q^2)\frac{dq}{d\zeta} - \frac{3}{2} q \frac{d}{d\zeta}(p^2 + q^2) = -\frac{3}{4}\frac{n_1}{C}\zeta + C_4' \quad \left(n_1 = \frac{c\, y_0}{A}\right)$$

For $c = 0$ and $C_4' = 0$ this last integral reduces to that of Goryačev (after interchanging the variables p and q). Hence a fourth integral of the equations of motion exists also in the case where $c \neq 0$, i.e. when the initial angular momentum of the body about the vertical through O is different from zero.

Further, assuming that the initial values $p_0 = q_0 = 0$, Mercalov gave an approximate expression for the component q of the angular velocity on the barycentric axis.

8.8. Center of mass lies in the characteristic plane. P. V. Myasnikov [1] proposed a new method for finding a number of integrable cases from a unified point of view, assuming that the center of mass G of the body lies in the so-called characteristic plane determined by the angular velocity vector and the angular momentum vector. This assumption yields a fourth algebraic integral

$$A\,p\,x_0 + B\,q\,y_0 + C\,r\,z_0 = m$$

where m is a constant.

If G lies simultaneously in the characteristic plane and on one of the principal axes of inertia at O, say, on the z-axis, i.e. $x_0 = y_0 = 0$, the cases of Euler and Poinsot, Lagrange and Poisson, and Steklov and Bobylev are obtained. When G is in the characteristic plane and in a principal plane of inertia at O, say, in $x_0 = 0$, then if $p \neq 0$, uniform rotations of Staude about the vertical, or the pendular motions of Mlodzeevskii about the horizontal x-axis, or the Euler and Poinsot case, or the spherical gyroscope are obtained. When G lies in the plane $z_0 = 0$, the same cases are possible. If, however, G lies in the plane $y_0 = 0$, a loxodromic pendulum is obtained. If G lies in the plane $x_0 = 0$, and $p = 0$, then in the case $m \neq 0$, $B - C \neq 0$ the Staude permanent rotations about the vertical follow, whereas in the case $m = 0$, $B - C = 0$ the pendular motions of Mlodzeevskii about the horizontal y-axis are obtained.

8.9. Case of N. Kowalewski. N. KOWALEWSKI [1] assumed that the center of mass G of the body lies on one of the principal axes of inertia at O, say, on the x-axis, i.e. $y_0 = z_0 = 0$, and sought all cases for which q^2 and r^2 can be expressed as polynomials of degree three in p. In doing so he arrived at the last three particular solutions of GORYAČEV, STEKLOV, and ČAPLYGIN and, in addition, at one new case.

Let

$$\frac{dx_i}{dt} = X_i + X_i' \qquad (i = 1, \ldots, n) \tag{8.65}$$

be a given system of differential equations to be compared with the system

$$\frac{dx_i}{dt} = X_i \qquad (i = 1, \ldots, n) \tag{8.66}$$

and let

$$F(x_1, \ldots, x_n; C_1, \ldots, C_n; t) = 0$$

be an integral of system (8.66). Then the basic equation of the LAGRANGE method of variation of constants (which is widely used in the perturbation theory) reads

$$\sum_{i=1}^{n} \left(\frac{\partial F}{\partial C_i} \frac{\partial C_i}{\partial t} + \frac{\partial F}{\partial x_i} X_i' \right) = 0 \tag{8.67}$$

While in the perturbation theory the motions of the celestial bodies are compared with those along conic sections, KOWALEWSKI chooses the EULER and POINSOT motion as the reference motion of a heavy rigid body with a fixed point. In the latter case there exist two time independent integrals

$$\frac{B-C}{A} q^2 - \frac{C-A}{B} p^2 = C_1 \tag{8.68}$$

$$\frac{B-C}{A} r^2 - \frac{A-B}{C} p^2 = C_2 \tag{8.69}$$

If we compare the EULER and POISSON differential equations of motion for $y_0 = z_0 = 0$ [corresponding to system (8.65)] with those in the case of EULER and POINSOT [corresponding to system (8.66)], and consider the terms $\frac{x_0}{B}\gamma$ and $-\frac{x_0}{C}\beta$ which occur in the first set as supplementary terms [X_i' in system (8.65)], then, following the method of variation of arbitrary constants, C_1 and C_2 must be considered as new variables which by virtue of (8.67) satisfy the equations

$$\gamma = \frac{B}{2x_0} \dot{C}_1 r, \quad \beta = -\frac{C}{2x_0} \dot{C}_2 q \tag{8.70}$$

where dots denote derivatives with respect to p.

Let us note that equations (8.70) can be obtained by differentiating equations (8.68) and (8.69) with respect to t. The meaning of the new variables C_1 and C_2 can be easily seen by considering the following two linear homogeneous combinations of them:

$$B\,C_1 + C\,C_2 = \frac{B-C}{A}\,(A\,p^2 + B\,q^2 + C\,r^2) \tag{8.71}$$

$$B^2\,C_1 + C^2\,C_2 = \frac{B-C}{A}\,(A^2\,p^2 + B^2\,q^2 + C^2\,r^2) \tag{8.72}$$

Hence, except for a constant factor, they give the kinetic energy of the body and the square of the modulus of the angular momentum vector respectively. On the other hand, C_1 and C_2 can be considered as the osculating elements since the motion of the endpoint of the angular velocity vector about the fixed point is the analog to the motion of a celestial body about the Sun. Therefore, we can speak about the osculating polhode in analogy to the osculating orbit.

From the energy integral and (8.71) it follows that

$$\alpha = \frac{1}{2x_0}\left[h - \frac{A}{B-C}\,(B\,C_1 + C\,C_2)\right] \tag{8.73}$$

where h denotes the energy constant.

Differentiating equations (8.70) with respect to t and observing the Euler and Poisson equations for $y_0 = z_0 = 0$ we obtain two second order differential equations for C_1 and C_2, not containing the direction cosines β and γ. If further q^2, r^2 and α are eliminated from these equations by means of equations (8.68), (8.69) and (8.73), and C_1, C_2 are replaced by new variables σ, τ defined by formulas

$$C_1 + \frac{C-A}{B}\,p^2 = \sigma$$

$$C_2 + \frac{A-B}{C}\,p^2 = \tau$$

then the new variables satisfy the equations

$$\begin{aligned}
&\tau\ddot{\sigma} + \frac{1}{2}\dot{\tau}\dot{\sigma} + \frac{A-2C}{B}\,p\,\dot{\tau} + \frac{A}{B-C}\,\sigma + \\
&\quad + \frac{1}{B}\left[\frac{A\,C}{B-C} - 2(C-A)\right]\tau + \frac{3A-2B}{B}\,p^2 - \frac{h}{B} = 0 \\
&\sigma\ddot{\tau} + \frac{1}{2}\dot{\sigma}\dot{\tau} + \frac{2B-A}{C}\,p\,\dot{\sigma} + \frac{A}{B-C}\,\tau + \\
&\quad + \frac{1}{C}\left[\frac{A\,B}{B-C} - 2(A-B)\right]\sigma + \frac{3A-2C}{C}\,p^2 - \frac{h}{C} = 0
\end{aligned} \tag{8.74}$$

while equations (8.68), (8.69), (8.70) and (8.73) assume the form

$$\left.\begin{aligned}
&\frac{B-C}{A} q^2 = \sigma \\
&\frac{B-C}{A} r^2 = \tau \\
&\alpha = \frac{1}{2x_0}\left[h - \frac{A}{B-C}(B\sigma + C\tau) - A p^2\right] \\
&\beta = -\frac{C}{2x_0}\left(\dot{\tau} - 2\frac{A-C}{C} p\right) q \\
&\gamma = \frac{B}{2x_0}\left(\dot{\sigma} - 2\frac{C-A}{B} p\right) r
\end{aligned}\right\} \tag{8.75}$$

To every solution $\sigma(p)$ and $\tau(p)$ of system (8.74) there correspond five time independent integrals (8.75) of the EULER and POISSON equations.

The sixth integral is given by the quadrature

$$t + \text{const.} = \int \frac{dp}{\sqrt{\sigma\tau}}$$

While searching for integrals of system (8.74) which in a neighborhood of $p = 0$ are expansible into power series of the form

$$\sigma = p^k \sum_{m=0}^{\infty} \alpha_m p^m, \qquad \tau = p^k \sum_{n=0}^{\infty} \beta_n p^n \tag{8.76}$$

KOWALEWSKI asked whether it is possible to select the initial values and the coefficients in such a way that the infinite series (8.76) reduce to polynomials. If in the expansions of σ, τ all terms up to the exponents d and g respectively are retained, then it turns out that for $d = g = 2$ the second case of STEKLOV is obtained, while for $d = 4$, $g = 2$ the second case of GORYAČEV results. If in the expansions of the form

$$\sum_0^{\infty} \lambda_m p^m + p^{2/3} \sum_0^{\infty} \mu_n p^n$$

where the exponents are roots of a certain system of algebraic equations with coefficients depending on A, B and C, only the first terms are considered, then the second case of ČAPLYGIN results.

Finally for $d = g = 3$ a new case is obtained provided that the coefficients α_i, β_i $(i = 0, 1, 2, 3)$ satisfy the above system of algebraic equations. If B, C and β_1 are considered as given, and the remaining quantities as unknown, then for $\beta_3 = 0$ the said system leads to well-defined expressions for α_0, α_1, α_2, α_3, β_0, β_2. The moments of inertia are subject to the relation

$$A = \frac{18B(B-C)}{9B - 10C}$$

The explicit expressions for the particular solution obtained by KOWALEWSKI are

$$\frac{B-C}{A} q^2 = \alpha_0' \beta_1^2 + \alpha_1' \beta_1 p + \alpha_2 p^2 + \alpha_3' \beta_1^{-1} p^3$$

$$\frac{B-C}{A} r^2 = \beta_0' \beta_1^2 + \beta_1 p + \beta_2 p^2$$

$$\alpha = \frac{1}{2x_0} \left\{ H - \frac{A}{B-C} [(B\alpha_1' + C) \beta_1 p + \right.$$

$$\left. + (B\alpha_2 + C\beta_2 + B - C) p^2 + B\alpha_3' \beta_1^{-1} p^3)] \right\}$$

$$\beta = -\frac{C}{2x_0} \left[\beta_1 + 2\left(\beta_2 - \frac{A-B}{C}\right) p \right] q$$

$$\gamma = \frac{B}{2x_0} \left[\alpha_1' \beta_1 + 2\left(\alpha_2 - \frac{C-A}{B}\right) p + 3\alpha_3' \beta_1^{-1} p^2 \right] r$$

where

$$H = h - \frac{A}{B-C} (B\alpha_0 + C\beta_0)$$

In order that $\alpha^2 + \beta^2 + \gamma^2$ be constantly equal to 1, and hence also for $p = 0$, then β_1 must satisfy the equation

$$\left[\frac{A}{B-C} (B^2 \alpha_1'^2 \beta_0' + C^2 \alpha_0') + H^2 \beta_1^{-4} \right] \beta_1^4 = 4x_0^2$$

where α_0', α_1', α_3', β_0' are certain expressions in terms of A, B, C.

Remark. Assuming that the center of mass G of the body lies on one of the principal axes of inertia at O, C. AGOSTINELLI [1] tried to determine all cases of motion for which p^2 and q^2 are polynomials of any degree in r. He showed that there are no other cases besides those given by V. A. STEKLOV [2], M. R. FABBRI [1—3], N. KOWALEWSKI [1] and D. N. GORYAČEV [1]. M. R. FABBRI's results were improved by Miss E. DE ANGELI [1].

8.10. Cases of Corliss and Field. The investigations of N. KOWALEWSKI have been continued by J. J. CORLISS [1]. In analogy with the KOWALEWSKI case, CORLISS assumes that (i) the center of mass G of the body lies on one of the principal axes of inertia at O, say, on the z-axis, i.e. $x_0 = y_0 = 0$, and (ii) the projection of the angular momentum upon the vertical is zero, i.e. $c = 0$. With these assumptions the EULER equations and the three classical integrals become

$$A\dot{p} + (C-B) q r = z_0 \beta \tag{8.77}$$

$$B\dot{q} + (A-C) r p = -z_0 \alpha \tag{8.78}$$

$$C\dot{r} + (B-A) p q = 0 \tag{8.79}$$

$$A p^2 + B q^2 + C r^2 = 2h_1 - 2z_0 \gamma \tag{8.80}$$

$$A p \alpha + B q \beta + C r \gamma = c \tag{8.81}$$

$$\alpha^2 + \beta^2 + \gamma^2 = 1 \tag{8.82}$$

If we eliminate α, β, γ and dt from equations (8.81) and (8.82) by means of equations (8.77) to (8.80), then we obtain the following two equations:

$$AB(A-B)\left[-p^2(2q\,dq)+q^2(2p\,dp)\right]+\left[(AC-2A^2)\,p^2+\right.$$
$$\left.+(BC-2B^2)\,q^2-C^2r^2+2C\,h_1\right]C\,r\,dr=2z_0\,c\,C\,dr \quad (8.83)$$

$$p^2[B(A-B)\,2q\,dq+C(A-C)\,2r\,dr]^2+q^2[A(A-B)\,2p\,dp+$$
$$+C(C-B)\,2r\,dr]^2=\{4z_0^2-[2h_1-(A\,p^2+B\,q^2+C\,r^2)]^2\}\,C^2dr^2 \quad (8.84)$$

If we change the variables according to the formulas

$$u=B(A-B)\,q^2+C(A-C)\,r^2 \quad (8.85)$$

$$v=A(A-B)\,p^2+C(C-B)\,r^2 \quad (8.86)$$

then

$$A\,v+B\,u=(A-B)\,(A^2p^2+B^2q^2+C^2r^2) \quad (8.87)$$

$$u+v=(A-B)\,(A\,p^2+B\,q^2+C\,r^2) \quad (8.88)$$

and equations (8.83) and (8.84) in terms of the new variables u, v become

$$-[v-C(C-B)\,r^2]\,du+[u-C(A-C\,r^2)]\,dv+$$
$$+[2h_1(A-B)-(u+v)]\,C^2r\,dr=2z_0\,c\,C(A-B)\,dr \quad (8.89)$$

$$B[v-C(C-B)\,r^2]\,du^2+A[u-C(A-C)\,r^2]\,dv^2+$$
$$+\frac{A\,B\,C^2}{A-B}\,[2h_1(A-B)-(u+v)]^2\,dr^2=4AB\,C^2(A-B)\,z_0^2\,dr^2 \quad (8.90)$$

The above equations may further be simplified by introducing two new variables ξ, η, defined by the relations

$$\xi=A^2p^2+B^2q^2+C^2r^2 \quad (8.91)$$

$$\eta=\frac{2h_1(A-B)-(u+v)}{A-B} \quad (8.92)$$

Then by virtue of (8.88) it follows that

$$\eta=2h_1-(A\,p^2+B\,q^2+C\,r^2) \quad (8.93)$$

and as a consequence equation (8.80) gives for γ the expression

$$\gamma=\frac{\eta}{2z_0} \quad (8.94)$$

Solving (8.91) and (8.92) for u and v we find that

$$u=-(A\,\eta+\xi-2A\,h_1) \quad (8.95)$$

$$v=B\,\eta+\xi-2B\,h_1 \quad (8.96)$$

Upon substitution of (8.95) and (8.96) into equations (8.89) and (8.90) and division by the common factor $A - B$, the equations

$$(2h_1 - \eta - C r^2)\, d\xi + (\xi - C^2 r^2)\, d\eta + C^2 \eta r\, dr = 2 z_0 c C\, dr \tag{8.97}$$

$$\begin{aligned}
&[-\xi - \eta(A + B) + 2h_1(A + B) + C(C - A - B) r^2]\, d\xi^2 + \\
&\quad + AB(\xi - C^2 r^2)\, d\eta^2 + 2AB(-\eta + 2h_1 - C r^2)\, d\xi\, d\eta + \\
&\quad + AB C^2 \eta^2\, dr^2 = 4 z_0^2 AB C^2\, dr^2
\end{aligned} \tag{8.98}$$

result.

Combining equations (8.97) and (8.98) we arrive at the equations

$$\frac{d\eta}{dr} - \frac{\dfrac{d\xi}{dr} - C^2 r}{\xi - C^2 r^2}\, \eta = \frac{(C r^2 - 2h_1)\dfrac{d\xi}{dr} + 2 z_0 c C}{\xi - C^2 r^2} \tag{8.99}$$

$$\begin{aligned}
&[-\xi - \eta(A + B) + 2h_1(A + B) + C(C - A - B) r^2]\, d\xi^2 + \\
&\quad + AB(\xi - C^2 r^2)\, d\eta^2 + 2AB(-\eta + 2h_1 - C r^2)\, d\xi\, d\eta + \\
&\quad + AB C^2 \eta^2\, dr^2 = 4 z_0^2 AB C^2\, dr^2
\end{aligned} \tag{8.100}$$

P. Stäckel [1—4] and O. Lazzarino [1—4] have shown that if instead of p, q, r the three scalars S, T and U are introduced, where

$$S = \vec{r}_0 \cdot \vec{I} = A p x_0 + B q y_0 + C r z_0$$

$$T = \tfrac{1}{2} \vec{\omega} \cdot \vec{I} = \tfrac{1}{2}(A p^2 + B q^2 + C r^2)$$

$$U = \tfrac{1}{2} \vec{I} \cdot \vec{I} = \tfrac{1}{2}(A^2 p^2 + B^2 q^2 + C^2 r^2),$$

then the Hess and Šiff [1] equations are equivalent to those of Euler and Poisson provided that S, T and U are all variable quantities. (See also R. Marcolongo [5], C. W. Oseen [1] and G. Hamel [1].) Since the variables $\xi = 2U$, $\eta = 2h_1 - 2T$ differ from U, T by constants, we may regard equations (8.99) and (8.100) as giving necessary and sufficient conditions in order that ξ and η (ξ, η and r being all variable quantities) satisfying these equations will lead to expressions of p, q, r satisfying the Euler equations (8.77) to (8.79).

In order to find a new particular solution J. J. Corliss assumes that

$$\xi = C^2(a r^2 + b) \tag{8.101}$$

$$\eta = a_1 r^2 + b_1 \tag{8.102}$$

where a, b, a_1, b_1 are constants to be determined. Substituting the above expressions for ξ, η into equations (8.99) and (8.100) and equating the coefficients of equal powers of r, he obtained the following expressions for h_1, a, b, a_1, b_1:

$$\left.\begin{aligned} &c = 0: \quad 2h_1 = \frac{(2AB - 2AC - 2BC + C^2)z_0}{(C-A)(C-B)} \\ &a = \frac{(C-A)(C-B)}{(C-2A)(C-2B)}, \quad b = \frac{[2AB - (A+B)C]z_0}{C(C-A)(C-B)} \\ &a_1 = -2aC, \quad b_1 = 2z_0 \end{aligned}\right\} \qquad (8.103)$$

The second condition of (8.103) imposes a restriction upon the total energy, i.e. the moments of inertia must be chosen in such a way that h_1 becomes positive.

Retracing the steps we find for p^2 and q^2 the expressions

$$\begin{aligned} p^2 &= \frac{(C-B)C}{(C-2A)(A-B)}\left[r^2 - \frac{z_0(C-2B)(C-2A)}{(C-A)(C-B)^2}\right] \\ q^2 &= -\frac{(C-A)C}{(C-2B)(A-B)}\left[r^2 - \frac{z_0(C-2B)(C-2A)}{(C-A)^2(C-B)}\right] \end{aligned} \qquad (8.104)$$

If we assume that $A > B$, then, for $A > C$, $2A > C > 2B$, the inequalities $A > C > 2B$ are the only restrictions to be imposed on the moments of inertia in order that the energy constant h_1, given by the second formula of (8.103), be positive, γ be less than or equal to one, and the time t be real.

The remaining two EULER angles are given by the formulas

$$\tan\Phi = \frac{(C-2A)p}{(C-2B)q}$$

and

$$\psi = \Phi + C_0$$

For convenience we may take $C_0 = 0$. The time is given as a function of r by means of an elliptic integral.

The second particular case obtained by J. J. CORLISS is a special case of S. KOVALEVSKAYA. The method for obtaining it is identical to that outlined above. The imposed restrictions are

$$A = C = 2B$$

$$c = C b \sqrt{\frac{n+1}{2}}$$

$$h_1 = \tfrac{1}{2} n C b^2$$

where n must be greater than or equal to one, and b^2 is given by the formula

$$b^2 = \frac{z_0}{C}\sqrt{\frac{2}{n+1}}$$

The expressions for p^2 and q^2 in terms of r are

$$\begin{aligned} p^2 &= -(r^2 - 2br - b^2) \\ q^2 &= -2(r^2 + 2br - nb^2) \end{aligned} \qquad (8.105)$$

In the two particular cases obtained by J. J. Corliss in his first paper the kinetic energy and the square of the modulus of the angular momentum vector are quadratic functions of r. In the second paper J. J. Corliss [2] proposed to find all cases in which the above quantities are polynomials in r.

The problem, as we shall see, can be reduced to the solution of two symmetric second order differential equations.

If we substitute into equations (8.77) and (8.78) for $1/dt$ its expression from equation (8.79) and introduce two new variables u, v, equations (8.77), (8.78) and the energy integral (8.80) become

$$q\,dv = 2z_0\,C\,\beta\,dr \tag{8.106}$$

$$p\,du = -2z_0\,C\,\alpha\,dr \tag{8.107}$$

$$2z_0\,\gamma = \frac{2h_1(A-B)-(u+v)}{A-B} = \eta \tag{8.108}$$

If expressions for α, β, γ in terms of u, v and η are substituted into equations (8.81) and (8.82), then we get

$$-A\,p^2\,du + B\,q^2\,dv + C^2\,\eta\,r\,dr = 2z_0\,c\,C\,dr \tag{8.109}$$

$$p^2(du)^2 + q^2(dv)^2 + C^2\eta^2(dr)^2 = 4z_0^2\,C^2(dr)^2 \tag{8.110}$$

Equations (8.109) and (8.110) are equivalent to the Euler equations as long as u, v, r are all variable quantities. On the other hand they are rather complicated and unsymmetric. Therefore two symmetric equations which are much easier to work with will be derived. However, the latter have the disadvantage of giving only necessary conditions contrary to equations (8.109) and (8.110) which give necessary and sufficient conditions in order that u, v lead to p, q, r, which satisfy the Euler equations (8.77) to (8.79). In order to obtain such symmetric equations we differentiate equations (8.109) and (8.110) with respect to r and substitute for dp^2/dr, dq^2/dr and $d\eta/dr$ their expressions obtained by differentiating equations (8.85), (8.86) and (8.108) with respect to r. It is easily seen that the equations so obtained are linear combinations of the following two symmetric equations

$$A(A-B)\,p^2\frac{d^2u}{dr^2} + \frac{1}{2}\frac{du}{dr}\frac{dv}{dr} + A\,C\,r\frac{dv}{dr} + C(B-C)\,r\frac{du}{dr} - \\ - A\,C^2\eta = 0 \tag{8.111}$$

$$B(A-B)\,q^2\frac{d^2v}{dr^2} + \frac{1}{2}\frac{du}{dr}\frac{dv}{dr} - B\,C\,r\frac{du}{dr} + C(C-A)\,r\frac{dv}{dr} - \\ - B\,C^2\eta = 0 \tag{8.112}$$

In actual computations it is advisable to use both pairs of equations, i.e. equations (8.109) and (8.110), and equations (8.111) and (8.112).

We now assume that u and v may be expressed in the form

$$u = B_0 + \sum_{i=1}^{m} B_i r^{b_i+1}$$

$$v = A_0 + \sum_{i=1}^{n} A_i r^{a_i+1}$$

From this assumption several general conclusions can be drawn: (i) that a_n and b_n cannot be fractions, (ii) that m must be taken equal to 2, (iii) that there does not exist a solution for $n \geqq 7$, and (iv) that A_6 and A_5 must be taken equal to zero, i.e. n must be taken $\leqq 4$.

The cases $n = 1$ and $n = 2$ lead to the two cases given by J. J. Corliss in his first paper, the case of P. Field [1] being a special case of $n = 2$. The case $n = 3$ corresponds to the case of N. Kowalewski. For $n = 4$ a new case is obtained. In this case

$$u = B_0 + B_1 r + B_2 r^2$$

$$v = A_0 + \sum_{i=1}^{4} A_i r^i$$

where the coefficients have the following expressions

$$A_0 = \frac{A\,C(64A^2 - 64A\,C + 15C^2)h_1}{(C - 2A)(9C^2 - 56A\,C + 64A^2)}$$

$$A_2 = \frac{C(4A - 3C)(64A^2 - 64A\,C + 15C^2)}{2(16A - 9C)(C - 2A)}$$

$$A_4 = \frac{C(4A - 3C)(16A^2 - 16A\,C + 3C^2)(9C^2 - 56A\,C + 64A^2)}{32(16A - 9C)(C - 2A)A\,h_1}$$

$$B_0 = -\frac{8A\,C(C - 2A)h_1}{(9C^2 - 56A\,C + 64A^2)}$$

$$B_2 = \frac{C(4A - 3C)}{2}$$

$$B_1 = A_1 = A_3 = A_4 = 0$$

The moments of inertia are subject to the relation

$$B = \frac{A(16A - 8C)}{16A - 9C}$$

and $2h_1$ is given by the formula

$$2h_1 = \frac{4z_0(C - 2A)(9C^2 - 56A\,C + 64A^2)}{(4A - 3C)(64A^2 - 64A\,C + 15C^2)}$$

The projection of the angular momentum vector on the vertical is zero, i.e. $c = 0$.

The value of $\tan\Phi$ is given by the formula

$$\tan\Phi = -\frac{p\dfrac{du}{dr}}{q\dfrac{dv}{dr}}$$

which follows from equations (8.106) and (8.107).

In order to find ψ and the time t in terms of r elliptic functions must be introduced.

In a second paper P. FIELD [2] has studied the case in which B approaches A, and C becomes small in such a way that $(A - B)/C$ approaches p.

8.11. Center of mass lies on one of the principal planes of inertia. Investigations which in a certain sense are parallel to those of N. KOWALEWSKI have been carried out by T. MANACORDA [2] and A. NADILE [1], who assumed that the center of mass G lies in one of the principal planes of inertia of the body at the fixed point O. In the terminology of P. STÄCKEL this is the so-called planar case of the motion of a rigid body.

MANACORDA sets up two invariant relations in terms of the components of the angular velocity vector as power series in s, s being a certain linear combination of these components. For only the zero value of a certain exponent does he arrive at the general solution depending upon five arbitrary constants. He also shows how the cases of HESS ($s = 0$), GRIOLI, and KOWALEWSKI fit into his theory.

NADILE finds two types of uniform rotations and certain other motions involving only quadratures and integration of a RICCATI equation. These latter motions depend upon five arbitrary constants of integration.

P. V. MYASNIKOV [2] studied the case where the angular momentum has constant projection on the line joining the fixed point O with the mass center G, the latter lying in one of the principal planes of inertia. All cases found are known.

8.12. Regular precessions about nonvertical axes. Consider a heavy rigid body which is free to rotate about a fixed point O different from its mass center G. Let E_G denote the central ellipsoid of inertia of the body, and let us consider the so-called asymmetric case, in which no two of the three principal axes of E_G are equal.

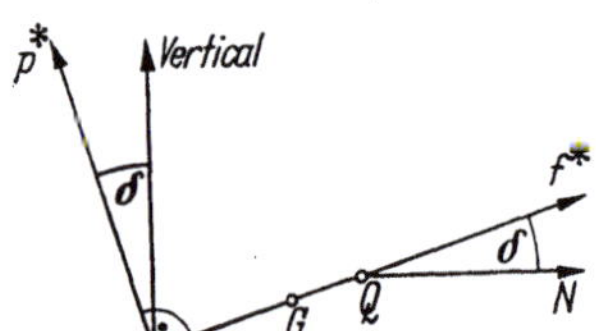

Fig. 11

E. J. ROUTH [1] showed that regular nondegenerate precessions of such a body, having for the axis of precession the vertical and for the figure axis one of the three principal axes of inertia at O, are dynamically impossible.

Let f^* be the barycentric line orthogonal to the plane of one of the circular cross sections of E_G, and let Q be one of the points of intersection of f^* with E_G. (Fig. 11) Further, let N be the normal to E_G at Q, and let δ be the acute angle between f^* and N.

Under the above assumptions G. GRIOLI [1, 3] showed that among the possible motions of a heavy asymmetric body there are ∞^2 regular nondegenerate precessions, having f^* for the figure axis. The period of each such precession is equal to the period of proper rotation, and the axis of precession p^* makes an angle δ with the vertical and is orthogonal to f^*. In addition, these are the only dynamically possible precessions. Hence regular precessions are among the possible motions of a heavy rigid body in the kinetically symmetric as well as asymmetric case.

If E_G is an ellipsoid of revolution, then the only dynamically possible regular precessions for a heavy rigid body are those for which the axis of precession is vertical and the figure axis is the axis of the body.

As G. Grioli remarked, we are faced here with a particular case of integrability of the Euler and Poisson equations, different from those considered by S. V. Kovalevskaya, S. A. Čaplygin and W. Hess. G. Grioli himself is not concerned with the actual integration of these equations. His proof is based on purely dynamical considerations without giving explicit expressions for the first integrals. For a characteristic property of regular precessions see G. Grioli [5]. Later M. P. Gulyaev [1] arrived at the Grioli solution by a straightforward integration of the Euler and Poisson equations.

Consider the motion of a heavy rigid body, fixed at the point for which the conditions

$$x_0 \sqrt{B-C} - z_0 \sqrt{A-B} = 0, \qquad y_0 = 0, \qquad A > B > C \quad (8.113)$$

hold.

Geometrically, conditions (8.113) imply that the center of mass G of the body lies on the perpendicular to the plane of one of the circular cross sections of the ellipsoid of inertia at the fixed point O. Then the Euler and Poisson equations reduce to

$$\left.\begin{aligned} &A\frac{dp}{dt} - (B-C)\,q\,r = \beta\, z_0 \\ &B\frac{dq}{dt} - (C-A)\,r\,p = \gamma\, x_0 - \alpha\, z_0 \\ &C\frac{dr}{dt} - (A-B)\,p\,q = -\beta\, x_0 \\ &\frac{d\alpha}{dt} = r\beta - q\gamma, \quad \frac{d\beta}{dt} = p\gamma - r\alpha, \quad \frac{d\gamma}{dt} = q\alpha - p\beta \end{aligned}\right\} \quad (8.114)$$

Rewrite conditions (8.113) in the form

$$\frac{A-B}{x_0^2} = \frac{B-C}{z_0^2} = \frac{A-C}{x_0^2+z_0^2} = k = \text{const.}, \qquad y_0 = 0, \qquad A > B > C \quad (8.115)$$

and let us try to obtain from system (8.114) an integrable combination. For this purpose multiply the first equation by x_0, the third by z_0 and add. Then by virtue of (8.115) it follows that

$$\frac{d}{dt}(A\, x_0\, p + C\, z_0\, r) = k\, x_0\, z_0\, q\,(x_0\, p + z_0\, r) \quad (8.116)$$

Further assume that the projection of the angular velocity of the body on the normal to the plane of the circular section of the ellipsoid of inertia through the center of mass G is constant (this property characterizes regular precessions of a symmetric body). Then

$$x_0 p + z_0 r = m = \text{const.} \tag{8.117}$$

By virtue of (8.116) and (8.117) system (8.114) reduces to the form

$$\left.\begin{aligned}
&\frac{dp}{dt} = \frac{z_0 m}{l^2} q \\
&\frac{dq}{dt} = \frac{C-A}{B} p r + \frac{1}{B}(x_0 \gamma - z_0 \alpha) \\
&\frac{dr}{dt} = -\frac{x_0 m}{l^2} q \\
&\frac{d\alpha}{dt} = \beta r - \gamma q, \quad \frac{d\beta}{dt} = \gamma p - \alpha r, \quad \frac{d\gamma}{dt} = \alpha q - \beta p
\end{aligned}\right\} \tag{8.118}$$

where $l^2 = x_0^2 + z_0^2$ is the square of the distance from the center of mass G to the fixed point O.

In order that systems (8.114) and (8.118) be equivalent it is necessary that the condition

$$(A x_0 p + C z_0 r) q - l^2 \beta = 0 \tag{8.119}$$

holds. This condition is obtained by comparing separately the first and the third equations in both systems.

Multiplying the first three equations of system (8.118) by p, q, r respectively, and adding we obtain that

$$\frac{d}{dt}(p^2 + q^2 + r^2) = \frac{2}{B l^2} q [B m (z_0 p - x_0 r) - (A - C) l^2 p r + l^2 (x_0 \gamma - z_0 \alpha)]$$

Since for regular precessions the quadratic integral

$$\omega^2 = p^2 + q^2 + r^2 = c = \text{const.} \tag{8.120}$$

holds, then by necessity

$$B m (z_0 p - x_0 r) - (A - C) l^2 p r + l^2 (x_0 \gamma - z_0 \alpha) = 0 \tag{8.121}$$

The other possible assumption $q = 0$ leads to a permanent rotation.

The existence of algebraic integrals (8.117), (8.119), (8.120) and (8.121) leads to a particular solution of system (8.118), and hence of system (8.114).

In fact, from (8.117) and (8.120) it follows that

$$q = \frac{1}{z_0} \sqrt{-l^2 p^2 + 2m x_0 p + z_0^2 c - m^2} \tag{8.122}$$

$$r = \frac{1}{z_0} (m - x_0 p) \tag{8.123}$$

Upon substitution of (8.122) into the first equation of system (8.118), an equation with separable variables results, namely,

$$dt = \frac{l^2 \, dp}{m \sqrt{-l^2 p^2 + 2m x_0 p + z_0^2 c - m^2}}$$

By integration we obtain the equation

$$\arcsin \frac{l^2 p - m x_0}{z_0 \sqrt{l^2 c - m^2}} = \frac{m}{l} (t - \tau)$$

where τ is a constant of integration. Solving the above equation for p we obtain

$$p = \frac{m}{l^2} \left[x_0 + z_0 \frac{\sqrt{l^2 c - m^2}}{m} \sin \left(\frac{m}{l} t - \frac{m}{l} \tau \right) \right]$$

or, after introduction of the notations

$$a = \frac{\sqrt{l^2 c - m^2}}{m}, \qquad n = \frac{m}{l}, \qquad \varepsilon = \frac{m}{l} \tau$$

$$p = \frac{n}{l} [x_0 + z_0 a \sin (n t - \varepsilon)] \tag{8.124}$$

The components q, r as functions of the time follow from formulas (8.122) and (8.123) without further integration. They give the following expressions

$$q = n a \cos (n t - \varepsilon) \tag{8.125}$$

$$r = \frac{n}{l} [z_0 - x_0 a \sin (n t - \varepsilon)]. \tag{8.126}$$

It remains to calculate the direction cosines α, β, γ. Since p, q, r are known, then β can be calculated from equation (8.119) by substituting for p, q, r their expressions (8.124), (8.125) and (8.126). Hence

$$\beta = \frac{n^2}{l^3} a [(A x_0^2 + C z_0^2) \cos (n t - \varepsilon) + \\ + (A - C) a x_0 z_0 \sin (n t - \varepsilon) \cos (n t - \varepsilon)] \tag{8.127}$$

For the calculation of α and γ as functions of the time t it is convenient to consider integral (8.121) and the classical energy integral

$$A p^2 + B q^2 + C r^2 + 2 (x_0 \alpha + z_0 \gamma) = 2h$$

h being the energy constant.

Substituting in these integrals for p, q, r their expressions (8.124), (8.125) and (8.126) and solving them for α and γ we obtain

$$\alpha = -\frac{n^2}{l^4}\left\{\frac{x_0}{2}\left[A\,x_0^2 + C\,z_0^2 + (A\,z_0^2 + C\,x_0^2)\,a^2 - \frac{2h\,l^2}{n^2}\right] - \right.$$
$$\left. - C\,l^2 z_0\,a\sin(n\,t-\varepsilon) + (A-C)\,x_0\,z_0^2[1 - a^2\sin^2(n\,t-\varepsilon)]\right\} \tag{8.128}$$

$$\gamma = -\frac{n^2}{l^4}\left\{\frac{z_0}{2}\left[A\,x_0^2 + C\,z_0^2 + (A\,z_0^2 + C\,x_0^2)\,a^2 - \frac{2h\,l^2}{n^2}\right] + \right.$$
$$\left. + A\,l^2 x_0\,a\sin(n\,t-\varepsilon) - (A-C)\,x_0^2\,z_0[1 - a^2\sin^2(n\,t-\varepsilon)]\right\} \tag{8.129}$$

It remains to verify that the obtained solution satisfies the system of equations (8.118). Substitution of formulas (8.124) to (8.129) into the first three equations of system (8.118) shows that they satisfy these equations.

Substitution of the same formulas into the last three equations of system (8.118) gives equations which determine a and h:

$$a = 1\,,\qquad h = \frac{n^2}{2}(A + C) \tag{8.130}$$

Hence the final expressions for p, q, r; α, β, γ are

$$\left.\begin{aligned}
p &= \frac{n}{l}[x_0 + z_0\sin(n\,t-\varepsilon)]\\
q &= n\cos(n\,t-\varepsilon)\\
r &= \frac{n}{l}[z_0 - x_0\sin(n\,t-\varepsilon)]\\
\alpha &= \frac{n^2}{l^2}[C\,z_0\sin(n\,t-\varepsilon) - (B-C)\,x_0\cos^2(n\,t-\varepsilon)]\\
\beta &= \frac{n^2}{l^3}[(A\,x_0^2 + C\,z_0^2)\cos(n\,t-\varepsilon) + \\
&\qquad + (A-C)\,x_0\,z_0\sin(n\,t-\varepsilon)\cos(n\,t-\varepsilon)]\\
\gamma &= -\frac{n^2}{l^2}[A\,x_0\sin(n\,t-\varepsilon) - (A-B)\,z_0\cos^2(n\,t-\varepsilon)]
\end{aligned}\right\} \tag{8.131}$$

This solution contains two constants ε and $n = m/l$, of which, as we shall see, only ε is arbitrary and n can be expressed in terms of the parameters of the body. In fact, consider the values of α, β, γ, calculated for $t_0 = \varepsilon/n$. Then

$$\alpha(t_0) = -\frac{n^2 x_0}{l^2}(B - C)$$

$$\beta(t_0) = \frac{n^2}{l^3}(A\,x_0^2 + C\,z_0^2)$$

$$\gamma(t_0) = \frac{n^2 z_0}{l^2}(A - B)$$

Substituting these expressions into the trigonometric identity

$$\alpha^2 + \beta^2 + \gamma^2 = 1$$

and solving it for n we obtain

$$n = \sqrt[4]{\frac{l^2}{(A-B)(B-C)+(A-B+C)^2}} \qquad (8.132)$$

On the other hand, the angular velocity of the body at the initial instant t_0 is given by the formula

$$\omega_0 = \sqrt{p_0^2 + q_0^2 + r_0^2} = \sqrt{c} = n\sqrt{2}$$

Hence, it follows from expression (8.132) that

$$\omega_0 = \sqrt[4]{\frac{4l^2}{(A-B)(B-C)+(A-B+C)^2}} \qquad (8.133)$$

This is the initial angular velocity to be given to the body rotating about the axis, whose position is determined by the ratios p_0/ω_0, q_0/ω_0, r_0/ω_0, in order that the desired motion be realized.

In order to make sure that the motion under consideration is a regular precession, it is sufficient to show that there exists an axis in the body and in space along which the projections of the angular velocity remain constant during the whole course of motion.

The axis f^* of the proper rotation coincides with the line which contains the center of mass and is perpendicular to a circular section of the ellipsoid of inertia at the fixed point. Dividing both sides of (8.117) by l, we obtain the projection of the angular velocity on this axis, namely

$$\omega_{f^*} = \frac{x_0}{l} p + \frac{z_0}{l} r = n = \text{const.} \qquad (8.134)$$

The position of the axis of precession p^* in fixed space can be determined as follows. Consider a fixed axis passing through the fixed point and having direction cosines α_1, α_2, α_3 relative to the moving coordinate trihedral. Then the latter satisfy the differential equations

$$\frac{d\alpha_1}{dt} = r\alpha_2 - q\alpha_3$$

$$\frac{d\alpha_2}{dt} = p\alpha_3 - r\alpha_1$$

$$\frac{d\alpha_3}{dt} = q\alpha_1 - p\alpha_2$$

which, after substituting for p, q, r their expressions as given by system (8.131), assume the form

$$\frac{d\alpha_1}{dt} = \frac{n}{l}[z_0 - x_0 \sin(nt - \varepsilon)]\,\alpha_2 - [n\cos(nt - \varepsilon)]\,\alpha_3$$

$$\frac{d\alpha_2}{dt} = \frac{n}{l}\{[x_0 + z_0 \sin(nt - \varepsilon)]\,\alpha_3 - [z_0 - x_0 \sin(nt - \varepsilon)]\,\alpha_1\}$$

$$\frac{d\alpha_3}{dt} = [n\cos(nt - \varepsilon)]\,\alpha_1 - \frac{n}{l}[x_0 + z_0 \sin(nt - \varepsilon)]\,\alpha_2$$

It is easy to see that one particular solution of this system, satisfying the algebraic integral

$$\vec{\omega}\cdot\vec{\alpha} = p\,\alpha_1 + q\,\alpha_2 + r\,\alpha_3 = d = \text{const.} \tag{8.135}$$

is

$$\left.\begin{aligned} \alpha_1 &= \frac{z_0}{l}\sin(nt - \varepsilon) \\ \alpha_2 &= \cos(nt - \varepsilon) \\ \alpha_3 &= -\frac{x_0}{l}\sin(nt - \varepsilon) \end{aligned}\right\} \tag{8.136}$$

Substitution of (8.136) into (8.135) shows that the projection of the angular velocity $\vec{\omega}$ on the axis whose position is determined by $(\alpha_1, \alpha_2, \alpha_3)$ is constant and equal to n, i.e.

$$\omega_{p^*} = \vec{\omega}\cdot\vec{\alpha} = d = n \tag{8.137}$$

Formulas (8.134) and (8.137) show that the velocity of precession is equal to that of proper rotation, i.e. $\omega_{f^*} = \omega_{p^*} = n = m/l$, and that the angle between both axes f^* and p^* is $\pi/2$ since

$$\frac{\vec{r}_0}{|\vec{r}_0|}\cdot\vec{\alpha} = \frac{x_0}{l}\alpha_1 + \frac{z_0}{l}\alpha_3 = 0$$

Hence

$$m = \omega_{p^*}\, l = n\, l$$

i.e. m is the linear velocity of the center of mass in the precessional motion.

A characteristic property of the precession under consideration is that the axis of precession p^* is not vertical but makes an angle δ with the latter. This angle depends on the shape of the ellipsoid of inertia and is determined by the formula

$$\cos\delta = \alpha\,\alpha_1 + \beta\,\alpha_2 + \gamma\,\alpha_3 = \frac{n^2}{l}(A - B + C)$$

In the case of a symmetric heavy rigid body ($A = B > C$) with the mass center G lying on the perpendicular to the plane of one of the circular sections of the ellipsoid of inertia at the fixed point, conditions (8.113) imply that $x_0 = 0$ and $z_0 = l$. In such a case formulas (8.131) reduce to the following ones

$$\begin{aligned} p &= n\sin(nt-\varepsilon), & \alpha &= \frac{n^2}{z_0} C\sin(nt-\varepsilon) \\ q &= n\cos(nt-\varepsilon), & \beta &= \frac{n^2}{z_0} C\cos(nt-\varepsilon) \\ r &= n = \text{const.}, & \gamma &= 0 \end{aligned}$$

where

$$n = \sqrt{\frac{z_0}{C}}$$

Further, it is easy to insure that

$$\alpha_1 = \alpha, \quad \alpha_2 = \beta, \quad \alpha_3 = \gamma$$

i.e. the axis of precession p^* is vertical. The axis of proper rotation f^* coincides with the perpendicular through the fixed point O to the plane of the circular section of the ellipsoid of inertia.

From the analytical point of view GRIOLI's solution is characterized by the presence of two quadratic integrals of a particular type.

Mrs. E. I. HARLAMOVA [1, 2] has investigated the question of the existence of new solutions of a similar type. For this purpose she has constructed a system of five algebraic equations for the determination of the parameters characterizing the motion. Two solutions of this system are obtained. One solution corresponds to GRIOLI's case, the other solution gives a new case of integrability. In the latter case the conditions imposed on the moments of inertia can be realized if the body possesses cavities filled with an ideal incompressible liquid.

G. COLOMBO [1] has derived the general expression for a moment vector acting on a gyro in order that merostatic motions be possible, i.e. regular precessions or uniform rotations about axes different from the gyro axis. From this expression he then derived a general stability criterion and applied it to another case studied by G. GRIOLI [2].

C. AGOSTINELLI [2, 4] proved the existence of a solution, depending upon five arbitrary constants. The integrals of motion are given by uniformly convergent power series in terms of a parameter s, which is proportional to the component of the angular momentum along the barycentric axis. Recursion formulas are given for the coefficients of these series in terms of four arbitrary constants. If no conditions are imposed upon the moments of inertia of the body, the only polynomial solution in s is obtained for ω_b = const., where ω_b is the component of the angular velocity $\vec{\omega}$ along the barycentric axis. The corresponding motions are the regular precessions obtained by GRIOLI. Another polynomial solution in s can be obtained under the assumption that ω_b is a linear function of s. The moments of inertia, then, have to satisfy a certain relationship. Concerning the stability of a particular periodic solution see C. AGOSTINELLI [3].

T. ZEULI [1] studied the motions of a heavy rigid body about a fixed point O under the assumption that the structure of the body is very close to that of a symmetric body. He expands the unknown functions p, q, r; α, β, γ in power series of a parameter, the coefficients being functions of the time. For the zero value of the parameter the problem reduces to that of a heavy symmetric body.

Miss B. GRISERI [1] studied solutions which are meromorphic in the plane of the complex variable t (time) with a pole at the origin. The existence of three such solutions is proved; although two of them do not impose any restrictions on the moments of inertia of the body, the third solution exists only if these structural constants satisfy a certain relation.

8.13. Case of Mrs. Harlamova. Mrs. E. I. HARLAMOVA [2] showed that under the assumptions

$$y_0 = 0$$

$$x_0(2C-A)\sqrt{A(C-B)(2C-A)} + z_0(C-2A)\sqrt{C(A-B)(C-2A)} = 0$$

$$C > 2A > 2B \qquad (x_0 < 0)$$

the EULER and POISSON equations (2.5) and (2.7) assume a particular solution

$$A\,p = \nu(\cos\varrho + \chi\sin\varrho\cos\tau)$$

$$q = \nu\sqrt{\frac{(C-2A)(2C-A)}{3AC[3AC-B(A+C)]}}\sin\tau$$

$$C\,r = \nu(\sin\varrho - \chi\cos\varrho\cos\tau)$$

$$\alpha = \frac{[3A(C-B)\sin^2\tau + B(A+C)]\cos\varrho + 3AC\chi\sin\varrho\cos\tau}{3AC - B(A+C)}$$

$$\beta = \frac{\sqrt{3AC(C-2A)(2C-A)}}{[3AC-B(A+C)]^{3/2}}\left\{B - 3\chi\sqrt{\frac{AC(A-B)(C-B)}{(C-2A)(2C-A)}}\cos\tau\right\}\sin\tau$$

$$\gamma = \frac{[3C(A-B)\sin^2\tau + B(A+C)]\sin\varrho - 3AC\chi\cos\varrho\cos\tau}{3AC - B(A+C)}$$

The variable τ as a function of the time t is given by the equation

$$\frac{d\tau}{dt} = \frac{\nu}{3AC\sqrt{3AC-B(A+C)}} \times$$
$$\times\left\{\sqrt{(C-2A)(2C-A)[3AC-2B(A+C)]} + \right.$$
$$\left. + (A+C)\sqrt{3(A-B)(C-B)}\cos\tau\right\}$$

while the constants ϱ, χ and ν are determined by the formulas

$$\cos\varrho = \sqrt{\frac{C(A-B)(C-2A)}{(C-A)[3AC-B(A+C)]}}$$

$$\sin\varrho = \sqrt{\frac{A(C-B)(2C-A)}{(C-A)[3AC-B(A+C)]}}$$

$$\chi = \sqrt{1 - \frac{2B(A+C)}{3AC}}$$

$$\nu = 3AC\sqrt[4]{\frac{(x_0^2+z_0^2)(C-A)}{[3AC-B(A+C)][A(C-B)(2C-A)^3 + C(A-B)(C-2A)^3]}}$$

For a rigid body the above requirement $C > 2A > 2B$ cannot be realized since it contradicts the well-known inequality $A + B > C$.

However, M. P. GULYAEV [4] showed that the above solution is physically realizable for a rotating body filled with a frictionless and incompressible fluid. The case under consideration represents the only regular precession about the vertical which is dynamically possible.

8.14. Linear integrals. S. A. ČAPLYGIN [2] concluded in 1898 his paper on the existence of linear integrals of the equations of motion of a heavy rigid body about a fixed point with the sentence that "the problem under consideration does not admit a particular linear integral in any other cases than those known up to the present." The fact, however, that some recently discovered particular solutions by G. GRIOLI [1] and Mrs. E. I. HARLAMOVA [1, 2] contain linear integrals stimulated P. V. HARLAMOV [1] to investigate the problem anew by starting from ČAPLYGIN's premises.

Denoting the components of the angular momentum vector $\vec{I}$ of the body with respect to the fixed point O by I_1, I_2, I_3, the coordinates of the mass center G by ξ_0, η_0, ζ_0 and retaining our earlier notations for the angular velocity $\vec{\omega}(p, q, r)$ and the unit vector $\vec{K}(\alpha, \beta, \gamma)$ along the positive Z-axis directed upward, we can rewrite the equations of motion in a body-fixed coordinate trihedral in the form

$$\left.\begin{aligned} \frac{dI_1}{dt} &= r I_2 - q I_3 + \zeta_0 \beta - \eta_0 \gamma \\ \frac{dI_2}{dt} &= p I_3 - r I_1 + \xi_0 \gamma - \zeta_0 \alpha \\ \frac{dI_3}{dt} &= q I_1 - p I_2 + \eta_0 \alpha - \xi_0 \beta \end{aligned}\right\} \tag{8.138}$$

and

$$\left.\begin{aligned} \frac{d\alpha}{dt} &= r\beta - q\gamma \\ \frac{d\beta}{dt} &= p\gamma - r\alpha \\ \frac{d\gamma}{dt} &= q\alpha - p\beta \end{aligned}\right\} \tag{8.139}$$

It is sufficient to establish conditions under which the above equations admit a single linear integral.

Two linear integrals exist only in the case of STEKLOV [1] and BOBYLEV [1] and that of a compound pendulum, while three linear integrals exist for the case of a uniform rotation of a body about a fixed axis.

By virtue of the fact that equations (8.138) and (8.139) are homogeneous in $I_1, I_2, I_3, \alpha, \beta, \gamma$ the sought linear integral must be of the form

$$A' I_1 + B' I_2 + C' I_3 = \text{const.}$$

where A', B', C' are certain constants. By choosing the coordinate trihedral properly this integral can be reduced to the form

$$I_3 = n = \text{const.} \tag{8.140}$$

By rotation of the coordinate trihedral about the axis corresponding to the component I_3 we can eliminate the term $I_1 I_2$ in the expression of the kinetic energy T so that T assumes the form

$$T = \tfrac{1}{2}(a_1 I_1^2 + a_2 I_2^2 + a_3 I_3^2) + (b_{13} I_1 + b_{23} I_2) I_3$$

Without any loss of generality we can assume that

$$b_{23} = \eta_0 = 0 \tag{8.141}$$

The components of the angular velocity and angular momentum are related by the relations

$$p = \frac{\partial T}{\partial I_1} = a_1 I_1 + b I_3, \quad q = \frac{\partial T}{\partial I_2} = a_2 I_2, \quad r = \frac{\partial T}{\partial I_3} = a_3 I_3 + b I_1$$
$$(b = b_{13})$$

Equation (8.140) implies that $\dfrac{d I_3}{dt} = \dfrac{d^2 I_3}{dt^2} = 0$. Hence

$$(a_2 - a_1) I_1 I_2 - n b I_2 - \xi_0 \beta = 0 \tag{8.142}$$

$$\begin{aligned}
&[(2a_1 - a_2) b I_1 + 2n \xi_0 b] \gamma + \\
&+ \{[(a_2 - a_1) \zeta_0 - b \xi_0] I_1 - n(b \zeta_0 + a_3 \xi_0)\} \alpha - (a_2 - a_1) \zeta_0 I_1 \beta - \\
&- (a_2 - a_1)(I_2^2 - I_1^2) b I_1 - n(a_2 - a_1)[(a_3 - a_2) I_2^2 + (a_1 - a_3) I_1^2] - \\
&- n b^2 I_1^2 - n^2 (a_2 + a_3 - 2a_1) b I_1 + n^3 b^2 = 0
\end{aligned} \tag{8.143}$$

The three known first integrals of systems (8.138) and (8.139) are

$$\tfrac{1}{2}(a_1 I_1^2 + a_2 I_2^2) + b n I_1 + \xi_0 \alpha + \zeta_0 \gamma = h \tag{8.144}$$

$$\alpha I_1 + \beta I_2 + \gamma n = m \tag{8.145}$$

$$\alpha^2 + \beta^2 + \gamma^2 = 1 \tag{8.146}$$

The problem consists in finding such values for the coefficients

$$a_1, a_2, a_3; \quad b, \xi_0, \zeta_0; \quad h, m, n \tag{8.147}$$

that equations (8.142) to (8.146) give values for I_1, I_2; α, β, γ which are constant and different from zero.

Substituting into equation (8.143) and (8.146) for α, β, γ their expressions in terms of I_1 and I_2 calculated from equations (8.142), (8.144) and (8.145) we obtain equations of the form

$$P_2 I_2^2 + P_4 = 0, \quad Q_2 I_2^4 + Q_4 I_2^2 + Q_6 = 0$$

where P_k, Q_k are polynomials in I_1 of degree k. Eliminating I_2 from the above relations we obtain the relation

$$Q_2 P_4^2 - Q_4 P_2 P_4 + Q_6 P_2^2 = 0 \tag{8.148}$$

which must be satisfied identically in I_1. Equating the coefficients of all powers in I_1 to zero we obtain the necessary number of equations for the determination of coefficients (8.147). The coefficient of I_1^{10} being

$$\xi_0^2 \zeta_0^2 (a_2 - a_1) [\tfrac{1}{4}(2a_1 - a_2)^2 \xi_0^2 + (a_2 - a_1)^2 \zeta_0^2] \times$$
$$\times \{[a_1(a_2 - a_1) + b^2] \xi_0 + 2b(a_2 - a_1) \zeta_0\}$$

this coefficient can be made zero by putting

$$\zeta_0 = -\frac{b^2 + a_1(a_2 - a_1)}{2b(a_2 - a_1)} \xi_0 \tag{8.149}$$

The coefficient of I_1^9 in (8.148) then assumes the form

$$\frac{n\,\xi_0^2(a_1^2 + b^2)}{32 b^5 (a_2 - a_1)} [b^2 + (a_2 - a_1)^2] [b^2 + a_1(a_2 - a_1)]^2 [b^2 - a_1(a_2 - a_1)] \times$$
$$\times [3b^2 - (a_2 - a_1)(a_1 - 2a_3)]$$

For

$$b^2 = \tfrac{1}{3}(a_2 - a_1)(a_1 - 2a_3) \tag{8.150}$$

the remaining coefficients vanish for

$$m = \frac{n^3 b}{3 a_2 \xi_0} (3a_2 - a_1 - a_3)$$
$$h = n^2 \left[\frac{a_1}{2} - \frac{(a_1 + a_3)^2}{9 a_2}\right]$$
$$n^4 = \frac{27 a_2^3 \xi_0^2}{(a_2 - a_1)(a_1 - 2a_3)(3a_2 - a_1 - a_3)^2}$$

while for

$$b^2 = a_1(a_2 - a_1) \tag{8.151}$$

the corresponding coefficients m, h, n^4 have the following expressions

$$m = \frac{n^3 b}{a_2^2 \xi_0} (a_2 - a_1 - a_3)^2$$
$$h = \frac{n^2}{2a_2} [(a_1 + a_3)^2 - a_2(a_1 + 2a_3)]$$
$$n^4 = \frac{a_2^4 \xi_0^2}{(a_2 - a_1)(a_2 - a_1 - a_3)^2 [(a_2 - a_1)(a_1 + a_3)^2 + a_1 a_2^2]}$$

Let $Oxyz$ be the principal inertia trihedral of the body at O, and let Φ be the angle through which this trihedral must be rotated about the Oy-axis until it coincides with the trihedral mentioned at the beginning of this Section. If A, B, C denote the principal moments of inertia of the body and $G(x_0, 0, z_0)$ the mass center, then

$$\begin{aligned}\xi_0 &= x_0 \cos\Phi - z_0 \sin\Phi \\ \zeta_0 &= x_0 \sin\Phi + z_0 \cos\Phi\end{aligned} \tag{8.152}$$

$$\left.\begin{aligned}&\frac{1}{A} = a_1 \cos^2\Phi + a_3 \sin^2\Phi + 2b \sin\Phi \cos\Phi \\ &\frac{1}{B} = a_2 \\ &\frac{1}{C} = a_1 \sin^2\Phi + a_3 \cos^2\Phi - 2b \sin\Phi \cos\Phi \\ &(a_3 - a_1)\sin 2\Phi + 2b \cos 2\Phi = 0\end{aligned}\right\} \tag{8.153}$$

Substituting into (8.149) for ξ_0, ζ_0, a_1, a_2, b their expressions calculated from (8.153), (8.150) and (8.152) we obtain the condition

$$x_0(2C - A)\sqrt{A(C-B)(2C-A)} + z_0(C-2A)\sqrt{C(A-B)(C-2A)} = 0$$

which characterizes the particular solution obtained by Mrs. E. I. Harlamova $(C > 2A > 2B)$. If condition (8.150) is replaced by (8.151), then the above substitution leads to the condition

$$x_0\sqrt{B-C} - z_0\sqrt{A-B} = 0$$

which characterizes the particular solution obtained by Grioli.

8.15. The principle of gyroscopic effect. As used here, the term refers to a method for obtaining approximations to solutions of problems concerning a rigid body rotating about a fixed point. For details see A. Signorini [1], F. Stoppelli [1, 2], L. Battaglia [1], D. Quilghini [1], S. Rionero [1, 2] and P. Levy [1].

Various properties of the familiar toy called the tippe-top have been discussed by P. Cor [1], G. Peretti [1], C. M. Braams [1], N. M. Hugenholtz [1], J. L. Synge [2], S. O'Brien and J. L. Synge [1], D. G. Parkin [1, 2], Miss L. S. Isaeva [1], W. Schallreuter [1], F. Schuh [1], A. P. Duvakin [1], and V. M. Morozov [1].

8.16. Intrinsic equations of motion. The cone, traced in space by the axis of a gyroscope, intersects the unit sphere, whose center is at the fixed point, in a curve Γ. Using the concepts of geodesic curvature and arc length of Γ, W. F. Osgood [1] introduced the so-called intrinsic equations of motion of a heavy symmetric rigid body about a fixed point. With the aid of Osgood's equations A. H. Copeland [1] described the properties of motion, exhibited their dependence upon the initial conditions, and showed how one type of motion changes into another as the initial conditions are varied continuously.

G. Grioli [7], using equations reminiscent of those due to Osgood, studied the existence of precessional motions of a heavy asymmetric rigid body. The motions obtained by him are special cases of those discovered by W. Hess. A. Bressan [1] showed that the motions discovered by Grioli are the only precessional motions of Hess if the spin axis coincides with the barycentric axis or if the angular momentum vector is horizontal.

8.17. Other cases of integrability. T. Manacorda [1] and A. Signorini [2] gave a geometrical approach to various matters connected with the motion of a rigid body about a fixed point O distinct from the mass center G.

F. DE SIMONI [2] studied the motion of a rigid body about a fixed point O, the distance d of which from the center of mass G is small enough so that d^n ($n \geqq 2$) can be neglected (quasi-barycentric suspension). He considered, in particular, the case when the angular momentum is parallel to the constant resultant force applied to the body.

K. STANGE [1, 2] studied the motion of a heavy symmetric rigid body assuming that the inclination of the axis of symmetry to the vertical is small.

H. TALLQVIST [1] pointed out several new integrable cases of the LAGRANGE-POISSON and KOVALEVSKAYA rigid bodies, when subjected to forces distinct from gravity.

P. UDESCHINI [1] discussed the motion of a heavy rigid body spinning on a horizontal plane subject to forces (in addition to those of gravity and reaction of the plane) consisting of a couple with respect to the center of mass G and of a horizontal force applied at G and intersecting a fixed vertical line.

G. ARRIGHI [1] considered the motion of a symmetric rigid body suspended frictionlessly from its center of mass G and subject to magnetostatic forces due to two dipoles: one fixed in space with its axis in line with G, the other located along the axis of symmetry of the body and rigidly connected with it. Such a problem is of some interest in atomic physics.

D. N. GORYAČEV [3—5] solved the problem of finding all cases when the EULER equations admit an integral which is a polynomial of degree three or four in p, q, r, and determined in each of these cases the corresponding force function U as a function of the position of the body. Under the further assumption that the projection of the angular momentum on the vertical is zero, i.e. $c = 0$, he obtained in each of the two cases mentioned above the solution in terms of four arbitrary constants. N. G. APYHTIN [1] determined permanent axes of rotation of a rigid body with a fixed point when the integrals of D. N. GORYAČEV [3, 5] exist. See also H. GEBELEIN [1], E. L. HILL [1], E. I. HARLAMOVA-ZABELINA [1], F. SIMONART [1] and F. SLUDSKII [1].

C. Application of Lie series to the Euler and Poisson equations

§ 9. Lie series and their application to the study of motion of a heavy rigid body about a fixed point

9.1. Definition of generalized Lie series. Let D denote a linear differential operator of the form

$$D = \sum_{k=1}^{n} \Theta_k(z_1, z_2, \ldots, z_n) \frac{\partial}{\partial z_k}$$

where the $\Theta_k(z_1, z_2, \ldots, z_n)$ are certain analytic functions of the complex variables $z_1, z_2, \ldots, z_n$. A series of the form

$$\sum_{\nu=0}^{\infty} \frac{t^\nu}{\nu!} D^\nu F(z_1, z_2, \ldots, z_n) \tag{9.1}$$

where $F(z_1, z_2, \ldots, z_n)$ is an analytic function and t a complex variable, is called a LIE series. Most complete results concerning these series are due to W. GRÖBNER [1].

In order to extend the range of applicability of such series A. N. FILATOV [1] has given the definition of generalized LIE series.

Let $\varphi_m(t, z_1, z_2, \ldots, z_n)$ $(m = 0, 1, \ldots, n)$ be certain analytic functions of the complex variables $t, z_1, z_2, \ldots, z_n$ defined in a domain G containing the origin of the space $Z_{n+1}\{t, z_1, z_2, \ldots, z_n\}$. Further, let $f(t, z_1, \ldots, z_n)$ be an analytic function defined in G. Consider the operator W of the form

$$W = \varphi_0(t, z_1, \ldots, z_n)\frac{\partial}{\partial t} + \sum_{k=1}^{n}\varphi_k(t, z_1, \ldots, z_n)\frac{\partial}{\partial z_k} \tag{9.2}$$

and define the powers of W by the equalities

$$W^0 f \equiv f, \qquad W^n f = W(W^{n-1} f)$$

Definition. A series of the form

$$\sum_{s=0}^{\infty}\frac{t^s}{s!}W^s f(t, z_1, \ldots, z_n) \tag{9.3}$$

is called a generalized LIE series.

A characteristic feature of series (9.3) is that the coefficients $W^s f$ of this series contain the variable t in powers of which the series is expanded. If $\varphi_0(t, z_1, \ldots, z_n) \equiv 0$ and the functions f and φ_k $(k = 1, \ldots, n)$ do not depend upon t, then the generalized LIE series reduces to the ordinary LIE series (9.1).

9.2. Convergence of generalized Lie series. By virtue of the above assumptions concerning the functions $f(t, z_1, \ldots, z_n)$ and $\varphi_m(t, z_1, \ldots, z_n)$ $(m = 0, 1, \ldots, n)$ in G they can be expanded in a neighborhood of the origin into infinite series,

$$f(t, z_1, \ldots, z_n) = \sum_{k_0 k_1 \ldots k_n}^{0,\infty} a_{k_0 k_1 \ldots k_n} t^{k_0} z_1^{k_1} \ldots z_n^{k_n}$$

$$\varphi_m(t, z_1, \ldots, z_n) = \sum_{k_0 k_1 \ldots k_n}^{0,\infty} a^m_{k_0 k_1 \ldots k_n} t^{k_0} z_1^{k_1} \ldots z_n^{k_n}$$

Consider in this neighborhood a closed domain defined by the inequalities

$$|t| \leqq \tau, \qquad |z_i| \leqq \alpha_i, \qquad (i = 1, 2, \ldots, n)$$

Then by virtue of the convergence of the series

$$\sum_{k_0 k_1 \ldots k_n}^{0,\infty} |a_{k_0 k_1 \ldots k_n}| \tau^{k_0} \alpha_1^{k_1} \ldots \alpha_n^{k_n}$$

$$\sum_{k_0 k_1 \ldots k_n}^{0,\infty} |a^m_{k_0 k_1 \ldots k_n}| \tau^{k_0} \alpha_1^{k_1} \ldots \alpha_n^{k_n}$$

there exist constants M and M_m such that

$$\left|a_{k_0 k_1 \ldots k_n}\right| \leqq \frac{M}{\tau^{k_0}\alpha_1^{k_1}\ldots\alpha_n^{k_n}} = b_{k_0 k_1 \ldots k_n}$$

$$\left|a^m_{k_0 k_1 \ldots k_n}\right| \leqq \frac{M_m}{\tau^{k_0}\alpha_1^{k_1}\ldots\alpha_n^{k_n}} = b^m_{k_0 k_1 \ldots k_n}$$

Consequently the functions of real variables $x_0, x_1, \ldots, x_n$ (where $|t| \leqq x_0 < \tau$, $|z_i| \leqq x_i < \alpha$)

$$\frac{M}{\left(1-\frac{x_0}{\tau}\right)\left(1-\frac{x_1}{\alpha_1}\right)\cdots\left(1-\frac{x_n}{\alpha_n}\right)} = \sum_{k_0 k_1 \ldots k_n}^{0,\infty} b_{k_0 k_1 \ldots k_n} x_0^{k_0} x_1^{k_1}\ldots x_n^{k_n}$$

$$\frac{M_m}{\left(1-\frac{x_0}{\tau}\right)\left(1-\frac{x_1}{\alpha_1}\right)\cdots\left(1-\frac{x_n}{\alpha_n}\right)} = \sum_{k_0 k_1 \ldots k_n}^{0,\infty} b^m_{k_0 k_1 \ldots k_n} x_0^{k_0} x_1^{k_1}\ldots x_n^{k_n}$$

are majorants of the functions $f(t, z_1, \ldots, z_n)$ and $\varphi_m(t, z_1, \ldots, z_n)$ respectively.

Let $\alpha = \min\limits_{1 \leqq i \leqq n}\{\tau, \alpha_i\}$. Then

$$\frac{1}{1-\frac{x_0}{\tau}} \leqq \frac{1}{1-\frac{x}{\alpha}}, \qquad \frac{1}{1-\frac{x_i}{\alpha_i}} \leqq \frac{1}{1-\frac{x}{\alpha}}$$

$$|t| \leqq x_0 \leqq x < \alpha, \qquad |z_i| \leqq x_i \leqq x < \alpha$$

and, consequently, the functions

$$F(x) = \frac{M}{\left(1-\frac{x}{\alpha}\right)^{n+1}}, \qquad \Phi_m(x) = \frac{M_m}{\left(1-\frac{x}{\alpha}\right)^{n+1}} \tag{9.4}$$

are also majorants of the functions $f(t, z_1, \ldots, z_n)$ and $\varphi_m(t, z_1, \ldots, z_n)$ respectively, and

$$\begin{aligned} &|f| \leqq F(x), && \left|\frac{\partial^k f}{\partial t^{k_0}\partial z_1^{k_1}\ldots\partial z_n^{k_n}}\right| \leqq \frac{d^k F}{dx^k} \\ &|\varphi_m| \leqq \Phi_m(x), && \left|\frac{\partial^k \varphi_m}{\partial t^{k_0}\partial z_1^{k_1}\ldots\partial z_n^{k_n}}\right| \leqq \frac{d^k \Phi_m}{dx^k} \quad (k_0 + \cdots + k_n = k) \end{aligned} \tag{9.5}$$

Furthermore, the function

$$\Phi(x) = \frac{N}{\left(1-\frac{x}{\alpha}\right)^{n+1}} \tag{9.6}$$

where

$$N = \sum_{m=0}^{n} M_m$$

is a majorant of all $\varphi_m(t, z_1, \ldots, z_n)$ $(m = 0, 1, \ldots, n)$. Let us introduce the operator

$$\overline{W} = \frac{N}{\left(1-\frac{x}{\alpha}\right)^{n+1}}\frac{d}{dx} \tag{9.7}$$

and prove the following lemmas.

Lemma 1. For $|t| \leqq x < \alpha$, $|z_k| \leqq x < \alpha$ $(k = 1, 2, \ldots, n)$ $s = 0, 1, 2, \ldots$ the inequality

$$|W^s f(t, z_1, \ldots, z_n)| \leqq \overline{W}^s F(x) \tag{9.8}$$

holds.

Proof by induction. For $s = 0$ inequality (9.8) holds by virtue of (9.5). For $s = 1$ we have that

$$|W f(t, z_1, \ldots, z_n)| = \left|\varphi_0 \frac{\partial f}{\partial t} + \sum_{k=1}^{n} \varphi_k \frac{\partial f}{\partial z_k}\right| \leqq$$

$$\leqq \frac{M_0}{\left(1 - \frac{x}{\alpha}\right)^{n+1}} \frac{dF}{dx} + \sum_{m=1}^{n} \frac{M_m}{\left(1 - \frac{x}{\alpha}\right)^{n+1}} \frac{dF}{dx} = \frac{N}{\left(1 - \frac{x}{\alpha}\right)^{n+1}} \frac{dF}{dx} = \overline{W} F(x)$$

Assume that inequality (9.8) holds for some s. Then

$$|W^{s+1} f| = |W(W^s f)| = \left|\varphi_0 \frac{\partial W^s f}{\partial t} + \sum_{k=1}^{n} \varphi_k \frac{\partial W^s f}{\partial z_k}\right| \leqq$$

$$\leqq \sum_{m=0}^{n} \frac{M_m}{\left(1 - \frac{x}{\alpha}\right)^{n+1}} \frac{d\overline{W}^s F}{dx} = \overline{W}^{s+1} F(x)$$

q.e.d.

Remark. Lemma 1 can be proved under less restrictive assumptions; namely, instead of $N = \sum_{m=0}^{n} M_m$ we can define $N = \max_{0 \leqq m \leqq n} \{M_m\}$.

In what follows both definitions of N may be used.

Lemma 2. For $s = 1, 2, \ldots$ the equality

$$\overline{W}^s F(x) = \frac{M\, s!}{\left(1 - \frac{x}{\alpha}\right)^{n+1}} \frac{N^s \prod_{k=1}^{s} \left(n + 2 - \frac{1}{k}\right)}{\alpha^s \left(1 - \frac{x}{\alpha}\right)^{s(n+2)}} \tag{9.9}$$

holds.

Proof by induction. For $s = 1$ the truth of (9.9) is obvious. Assume that (9.9) holds for some s. Then

$$\overline{W}^{s+1} F(x) = \overline{W}(\overline{W}^s F(x))$$

$$= \frac{N}{\left(1 - \frac{x}{\alpha}\right)^{n+1}} \frac{d}{dx} \left[\frac{M\, s!}{\left(1 - \frac{x}{\alpha}\right)^{n+1}} \frac{N^s \prod_{k=1}^{s} \left(n + 2 - \frac{1}{k}\right)}{\alpha^s \left(1 - \frac{x}{\alpha}\right)^{s(n+2)}}\right]$$

$$= \frac{M(s+1)!}{\left(1 - \frac{x}{\alpha}\right)^{n+1}} \frac{N^{s+1} \prod_{k=1}^{s+1} \left(n + 2 - \frac{1}{k}\right)}{\alpha^{s+1} \left(1 - \frac{x}{\alpha}\right)^{(s+1)(n+2)}}$$

q.e.d.

Theorem 1

The generalized LIE series converges absolutely for

$$|t| < \min\left\{x, \frac{\alpha}{N(n+2)}\left(1-\frac{x}{\alpha}\right)^{n+2}\right\} \qquad (9.10)$$
$$|z_k| \leqq x < \alpha$$

Proof. By virtue of Lemmas 1 and 2 we have that

$$\sum_{s=0}^{\infty} \frac{|t|^s}{s!} |W^s f| \leqq \frac{M}{\left(1-\frac{x}{\alpha}\right)^{n+1}} + \sum_{s=1}^{\infty} \frac{|t|^s}{s!} \overline{W}^s F$$

$$= \frac{M}{\left(1-\frac{x}{\alpha}\right)^{n+1}} + \sum_{s=1}^{\infty} \frac{|t|^s M}{\left(1-\frac{x}{\alpha}\right)^{n+1}} \frac{N^s \prod_{k=1}^{s}\left(n+2-\frac{1}{k}\right)}{\alpha^s\left(1-\frac{x}{\alpha}\right)^{s(n+2)}} \leqq$$

$$\leqq \frac{M}{\left(1-\frac{x}{\alpha}\right)^{n+1}} \left\{1 + \sum_{s=1}^{\infty}\left[\frac{|t| N(n+2)}{\alpha\left(1-\frac{x}{\alpha}\right)^{n+2}}\right]^s\right\}$$

The series in curled brackets converges for

$$|t| < \frac{\alpha}{N(n+2)}\left(1-\frac{x}{\alpha}\right)^{n+2}$$

Since, in addition, $|t| \leqq x < \alpha$, then for

$$|t| < \min\left\{x, \frac{\alpha}{N(n+2)}\left(1-\frac{x}{\alpha}\right)^{n+2}\right\}$$

the generalized LIE series converges absolutely.

Theorem 2

If $\bar{G}_{n+1} = \bar{G}_n \times [|t| \leqq l]$ [1] is a finite closed domain of the space Z_{n+1} in which the operator W and the function $f(t, z_1, \ldots, z_n)$ are analytic[2], then there always exists a positive constant T such that the generalized LIE series

$$\sum_{s=0}^{\infty} \frac{t^s}{s!} W^s f(t, z_1, \ldots, z_n)$$

converges absolutely and uniformly for $|t| \leqq T$ in the whole domain $\bar{G}_n \times [|t| \leqq T]$ and represents an analytic function of the complex variables $t, z_1, \ldots, z_n$.

[1] $\bar{G}_n \times [|t| \leqq l]$ denotes the direct product of the closed domain $\bar{G}_n$ of the space $Z_n \{z_1, \ldots, z_n\}$ and the domain $|t| \leqq l$.

[2] W is said to be analytic if the functions $\varphi_s(t, z_1, \ldots, z_n)$ $(s = 0, 1, \ldots, n)$ are analytic.

Inside the domain $\bar{G}_n \times [|t| \leqq T]$ the generalized LIE series can be differentiated term-by-term an arbitrary number of times. The following equalities

$$\frac{\partial}{\partial t}\left(\sum_{s=0}^{\infty} \frac{t^s}{s!} W^s f\right) = \sum_{s=0}^{\infty} \frac{t^s}{s!} W^{s+1} f + \sum_{s=0}^{\infty} \frac{t^s}{s!} \frac{\partial W^s f}{\partial t}$$

$$\frac{\partial}{\partial z_k}\left(\sum_{s=0}^{\infty} \frac{t^s}{s!} W^s f\right) = \sum_{s=0}^{\infty} \frac{t^s}{s!} \frac{\partial W^s f}{\partial z_k}$$

hold.

Proof. By virtue of Theorem 1 series (9.3) converges absolutely for

$$|t| < \min\left\{x, \frac{\alpha}{N(n+2)}\left(1 - \frac{x}{\alpha}\right)^{n+2}\right\}$$

$$|z_k| \leqq x < \alpha$$

Let us assign to x a fixed value, and let us put $x = \Theta\alpha$, $0 < \Theta < 1$. Then for

$$|t| \leqq \tau_0 < \min\left\{\Theta\alpha, \frac{\alpha(1-\Theta)^{n+2}}{N(n+2)}\right\}$$

series (9.3) converges absolutely and uniformly in the domain $\bar{Q}_n$ defined by the inequalities

$$|z_k| \leqq \alpha\Theta$$

and represents an analytic function of the variables $t, z_1, \ldots, z_n$.

The above argument refers to a neighborhood of the origin $t = z_1 = \cdots = z_n = 0$. Obviously the same argument can be repeated for any point of the domain $\bar{G}_{n+1}$. Thus for each point of the domain $\bar{G}_{n+1}$ there exists a number τ_0 and a domain Q_n such that inside this domain series (9.3) converges absolutely and uniformly. The set of domains Q_n represents a covering of the domain $\bar{G}_n$. By the HEINE-BOREL theorem there exists a finite number of domains Q_n which cover the domain $\bar{G}_n$.

To each such domain Q_n of the finite covering of the domain $\bar{G}_n$ there corresponds a number τ_0. Select the smallest of these and denote it by T. Then for $|t| \leqq T$ series (9.3) will converge absolutely and uniformly in the whole domain $\bar{G}_n$.

Consider the series

$$\sum_{s=0}^{\infty} \frac{t^s}{s!} \frac{\partial W^s f}{\partial z_k}$$

obtained from series (9.3) by differentiating the latter term-by-term. Let us prove that also this series converges in the same domain as series (9.3). In fact, since $\overline{W}^s F$ is a majorant of $W^s f$, then

$$\left|\frac{\partial W^s f}{\partial z_k}\right| \leqq \frac{d\overline{W}^s F}{dx}, \quad |t| \leqq x < \alpha, \quad |z_i| \leqq x < \alpha$$

Therefore

$$\sum_{s=0}^{\infty} \frac{|t|^s}{s!} \left| \frac{\partial W^s f}{\partial z_k} \right| \leqq \sum_{s=0}^{\infty} \frac{|t|^s}{s!} \frac{d \overline{W}^s F}{dx} \leqq$$

$$\leqq \frac{M(n+1)}{\alpha\left(1-\frac{x}{\alpha}\right)^{n+2}} + \sum_{s=1}^{\infty} \frac{|t|^s M(s+1) N^s \prod_{k=1}^{s+1}\left(n+2-\frac{1}{k}\right)}{\alpha^{s+1}\left(1-\frac{x}{\alpha}\right)^{(s+1)(n+2)}} \leqq$$

$$\leqq \frac{M(n+2)}{\alpha\left(1-\frac{x}{\alpha}\right)^{n+2}} \left\{1 + \sum_{s=1}^{\infty} \left[\frac{|t| N(n-2)}{\alpha\left(1-\frac{x}{\alpha}\right)^{n+2}}\right]^s (s+1)\right\}$$

The last series converges for

$$|t| < \min\left\{x, \frac{\alpha}{N(n+2)}\left(1-\frac{x}{\alpha}\right)^{n+2}\right\}$$

The further reasoning is obvious. This completes the proof of the theorem.

9.3. Operations with generalized Lie series. Let us consider some of the elementary operations, such as addition, subtraction, multiplication, division and permutability of the generalized Lie series.

Assume that $F_1(t, z_1, \ldots, z_n)$ and $F_2(t, z_1, \ldots, z_n)$ are certain analytic functions of the variables $t, z_1, \ldots, z_n$ and c_1, c_2 are arbitrary constants. Then the following equalities are obvious:

$$W^s\{c_1 F_1 \pm c_2 F_2\} = c_1 W^s F_1 \pm c_2 W^s F_2 \tag{9.11}$$

$$W^s\{F_1 F_2\} = \sum_{m=0}^{\infty} \frac{s!}{m!(s-m)!} W^{s-m} F_1 W^m F_2 \tag{9.12}$$

1. Addition and subtraction. The equality

$$c_1 \sum_{s=0}^{\infty} \frac{t^s}{s!} W^s F_1 \pm c_2 \sum_{s=0}^{\infty} \frac{t^s}{s!} W^s F_2 = \sum_{s=0}^{\infty} \frac{t^s}{s!} W^s\{c_1 F_1 \pm c_2 F_2\}$$

holds. The proof is obvious due to the first of the above equalities.

2. Multiplication. The equality

$$\left(\sum_{s=0}^{\infty} \frac{t^s}{s!} W^s F_1\right)\left(\sum_{s=0}^{\infty} \frac{t^s}{s!} W^s F_2\right) = \sum_{s=0}^{\infty} \frac{t^s}{s!} W^s\{F_1 F_2\} \tag{9.13}$$

holds.

Proof. By virtue of formula (9.12) we can write

$$\sum_{s=0}^{\infty} \frac{t^s}{s!} W^s\{F_1 F_2\} = \sum_{s=0}^{\infty} \sum_{m=0}^{s} \frac{t^s}{m!(s-m)!} W^{s-m} F_1 W^m F_2$$

Let

$s - m = k$, $s = m + k$. Then

$$\sum_{s=0}^{\infty} \sum_{m=0}^{s} \frac{t^s}{m!(s-m)!} W^{s-m} F_1 W^m F_2 = \sum_{k=0}^{\infty} \sum_{m=0}^{\infty} \frac{t^{m+k}}{m!\,k!} W^k F_1 W^m F_2$$

$$= \left(\sum_{s=0}^{\infty} \frac{t^s}{s!} W^s F_1\right)\left(\sum_{s=0}^{\infty} \frac{t^s}{s!} W^s F_2\right)$$

q.e.d.

From formula (9.13) follows, in particular, that

$$\left(\sum_{s=0}^{\infty} \frac{t^s}{s!} W^s F\right)^m = \sum_{s=0}^{\infty} \frac{t^s}{s!} W^s\{F^m\} \qquad (m = 0, 1, 2, \ldots)$$

3. *Division.* The formula

$$\frac{\sum_{s=0}^{\infty} \frac{t^s}{s!} W^s F_1}{\sum_{s=0}^{\infty} \frac{t^s}{s!} W^s F_2} = \sum_{s=0}^{\infty} \frac{t^s}{s!} W^s \left\{\frac{F_1}{F_2}\right\}$$

holds provided that the function $F_2(t, z_1, \ldots, z_n)$ and the series $\sum_{s=0}^{\infty} \frac{t^s}{s!} W^s F_2$ do not have roots in a certain common domain of their existence.

Proof. Since

$$\sum_{s=0}^{\infty} \frac{t^s}{s!} W^s F_1 = \sum_{s=0}^{\infty} \frac{t^s}{s!} W^s \left\{\frac{F_1}{F_2} F_2\right\} = \left(\sum_{s=0}^{\infty} \frac{t^s}{s!} W^s \frac{F_1}{F_2}\right)\left(\sum_{s=0}^{\infty} \frac{t^s}{s!} W^s F_2\right)$$

then

$$\sum_{s=0}^{\infty} \frac{t^s}{s!} W^s \left\{\frac{F_1}{F_2}\right\} = \frac{\sum_{s=0}^{\infty} \frac{t^s}{s!} W^s F_1}{\sum_{s=0}^{\infty} \frac{t^s}{s!} W^s F_2}$$

q.e.d.

4. *Permutability.* Let

$$\sum_{s=0}^{\infty} \frac{t^s}{s!} W^s z_i = u_i, \qquad \sum_{s=0}^{\infty} \frac{t^s}{s!} W^s t = u_0$$

Then the following theorem holds.

Theorem 3

Let $F(t, z_1, \ldots, z_n)$ be an arbitrary function analytic in a neighborhood of the origin $t = z_1 = \cdots = z_n = 0$, the power series expansion of which in this neighborhood converges at the point $\{u_0, u_1, \ldots, u_n\}$. Then the equality

$$\sum_{s=0}^{\infty} \frac{t^s}{s!} W^s F(t, z_1, \ldots, z_n) = F\left(\sum_{s=0}^{\infty} \frac{t^s}{s!} W^s t, \sum_{s=0}^{\infty} \frac{t^s}{s!} W^s z_1, \ldots, \sum_{s=0}^{\infty} \frac{t^s}{s!} W^s z_n\right)$$

holds.

9.4. First integrals of a system of ordinary differential equations. Consider the system of differential equations

$$\frac{dz_i}{dt} = f_i(t, z_1, \ldots, z_n) \qquad (i = 1, 2, \ldots, n) \qquad (9.14)$$

where the functions $f_i(t, z_1, \ldots, z_n)$ are analytic in the domain $\bar{G}$ defined by the inequalities

$$|t| \leqq \tau, \qquad |z_i| \leqq \alpha_i \qquad (9.15)$$

Let

$$M = \max_{\bar{G}} |f_i|, \qquad \alpha = \min_{1 \leq i \leq n} (\tau, \alpha_i), \qquad 0 < \mu < 1$$

Theorem 4

If the functions $f_i(t, z_1, \ldots, z_n)$ of system (9.14) are analytic in the domain $\bar{G}$ defined by (9.15), then n independent first integrals of system (9.14) have the form

$$\sum_{s=0}^{\infty} \frac{(-1)^s t^s}{s!} W^s\{z_i\} = c_i \qquad (i = 1, 2, \ldots, n) \tag{9.16}$$

where the c_i are arbitrary constants and W is an operator of the form

$$W = \frac{\partial}{\partial t} + \sum_{k=1}^{n} f_k(t, z_1, \ldots, z_n) \frac{\partial}{\partial z_k} \tag{9.17}$$

The series (9.16) converge absolutely and uniformly in the domain $\bar{Q}$ defined by the inequalities

$$\begin{aligned} &|t| \leq T < \min\left\{\mu\alpha, \frac{\alpha(1-\mu)^{n+2}}{M(n+2)}\right\} \\ &|z_i| \leq \mu\alpha, \qquad 0 < \mu < 1 \end{aligned} \tag{9.18}$$

Proof. In order to prove that the series (9.16) are first integrals of system (9.14) it is sufficient to show that the total derivative of (9.16) by virtue of (9.14) is identically zero.

We have that

$$\begin{aligned} &\frac{d}{dt} \sum_{s=0}^{\infty} \frac{(-1)^s t^s}{s!} W^s z_i \\ &= \sum_{s=1}^{\infty} \frac{(-1)^s t^{s-1}}{(s-1)!} W^s z_i + \sum_{s=0}^{\infty} \frac{(-1)^s t^s}{s!} \cdot \left(\frac{\partial W^s z_i}{\partial t} + \sum_{k=1}^{n} \frac{\partial W^s z_i}{\partial z_k} \frac{dz_k}{dt}\right) \\ &= -\sum_{s=0}^{\infty} \frac{(-1)^s t^s}{s!} W^{s+1} z_i + \sum_{s=0}^{\infty} \frac{(-1)^s t^s}{s!} \left(\frac{\partial W^s z_i}{\partial t} + \sum_{k=1}^{n} f_k(t, z_1, \ldots, z_n) \frac{\partial W^s z_i}{\partial z_k}\right) \\ &= -\sum_{s=0}^{\infty} \frac{(-1)^s t^s}{s!} W^{s+1} z_i + \sum_{s=0}^{\infty} \frac{(-1)^s t^s}{s!} W(W^s z_i) \equiv 0 \end{aligned}$$

Series (9.16) being generalized LIE series, their convergence can be proved in the same way as before.

The independence of the first integrals (9.16) can be proved as follows. Assume that the k-th integral is a combination of the remaining integrals, i.e. let

$$\begin{aligned} \sum_{s=0}^{\infty} \frac{(-1)^s t^s}{s!} W^s z_k = \varphi\Bigg(&\sum_{s=0}^{\infty} \frac{(-1)^s t^s}{s!} W^s z_1, \ldots, \sum_{s=0}^{\infty} \frac{(-1)^s t^s}{s!} W^s z_{k-1}, \\ &\sum_{s=0}^{\infty} \frac{(-1)^s t^s}{s!} W^s z_{k+1}, \ldots, \sum_{s=0}^{\infty} \frac{(-1)^s t^s}{s!} W^s z_n\Bigg) \end{aligned}$$

where φ is some analytic function. Applying the permutability property we find that

$$\sum_{s=0}^{\infty} \frac{(-1)^s t^s}{s!} W^s z_k = \sum_{s=0}^{\infty} \frac{(-1)^s t^s}{s!} W^s \varphi(z_1, \ldots, z_{k-1}, z_{k+1}, \ldots, z_n)$$

or

$$\sum_{s=0}^{\infty} \frac{(-1)^s t^s}{s!} W^s\{z_k - \varphi(z_1, \ldots, z_{k-1}, z_{k+1}, \ldots, z_n)\} = 0$$

Since this equality must be satisfied for any t in the domain of convergence, then by putting $t = 0$ we obtain that

$$z_k - \varphi(z_1, \ldots, z_{k-1}, z_{k+1}, \ldots, z_n) = 0$$

This contradicts the assumption that the variables z_i are independent. Hence the truth of the theorem is proved.

In practice the integrals found are often combinations of integrals (9.16). Let F denote an analytic function of n variables, then

$$F\left(\sum_{s=0}^{\infty} \frac{(-1)^s t^s}{s!} W^s z_1, \ldots, \sum_{s=0}^{\infty} \frac{(-1)^s t^s}{s!} W^s z_n\right) = \text{const.}$$

is also an integral of system (9.14). By the permutability property this integral can be rewritten in the form

$$\sum_{s=0}^{\infty} \frac{(-1)^s t^s}{s!} W^s F(z_1, \ldots, z_n) = \text{const.} \tag{9.19}$$

Example:

$$\frac{dx}{dt} = y, \quad \frac{dy}{dt} = -x \tag{9.20}$$

Introducing the operator $W = y\frac{\partial}{\partial x} - x\frac{\partial}{\partial y}$, the first integrals of the above system are

$$\sum_{s=0}^{\infty} \frac{(-1)^s t^s}{s!} W^s x = c_1, \quad \sum_{s=0}^{\infty} \frac{(-1)^s t^s}{s!} W^s y = c_2 \tag{9.21}$$

Let $F(x, y)$ be an analytic function. Then

$$\sum_{s=0}^{\infty} \frac{(-1)^s t^s}{s!} W^s F(x, y) = \text{const.} \tag{9.22}$$

is a first integral of the above system. If, in particular, $F(x, y) = x^2 + y^2$, then

$$\sum_{s=0}^{\infty} \frac{(-1)^s t^s}{s!} W^s(x^2 + y^2) = x^2 + y^2 = \text{const.} \tag{9.23}$$

is an integral of our system.

Since

$$W^s x = \begin{cases} (-1)^{n-1} y, & s = 2n - 1 \\ (-1)^n x, & s = 2n \end{cases} \qquad (n = 1, 2, \ldots)$$

$$W^s y = \begin{cases} (-1)^n x, & s = 2n - 1 \\ (-1)^n y, & s = 2n \end{cases} \qquad (n = 1, 2, \ldots)$$

then integrals (9.21) assume the form

$$x \cos t - y \sin t = c_1, \quad x \sin t + y \cos t = c_2$$

Corollary. If the given system is autonomous, i.e. it has the form

$$\frac{dz_i}{dt} = f_i(z_1, \ldots, z_n) \tag{9.24}$$

then its first integrals are of the form

$$\sum_{s=0}^{\infty} \frac{(-1)^s t^s}{s!} U^s z_i = c_i \tag{9.25}$$

where

$$U = \sum_{k=1}^{n} f_k(z_1, \ldots, z_n) \frac{\partial}{\partial z_k} \tag{9.26}$$

9.5. First integrals of canonical equations. Consider the system of canonical equations

$$\dot{q}_i = \frac{\partial H}{\partial p_i}, \qquad \dot{p}_i = -\frac{\partial H}{\partial q_i} \qquad (i = 1, 2, \ldots, n) \tag{9.27}$$

where $H = H(t, q_i, p_i)$ is the HAMILTONIAN function. Assuming that the right-hand sides of (9.27) are analytic and introducing the operator

$$\Gamma = \frac{\partial}{\partial t} + \sum_{k=1}^{n} \left(\frac{\partial H}{\partial p_k} \frac{\partial}{\partial q_k} - \frac{\partial H}{\partial q_k} \frac{\partial}{\partial p_k} \right) \tag{9.28}$$

the first integrals of canonical system (9.27) are

$$\begin{aligned} &\sum_{s=0}^{\infty} \frac{(-1)^s t^s}{s!} \Gamma^s q_i = a_i \\ &\sum_{s=0}^{\infty} \frac{(-1)^s t^s}{s!} \Gamma^s p_i = b_i \end{aligned} \tag{9.29}$$

The integrals found in particular cases are not of the form (9.29) but they represent certain combinations of integrals (9.29). Let F be an analytic function of $2n$ variables. Then

$$F\left(\sum_{s=0}^{\infty} \frac{(-1)^s t^s}{s!} \Gamma^s q_1, \ldots, \sum_{s=0}^{\infty} \frac{(-1)^s t^s}{s!} \Gamma^s q_n; \sum_{s=0}^{\infty} \frac{(-1)^s t^s}{s!} \Gamma^s p_1, \ldots, \sum_{s=0}^{\infty} \frac{(-1)^s t^s}{s!} \Gamma^s p_n \right) = \text{const.}$$

is also an integral of system (9.27), which by virtue of the permutability property, can be written in the form

$$\sum_{s=0}^{\infty} \frac{(-1)^s t^s}{s!} \Gamma^s \{F(q_1, \ldots, q_n; p_1, \ldots, p_n)\} = \text{const.} \tag{9.30}$$

If the HAMILTONIAN function H does not contain the time t explicitly, then system (9.27) assumes the integral

$$H(q_1, \ldots, q_n; p_1, \ldots, p_n) = h = \text{const.} \tag{9.31}$$

which can be obtained from integrals (9.29) as follows. We have that

$$h = H(a_1, \ldots, a_n;\ b_1, \ldots, b_n)$$

$$= H\left(\sum_{s=0}^{\infty} \frac{(-1)^s t^s}{s!} \Gamma^s q_1, \ldots, \sum_{s=0}^{\infty} \frac{(-1)^s t^s}{s!} \Gamma^s q_n;\ \sum_{s=0}^{\infty} \frac{(-1)^s t^s}{s!} \Gamma^s p_1, \ldots, \sum_{s=0}^{\infty} \frac{(-1)^s t^s}{s!} \Gamma^s p_n\right)$$

$$= \sum_{s=0}^{\infty} \frac{(-1)^s t^s}{s!} \Gamma^s \{H(q_1, \ldots, q_n;\ p_1, \ldots, p_n)\} = H(q_1, \ldots, q_n;\ p_1, \ldots, p_n) - \frac{t}{1!} \Gamma H + \cdots$$

Since

$$\Gamma H = \sum_{k=1}^{n} \frac{\partial H}{\partial p_k} \frac{\partial H}{\partial q_k} - \sum_{k=1}^{n} \frac{\partial H}{\partial q_k} \frac{\partial H}{\partial p_k} \equiv 0$$

then

$$H(q_1, \ldots, q_n;\ p_1, \ldots, p_n) = h = \text{const.}$$

Integral (9.31) follows from (9.30) by means of the substitution

$$F(q_1, \ldots, q_n;\ p_1, \ldots, p_n) = H(q_1, \ldots, q_n;\ p_1, \ldots, p_n)$$

9.6. First integrals in the problem of motion of a heavy rigid body about a fixed point. The equations of motion can be rewritten in the form

$$\left.\begin{aligned}
\frac{dp}{dt} &= \frac{B-C}{A} q r + \frac{m g}{A} (\beta z_0 - \gamma y_0) \\
\frac{dq}{dt} &= \frac{C-A}{B} r p + \frac{m g}{B} (\gamma x_0 - \alpha z_0) \\
\frac{dr}{dt} &= \frac{A-B}{C} p q + \frac{m g}{C} (\alpha y_0 - \beta x_0) \\
\frac{d\alpha}{dt} &= \beta r - \gamma q, \quad \frac{d\beta}{dt} = \gamma p - \alpha r, \quad \frac{d\gamma}{dt} = \alpha q - \beta p
\end{aligned}\right\} \tag{9.32}$$

In the general case system (9.32) has three first integrals:

$$A p^2 + B q^2 + C r^2 + 2 m g (x_0 \alpha + y_0 \beta + z_0 \gamma) = c_1 \quad (c_1 = 2 h_1) \tag{9.33}$$

$$A p \alpha + B q \beta + C r \gamma = c_2 \quad (c_2 = c) \tag{9.34}$$

$$\alpha^2 + \beta^2 + \gamma^2 = c_3 = 1 \tag{9.35}$$

Introduce the operator

$$\begin{aligned}
E = &\left\{\frac{B-C}{A} q r + \frac{m g}{A} (\beta z_0 - \gamma y_0)\right\} \frac{\partial}{\partial p} + \\
&+ \left\{\frac{C-A}{B} r p + \frac{m g}{B} (\gamma x_0 - \alpha z_0)\right\} \frac{\partial}{\partial q} + \\
&+ \left\{\frac{A-B}{C} p q + \frac{m g}{C} (\alpha y_0 - \beta x_0)\right\} \frac{\partial}{\partial r} + \\
&+ (\beta r - \gamma q) \frac{\partial}{\partial \alpha} + (\gamma p - \alpha r) \frac{\partial}{\partial \beta} + (\alpha q - \beta p) \frac{\partial}{\partial \gamma}
\end{aligned} \tag{9.36}$$

Then the complete system of first integrals of system (9.32) is

$$\sum_{s=0}^{\infty} \frac{(-1)^s t^s}{s!} E^s p = a_1, \quad \sum_{s=0}^{\infty} \frac{(-1)^s t^s}{s!} E^s q = a_2, \quad \sum_{s=0}^{\infty} \frac{(-1)^s t^s}{s!} E^s r = a_3$$

$$\sum_{s=0}^{\infty} \frac{(-1)^s t^s}{s!} E^s \alpha = a_4, \quad \sum_{s=0}^{\infty} \frac{(-1)^s t^s}{s!} E^s \beta = a_5, \quad \sum_{s=0}^{\infty} \frac{(-1)^s t^s}{s!} E^s \gamma = a_6 \tag{9.37}$$

Also let us note that if F is an analytic function of $p, q, r; \alpha, \beta, \gamma$, then

$$\sum_{s=0}^{\infty} \frac{(-1)^s t^s}{s!} E^s \{F(p, q, r, \alpha, \beta, \gamma)\} = \text{const.} \tag{9.38}$$

is an integral of system (9.32). Thus given F, we shall be able to obtain integrals of system (9.32).

For example, from integral (9.38) follows the energy integral (9.33) for

$$F = A p^2 + B q^2 + C r^2 + 2m g (x_0 \alpha + y_0 \beta + z_0 \gamma)$$

the angular momentum integral (9.34) for

$$F = A p \alpha + B q \beta + C r \gamma$$

and integral (9.35) for

$$F = \alpha^2 + \beta^2 + \gamma^2$$

Let us pass now to the consideration of cases where the fourth integral of system (9.32) is known.

Case of Euler and Poinsot ($x_0 = y_0 = z_0 = 0$). In this case, in addition to integrals (9.33), (9.34) and (9.35), system (9.32) assumes the fourth integral

$$A^2 p^2 + B^2 q^2 + C^2 r^2 = c_4$$

This integral can be obtained from system (9.37) of first integrals as follows. We have that

$$A^2 a_1^2 + B^2 a_2^2 + C^2 a_3^2 = \sum_{s=0}^{\infty} \frac{(-1)^s t^s}{s!} E^s \{A^2 p^2 + B^2 q^2 + C^2 r^2\}$$

$$= A^2 p^2 + B^2 q^2 + C^2 r^2 + \sum_{s=1}^{\infty} \frac{(-1)^s t^s}{s!} E^s \{A^2 p^2 + B^2 q^2 + C^2 r^2\}$$

Further, we find that

$$\begin{aligned} E\{A^2 p^2 + B^2 q^2 + C^2 r^2\} &= 2A(B - C) p q r + 2A m g p (\beta z_0 - \gamma y_0) + \\ &+ 2B(C - A) p q r + 2B m g q (\gamma x_0 - \alpha z_0) + \\ &+ 2C(A - B) p q r + 2C m g r (\alpha y_0 - \beta x_0) = 0 \end{aligned}$$

Consequently,

$$A^2 p^2 + B^2 q^2 + C^2 r^2 = A^2 a_1^2 + B^2 a_2^2 + C^2 a_3^2 = \text{const.} = c_4$$

This integral follows from (9.38) for

$$F = A^2 p^2 + B^2 q^2 + C^2 r^2$$

Case of Lagrange and Poisson $(A = B,\ x_0 = y_0 = 0)$. The fourth integral

$$r = c_4$$

can be obtained from (9.37) as follows. We have that

$$a_3 = \sum_{s=0}^{\infty} \frac{(-1)^s t^s}{s!} E^s r = r + \sum_{s=1}^{\infty} \frac{(-1)^s t^s}{s!} E^s r$$

Since

$$E r = \frac{A-B}{C} p q + \frac{m g}{C} (\alpha y_0 - \beta x_0) = 0$$

then

$$r = a_3 = \text{const.} = c_4$$

Case of kinetic symmetry $(A = B = C)$. In this case the fourth integral is

$$x_0 p + y_0 q + z_0 r = c_4$$

Let us derive this integral from system (9.37) of first integrals. We have that

$$x_0 a_1 + y_0 a_2 + z_0 a_3 = \sum_{s=0}^{\infty} \frac{(-1)^s t^s}{s!} E^s \{x_0 p + y_0 q + z_0 r\}$$

$$= x_0 p + y_0 q + z_0 r + \sum_{s=1}^{\infty} \frac{(-1)^s t^s}{s!} E^s \{x_0 p + y_0 q + z_0 r\}$$

On the other hand,

$$E \{x_0 p + y_0 q + z_0 r\} = \frac{m g x_0}{A} (\beta z_0 - \gamma y_0) + $$

$$+ \frac{m g y_0}{A} (\gamma x_0 - \alpha z_0) + \frac{m g z_0}{A} (\alpha y_0 - \beta x_0) = 0$$

Consequently,

$$x_0 p + y_0 q + z_0 r = x_0 a_1 + y_0 a_2 + z_0 a_3 = \text{const.} = c_4$$

Case of Kovalevskaya $(A = B = 2C,\ y_0 = z_0 = 0)$. The fourth integral is

$$\left(p^2 - q^2 - \frac{m g x_0}{C} \alpha\right)^2 + \left(2 p q - \frac{m g x_0}{C} \beta\right)^2 = c_4$$

which can be derived from system (9.37) of first integrals. We have

$$\left(a_1^2 - a_2^2 - \frac{m g x_0}{C} a_4\right)^2 + \left(2 a_1 a_2 - \frac{m g x_0}{C} a_5\right)^2$$

$$= \sum_{s=0}^{\infty} \frac{(-1)^s t^s}{s!} E^s \left\{\left(p^2 - q^2 - \frac{m g x_0}{C} \alpha\right)^2 + \left(2 p q - \frac{m g x_0}{C} \beta\right)^2\right\}$$

$$= \left(p^2 - q^2 - \frac{m g x_0}{C} \alpha\right)^2 + \left(2 p q - \frac{m g x_0}{C} \beta\right)^2$$

$$+ \sum_{s=1}^{\infty} \frac{(-1)^s t^s}{s!} E^s \left\{\left(p^2 - q^2 - \frac{m g x_0}{C} \alpha\right)^2 + \left(2 p q - \frac{m g x_0}{C} \beta\right)^2\right\}$$

As in the previous cases it can be proved that

$$E \left\{\left(p^2 - q^2 - \frac{m g x_0}{C} \alpha\right)^2 + \left(2 p q - \frac{m g x_0}{C} \beta\right)^2\right\} = 0$$

Therefore

$$\left(p^2 - q^2 - \frac{m g x_0}{C} \alpha\right)^2 + \left(2 p q - \frac{m g x_0}{C} \beta\right)^2$$

$$= \left(a_1^2 - a_2^2 - \frac{m g x_0}{C} a_4\right)^2 + \left(2 a_1 a_2 - \frac{m g x_0}{C} a_5\right)^2 = \text{const.} = c_4$$

Case of Goryačev and Čaplygin ($A = B = 4C$, $y_0 = z_0 = 0$). The fourth integral reads

$$r(p^2 + q^2) - \frac{m g x_0}{C} p \gamma = c_4$$

It can be derived from system (9.37). In fact,

$$a_3 (a_1^2 + a_2^2) - \frac{m g x_0}{C} a_1 a_6 = \sum_{s=0}^{\infty} \frac{(-1)^s t^s}{s!} E^s \left\{r(p^2 + q^2) - \frac{m g x_0}{C} p \gamma\right\}$$

Since

$$E \left\{r(p^2 + q^2) - \frac{m g x_0}{C} p \gamma\right\} \equiv 0$$

then

$$r(p^2 + q^2) - \frac{m g x_0}{C} p \gamma = a_3 (a_1^2 + a_2^2) - \frac{m g x_0}{C} a_1 a_6 = \text{const.} = c_4$$

Case of a loxodromic pendulum (Hess and Appel'rot) ($y_0 = 0$, $x_0 \sqrt{A(B - C)} + z_0 \sqrt{C(A - B)} = 0$). The fourth integral

$$A x_0 p + C z_0 r = 0$$

can again be derived from system (9.37). In fact, consider the expression

$$A x_0 a_1 + C z_0 a_3 = \sum_{s=0}^{\infty} \frac{(-1)^s t^s}{s!} E^s \{A x_0 p + C z_0 r\}$$

It can be shown that

$$E\{A\,x_0\,p + C\,z_0\,r\} \equiv 0$$

Therefore

$$A\,x_0\,p + C\,z_0\,r = A\,x_0\,a_1 + C\,z_0\,a_3$$

Selecting the constants a_1 and a_3 in such a way that $A\,x_0\,a_1 + C\,z_0\,a_3 = 0$, we obtain that

$$A\,x_0\,p + C\,z_0\,r = 0$$

Other integrable cases can be discussed in a similar way.

Chapter II

Self-excited rigid body

§ 10. Self-excited symmetric rigid body

10.1. Introduction. From the mathematical point of view EULER considered the simplest case of motion of a rigid body about a fixed point; namely, when the resultant external torque acting on the body is zero.

The next simplest case we can think of is that of the motion of a rigid body about a fixed point subject to forces with a fixed torque direction in the body. To say that in the EULER equations the torque direction remains constant means that the torque changes its direction as the body continues to rotate about a fixed point O. This is so since the EULER equations refer to the principal axes of inertia of the body at the fixed point O which are rigidly attached to the body. Consequently the torque as well is rigidly attached to the body.

R. GRAMMEL [7—9] defines a rigid body which is free to rotate about a point fixed in the body and in space as "self-excited" when it is acted on by a torque vector $\vec{\mathfrak{M}}$ arising from internal reactions which, however, are not capable of changing appreciably its mass and its mass distribution. Two cases are possible, depending on whether $\vec{\mathfrak{M}}$ is fixed in the body or moves in it in a prescribed manner. We shall refer to these two cases as self-excitements with fixed or moving direction within the body respectively. In each of the above cases two subcases are to be distinguished, depending on whether the modulus of $\vec{\mathfrak{M}}$ changes with the time or not. These are the time-dependent or the time-independent self-excitements.

U. T. BÖDEWADT [1] considered the integration of the EULER equations of motion of a symmetric rigid body ($A = B$) subject to a time-independent self-excitement in a fixed direction in the body. The solution is given in terms of integrals of the FRESNEL type. The integration becomes possible due to a partial uncoupling of the EULER equations.

10.2 The angular velocity of a rigid body subject to a time-independent self-excitement with a fixed direction in the body. The

EULER equations of motion are

$$\left.\begin{aligned} A\dot{p} - (A - C)\,q\,r &= \mathcal{P} \\ A\dot{q} + (A - C)\,p\,r &= \mathcal{Q} \\ C\dot{r} \qquad\qquad &= \mathcal{R} \end{aligned}\right\} \tag{10.1}$$

ere $\mathcal{P}$, $\mathcal{Q}$, $\mathcal{R}$ represent the constant components of $\vec{\mathfrak{M}}$.

We shall introduce the notations

$$1 - \frac{C}{A} = h, \qquad \frac{\mathcal{P}}{A} = P, \qquad \frac{\mathcal{Q}}{A} = Q, \qquad \frac{\mathcal{R}}{C} = R \tag{10.2}$$

assuming first that $A \neq C$, i.e. $h \neq 0$. In the case of kinetic symmetry we have $h = 0$, and the rotations about the three axes are completely independent of each other. In terms of notations (10.2) equations (10.1) assume the form

$$\left.\begin{aligned} \dot{p} - h\,q\,r &= P \\ \dot{q} + h\,p\,r &= Q \\ \dot{r} \qquad &= R \end{aligned}\right\} \tag{10.3}$$

The last equation can be immediately integrated to give for r the expression

$$r = r_1 + R(t - t_1) \tag{10.4}$$

where the subscript 1 here and in what follows denotes the value of the corresponding quantity for $t = t_1$.

If we introduce the complex variables

$$\pi = p + i\,q, \qquad \pi_1 = p_1 + i\,q_1 \tag{10.5}$$

and multiply the second equation of system (10.3) by i, and add it to the first equation of the same set, we obtain

$$\dot{\pi} + i\,h\,r\,\pi = \Pi \tag{10.6}$$

where

$$\Pi = P + i\,Q \tag{10.7}$$

Further, if we use the notations

$$\operatorname{cis} x = \exp i\,x = \cos x + i \sin x \tag{10.8}$$

$$\zeta(t) = \int_{t_1}^{t} h\,r(\tau)\,d\tau \tag{10.9}$$

$$\Omega(t) = \int_{t_1}^{t} \operatorname{cis}\zeta(\tau)\,d\tau \tag{10.10}$$

equation (10.6) has the solution

$$\pi(t) = \operatorname{cis}(-\zeta)\,(\pi_1 + \Pi\,\Omega) \tag{10.11}$$

In what follows we shall distinguish four cases depending on the values of R, r_1, and h.

Case 1: $h \neq 0$, $R = 0$, $r_1 = 0$. In this case it follows from equations (10.4), (10.9) and (10.10) that $r = 0$,

$$\zeta = 0 \tag{10.12}$$

$$\Omega = t - t_1 \tag{10.13}$$

and hence by virtue of equation (10.11),

$$\pi = \pi_1 + \Pi(t - t_1) \tag{10.14}$$

or

$$\begin{aligned} p &= p_1 + P(t - t_1) \\ q &= q_1 + Q(t - t_1) \end{aligned} \tag{10.15}$$

Case 2: $h \neq 0$, $R = 0$, $r_1 \neq 0$. It follows from equations (10.4), (10.9), (10.10) and (10.11) that

$$r = r_1 \neq 0 \tag{10.16}$$

$$\zeta = h\,r(t - t_1) \tag{10.17}$$

$$\Omega = \frac{1}{i\,h\,r}\,[\operatorname{cis} h\,r(t - t_1) - 1] = \frac{1}{i\,h\,r}\,(\operatorname{cis}\zeta - 1) \tag{10.18}$$

$$\pi = \frac{\Pi}{i\,h\,r} + \left(\pi_1 - \frac{\Pi}{i\,h\,r}\right)\operatorname{cis}(-\zeta) \tag{10.19}$$

$$\begin{aligned} p &= p_1\cos\zeta + q_1\sin\zeta + (h\,r)^{-1}[P\sin\zeta + Q(1 - \cos\zeta)] \\ q &= q_1\cos\zeta - p_1\sin\zeta + (h\,r)^{-1}[Q\sin\zeta - P(1 - \cos\zeta)] \end{aligned} \tag{10.20}$$

Case 3: $h \neq 0$, $R \neq 0$. If we introduce the constant

$$t_0 = t_1 - \frac{r_1}{R} \tag{10.21}$$

equation (10.4) assumes the form

$$r = R(t - t_0) \tag{10.22}$$

From equation (10.9) we obtain that

$$\zeta(t) = \tfrac{1}{2}\,hR(t - t_1)\,(t + t_1 - 2t_0) = \eta(t) - \eta_1 \tag{10.23}$$

where

$$\eta(t) = \tfrac{1}{2}\,hR(t - t_0)^2, \quad \eta_1 = \tfrac{1}{2}\,hR(t_1 - t_0)^2 \tag{10.24}$$

The quantity Ω can now be expressed in terms of the FRESNEL integrals

$$C(x) = \int_0^x \frac{\cos y}{\sqrt{y}}\,dy, \qquad S(x) = \int_0^x \frac{\sin y}{\sqrt{y}}\,dy \tag{10.25}$$

For this purpose introduce $y = |\eta(\tau)|$ in equation (10.10) as the variable of integration. If σ denotes the sign of $h\,R$, then y can be rewritten in the form $y = \sigma\eta(\tau)$. Since η vanishes for $\tau = t_0$, we must distinguish three cases:

$$(3\text{a})\quad t_0 \leqq t_1 \leqq t; \qquad (3\text{b})\quad t_1 \leqq t_0 \leqq t; \qquad (3\text{c})\quad t_1 \leqq t \leqq t_0$$

Using formula (10.23) the integrand of (10.10) can be rewritten in the form $\operatorname{cis}(-\eta_1)\operatorname{cis}\eta$, and by virtue of the formula

$$\operatorname{cis}\eta = \cos\eta + i\sin\eta = \cos\sigma\eta + i\sigma\sin\sigma\eta = \cos y + i\sigma\sin y$$

we need consider only the integrals of $\cos y$ and $\sin y$.

From

$$y = \sigma\eta(\tau) = \tfrac{1}{2}\sigma h R(\tau - t_0)^2 \tag{10.26}$$

it follows that

$$|\tau - t_0| = \frac{2}{\sqrt{2|hR|}}\sqrt{y}$$

where on the right the positive sign is to be taken.

Putting

$$(2|hR|)^{-\frac{1}{2}} = k \tag{10.27}$$

we obtain

$$\tau = t_0 + \operatorname{sgn}(\tau - t_0)\, 2k\sqrt{y} \tag{10.28}$$

and hence

$$d\tau = \operatorname{sgn}(\tau - t_0)\, k\frac{dy}{\sqrt{y}} \tag{10.29}$$

In case (3 a) we can write the integral of $\cos y$ as follows:

$$\int_{t_1}^{t} \cos y\, d\tau = \left(\int_{t_0}^{t} - \int_{t_0}^{t_1}\right)\cos y\, d\tau = k[C(\sigma\eta) - C(\sigma\eta_1)]$$

In case (3 b) we have

$$\int_{t_1}^{t} \cos y\, d\tau = \left(\int_{t_0}^{t} + \int_{t_1}^{t_0}\right)\cos y\, d\tau = k[C(\sigma\eta) + C(\sigma\eta_1)]$$

and in case (3 c)

$$\int_{t_1}^{t} \cos y\, d\tau = \left(\int_{t_1}^{t_0} - \int_{t}^{t_0}\right)\cos y\, d\tau = k[C(\sigma\eta_1) - C(\sigma\eta)]$$

If we introduce the notations

$$\varepsilon = \operatorname{sgn}(t - t_0); \qquad \varepsilon_1 = \operatorname{sgn}(t_1 - t_0) \tag{10.30}$$

the above three formulas can be combined into a single formula. Finally, using the notation $C(\eta_1, \eta)$ we have,

$$\int_{t_1}^{t} \cos y\, d\tau = k[\varepsilon\, C(\sigma\eta) - \varepsilon_1\, C(\sigma\eta_1)] = k\, C(\eta_1, \eta) \tag{10.31}$$

Similarly we can derive the formula

$$\int_{t_1}^{t} \sin y \, d\tau = k[\varepsilon\, S(\sigma\,\eta) - \varepsilon_1\, S(\sigma\,\eta_1)] = k\, S(\eta_1, \eta) \qquad (10.31')$$

where a new notation $S(\eta_1, \eta)$ is introduced.

With these notations Ω assumes the form

$$\Omega = \operatorname{cis}(-\eta_1)\, k[C(\eta_1, \eta) + i\,\sigma\, S(\eta_1, \eta)] \qquad (10.32)$$

Substituting this formula into (10.11) we obtain

$$\pi = \pi_1 \operatorname{cis}(-\zeta) + \Pi\, k \operatorname{cis}(-\eta)\,[C(\eta_1, \eta) + i\,\sigma\, S(\eta_1, \eta)] \qquad (10.33)$$

or

$$\left.\begin{aligned} p &= p_1 \cos\zeta + q_1 \sin\zeta + P\,k(C\cos\eta + \sigma\, S \sin\eta) + \\ &\quad + Q\,k(C \sin\eta - \sigma\, S\cos\eta) \\ q &= q_1 \cos\zeta - p_1 \sin\zeta + Q\,k(C\cos\eta + \sigma\, S\sin\eta) - \\ &\quad - Pk\,(C\sin\eta - \sigma\, S\cos\eta) \end{aligned}\right\} \qquad (10.34)$$

Case 4: $h = 0$. In the case of kinetic symmetry it follows directly from formula (10.6) by virtue of formula (10.4) that

$$\pi = \pi_1 + \Pi(t - t_1) \qquad (10.35)$$

Hence

$$\left.\begin{aligned} p &= p_1 + P(t - t_1) \\ q &= q_1 + Q(t - t_1) \\ r &= r_1 + R(t - t_1) \end{aligned}\right\} \qquad (10.36)$$

10.3. Formulas describing rotations. So far we have obtained the angular velocity components p, q, r with respect to a trihedral fixed in the body. Let us consider now the change of the position of the body with respect to a coordinate trihedral fixed in space.

If we denote by A the tensor which transforms the fixed trihedral $O\,X\,Y\,Z$ into the moving trihedral $O\,x\,y\,z$, then with respect to the trihedral $O\,X\,Y\,Z$ the elements α_{mn} $(m, n = 1, 2, 3)$ of the orthogonal matrix (α_{mn}) associated with A have the expressions

$$\left.\begin{aligned} \alpha_{11} &= \cos\psi\cos\Phi - \cos\Theta\sin\psi\sin\Phi \\ \alpha_{12} &= -\cos\psi\sin\Phi - \cos\Theta\sin\psi\cos\Phi \\ \alpha_{13} &= \sin\Theta\sin\psi \\ \alpha_{21} &= \sin\psi\cos\Phi + \cos\Theta\cos\psi\sin\Phi \\ \alpha_{22} &= -\sin\psi\sin\Phi + \cos\Theta\cos\psi\cos\Phi \\ \alpha_{23} &= -\sin\Theta\cos\psi \\ \alpha_{31} &= \sin\Theta\sin\Phi \\ \alpha_{32} &= \sin\Theta\cos\Phi \\ \alpha_{33} &= \cos\Theta \end{aligned}\right\} \qquad (10.37)$$

By formula (1.6) a vector $\vec{x}$ has the same components with respect to a moving trihedral as the vector $\vec{x}' = A^{-1}\vec{x}$ with respect to the fixed trihedral. To say that the vector $\vec{x}'$ is constant means that $\vec{x}$ has constant components with respect to the moving trihedral. The change of the vector

$$\vec{x} = A\,\vec{x}' \tag{10.38}$$

is implied by the change of the tensor A which reflects the rotation of the body. Since A gives the instantaneous position of the body with respect to the fixed trihedral, the derivative of $\vec{x}$ with respect to the time t can be represented on the one hand by $\dot{A}\,\vec{x}'$, and on the other hand by the vector product $\vec{\omega} \times \vec{x}$, $\vec{\omega}(\omega_1, \omega_2, \omega_3)$ being the vector of rotation, or by the product $W\vec{x}$ of the affinor of rotation W and $\vec{x}$. The components of W with respect to the fixed trihedral are

$$W_{11} = W_{22} = W_{33} = 0$$

$$W_{32} = -W_{23} = \omega_1, \qquad W_{13} = -W_{31} = \omega_2, \qquad W_{21} = -W_{12} = \omega_3$$

Hence

$$W = \begin{pmatrix} 0 & -\omega_3 & \omega_2 \\ \omega_3 & 0 & -\omega_1 \\ -\omega_2 & \omega_1 & 0 \end{pmatrix}$$

Consequently we can write that

$$\dot{\vec{x}} = \vec{\omega} \times \vec{x} = W\vec{x} = \dot{A}\,\vec{x}' \tag{10.39}$$

The angular velocity vector $\vec{\omega}$ and the vector $W\vec{x}$ transform according to the same law (10.38), i.e.

$$\vec{\omega} = A\,\vec{\omega}' \tag{10.40}$$

$$W\vec{x} = A\,W'\vec{x}' \tag{10.41}$$

The vector $\vec{\omega}' = A^{-1}\vec{\omega}$ is obtained from $\vec{\omega}$ by making the moving trihedral coincide with the fixed trihedral and keeping p, q, r constant. Hence $\vec{\omega}'$ varies only in as much as the components p, q, r vary.

From formula (10.41) and $\vec{x}' = A^{-1}\vec{x}$ we obtain the following law of transformation for the affinor of rotation W, namely

$$W = A\,W'A^{-1} \tag{10.42}$$

In addition, by virtue of the above transformation formulas we have that

$$W'\vec{x}' = \vec{\omega}' \times \vec{x}'$$

From formulas (10.39), (10.40), and (10.42) follow the relations

$$\begin{aligned} WA &= A\,W' = \dot{A} \\ W &= \dot{A}\,A^{-1}, \qquad W' = A^{-1}\dot{A} \end{aligned} \tag{10.43}$$

The affinor of rotation

$$W' = \begin{pmatrix} 0 & -r & q \\ r & 0 & -p \\ -q & p & 0 \end{pmatrix}$$

is determined by $\vec{\omega}'$, the components of which ($p = \operatorname{Re}\pi$, $q = \operatorname{Im}\pi$, r) have been calculated in Sec. 10.2.

By integration introduce (i) the vector

$$\vec{u} = \int_{t_1}^{t} \vec{\omega}'\,dt = \vec{i}\int_{t_1}^{t} p\,dt + \vec{j}\int_{t_1}^{t} q\,dt + \vec{k}\int_{t_1}^{t} r\,dt \tag{10.44}$$

and (ii) the antisymmetric affinor

$$U = \int_{t_1}^{t} W'\,dt \tag{10.45}$$

The vector $\vec{u}$ can be interpreted as a rotation generated by three rotations with velocities p, q, r about the corresponding principal axes of inertia under the assumption that these rotations are independent of each other. Therefore the components of $\vec{u}$ with respect to the moving trihedral are the corresponding angles of rotation

$$\alpha = \int_{t_1}^{t} p\,dt, \qquad \beta = \int_{t_1}^{t} q\,dt, \qquad \gamma = \int_{t_1}^{t} r\,dt \tag{10.44'}$$

about the principal axes of inertia. Hence

$$\vec{u}^2 = \alpha^2 + \beta^2 + \gamma^2 = \lambda^2 \tag{10.46}$$

The antisymmetric affinor U then assumes the form

$$U = \begin{pmatrix} 0 & -\gamma & \beta \\ \gamma & 0 & -\alpha \\ -\beta & \alpha & 0 \end{pmatrix}$$

Integrating the last formula in (10.43) we obtain for A, which characterizes the instantaneous position of the body, the expression

$$A = A_1\,C \tag{10.47}$$

where

$$A_1 = A(t_1)$$

and

$$C = e^{U}$$

Instead of formula (10.47) we can also write

$$A = DA_1$$

where

$$D = A_1 C A_1^{-1} \tag{10.47'}$$

The first few powers of U have the expressions

$$\begin{aligned} U^2 &= \vec{u}\,\vec{u} - \vec{u}^2 I = \vec{u}\,\vec{u} - \lambda^2 I \\ U^3 &= -\lambda^2 U, \qquad U^4 = -\lambda^2 U^2, \ \ldots \end{aligned} \tag{10.48}$$

where the symbol $\vec{u}\,\vec{u}$ denotes a dyad and I is the unit tensor.

If we take formulas (10.48) into account the formula $C = e^U$ assumes the form

$$\begin{aligned} C = e^U &= I + U \sum_{n=0}^{\infty} \frac{(-\lambda^2)^n}{(2n+1)!} + U^2 \sum_{n=1}^{\infty} \frac{(-\lambda^2)^{n-1}}{(2n)!} \\ &= I + U \frac{\sin\lambda}{\lambda} + U^2 \frac{1-\cos\lambda}{\lambda^2} \\ &= I \cos\lambda + U \frac{\sin\lambda}{\lambda} + \vec{u}\,\vec{u}\,\frac{1-\cos\lambda}{\lambda^2} \end{aligned} \tag{10.49}$$

and hence, by virtue of formula (10.47), we obtain for A the expression

$$A = A_1 \cos\lambda + A_1 U \frac{\sin\lambda}{\lambda} + A_1 \vec{u}\,\vec{u}\,\frac{1-\cos\lambda}{\lambda^2} \tag{10.50}$$

If we specify the fixed coordinate trihedral in such a way that for $t = t_1$ it coincides with the principal axes of inertia, then $A(t_1) = A_1 = I$ and $A = D = C$. The matrix C being an orthogonal matrix can be represented by means of the elements given by formulas (10.37) in terms of the EULER angles ψ, Θ, Φ, where

$$\tan\Phi = \frac{C_{31}}{C_{32}}, \qquad \cos\Theta = C_{33}, \qquad \tan\psi = -\frac{C_{13}}{C_{23}} \tag{10.51}$$

By virtue of formula (10.49) the above components of C have the expressions

$$\left.\begin{aligned} C_{13} &= \beta \frac{\sin\lambda}{\lambda} + \alpha\gamma \frac{1-\cos\lambda}{\lambda^2} \\ C_{23} &= -\alpha \frac{\sin\lambda}{\lambda} + \beta\gamma \frac{1-\cos\lambda}{\lambda^2} \\ C_{31} &= -\beta \frac{\sin\lambda}{\lambda} + \alpha\gamma \frac{1-\cos\lambda}{\lambda^2} \\ C_{32} &= \alpha \frac{\sin\lambda}{\lambda} + \beta\gamma \frac{1-\cos\lambda}{\lambda^2} \\ C_{33} &= \cos\lambda + \gamma^2 \frac{1-\cos\lambda}{\lambda^2} \end{aligned}\right\} \tag{10.52}$$

10.4. The angles of rotation of a rigid body subject to a time-independent self-excitement with a fixed direction in the body. The formulas obtained in Sec. 10.3 are valid for the four cases considered in Sec. 10.2 as well as for an unsymmetric rigid body. Let us now calculate the angles α, β, γ for each of the above four cases. By virtue of formulas (10.44′) and (10.5) we have the relations

$$\alpha + i\beta = \int_{t_1}^{t} \pi(\tau)\, d\tau, \qquad \gamma = \int_{t_1}^{t} r(\tau)\, d\tau \tag{10.53}$$

After the angles α, β, γ have been calculated the angle λ is given by formula (10.46).

Case 1: $h \neq 0$, $R = 0$, $r_1 = 0$.

$$\left.\begin{aligned} \alpha + i\beta &= \pi_1 (t - t_1) + \tfrac{1}{2} \Pi (t - t_1)^2 \\ \alpha &= p_1 (t - t_1) + \tfrac{1}{2} P (t - t_1)^2 \\ \beta &= q_1 (t - t_1) + \tfrac{1}{2} Q (t - t_1)^2 \\ \gamma &= 0 \end{aligned}\right\} \tag{10.54}$$

Case 2: $h \neq 0$, $R = 0$, $r = r_1 \neq 0$

$$\left.\begin{aligned} \alpha + i\beta &= \frac{\Pi}{i h r} (t - t_1) + \left(\frac{i \pi_1}{h r} - \frac{\Pi}{h^2 r^2}\right) [\operatorname{cis}(-\zeta) - 1] \\ \alpha &= \frac{Q}{h r} (t - t_1) + (1 - \cos\zeta) \left(\frac{q_1}{h r} + \frac{P}{h^2 r^2}\right) + \sin\zeta \left(\frac{p_1}{h r} - \frac{Q}{h^2 r^2}\right) \\ \beta &= -\frac{P}{h r} (t - t_1) + (1 - \cos\zeta) \left(\frac{-p_1}{h r} + \frac{Q}{h^2 r^2}\right) + \\ &\quad + \sin\zeta \left(\frac{q_1}{h r} + \frac{P}{h^2 r^2}\right) \\ \gamma &= r (t - t_1) \end{aligned}\right\} \tag{10.55}$$

Case 3: $h \neq 0$, $R \neq 0$.
Using the formulas

$$\left.\begin{aligned} &\int_0^x [\cos u\, C(u) + \sin u\, S(u)] \frac{du}{\sqrt{u}} = \frac{1}{2} [C^2(x) + S^2(x)] \\ &\int_0^x [\cos u\, S(u) - \sin u\, C(u)] \frac{du}{\sqrt{u}} = \cos x\, C(x) + \sin x\, S(x) - 2\sqrt{x} \end{aligned}\right. \tag{10.56}$$

we obtain from formulas (10.53) and (10.33) the equation

$$\begin{aligned} \alpha + i\beta = \{\pi_1 \operatorname{cis}\eta_1 - \Pi k \varepsilon_1 [C(\sigma \eta_1) + i \sigma S(\sigma \eta_1)]\} \int_{t_1}^{t} \operatorname{cis}(-\zeta)\, d\tau + \\ + \Pi k \varepsilon \int_{t_1}^{t} \operatorname{cis}(-\zeta) [C(y) + i \sigma S(y)]\, d\tau \end{aligned} \tag{10.57}$$

which by virtue of formulas (10.26), (10.29), (10.31), (10.31') and the formula

$$\operatorname{cis}(-\eta) = \cos y - i\,\sigma \sin y$$

reduces to

$$\begin{aligned}\alpha + i\beta = &\{\pi_1 k \operatorname{cis}\eta_1 - \Pi k^2 \varepsilon_1 [C(\sigma\eta_1) + i\sigma S(\sigma\eta_1)]\} [C(\eta_1, \eta) - i\sigma S(\eta_1, \eta)] + \\ &+ \Pi k^2 \varepsilon \left\{\frac{\varepsilon}{2} [C^2(\sigma\eta) + S^2(\sigma\eta)] - \frac{\varepsilon_1}{2} [C^2(\sigma\eta_1) + S^2(\sigma\eta_1)] + \right. \\ &+ i\sigma [\varepsilon \cos\eta\, C(\sigma\eta) + \varepsilon \sin\eta\, S(\sigma\eta) - \varepsilon_1 \cos\eta_1\, C(\sigma\eta_1) - \\ &\left. - \varepsilon_1 \sin\eta_1\, S(\sigma\eta_1) - 2\varepsilon\sqrt{|\eta|} + 2\varepsilon_1\sqrt{|\eta_1|}]\right\}\end{aligned}$$

Finally, after some simplifications, we obtain that

$$\begin{aligned}\alpha + i\beta = &\pi_1 k \operatorname{cis}\eta_1 [C(\eta_1, \eta) - i\sigma S(\eta_1, \eta)] + \\ &+ \tfrac{1}{2}\Pi k^2 \{C^2(\eta_1, \eta) + S^2(\eta_1, \eta) + (1 - \varepsilon\varepsilon_1) [C^2(\sigma\eta_1) + S^2(\sigma\eta_1)]\} + \\ &+ i\sigma \Pi k^2 \{S(\sigma\eta) [\sin\eta + \varepsilon\varepsilon_1 C(\sigma\eta_1)] + C(\sigma\eta) [\cos\eta - \varepsilon\varepsilon_1 S(\sigma\eta_1)] - \\ &- \varepsilon\varepsilon_1 [\cos\eta_1\, C(\sigma\eta_1) + \sin\eta_1\, S(\sigma\eta_1)] - 2(\sqrt{|\eta|} - \varepsilon\varepsilon_1\sqrt{|\eta_1|})\}\end{aligned} \tag{10.58}$$

Separating the real and imaginary parts we obtain

$$\left.\begin{aligned}\alpha = &p_1 k [\cos\eta_1\, C(\eta_1, \eta) + \sigma \sin\eta_1\, S(\eta_1, \eta)] - \\ &- q_1 k [\sin\eta_1\, C(\eta_1, \eta) - \sigma \cos\eta_1\, S(\eta_1, \eta)] + \\ &+ \tfrac{1}{2} P k^2 \{C^2(\eta_1, \eta) + S^2(\eta_1, \eta) + (1 - \varepsilon\varepsilon_1) [C^2(\sigma\eta_1) + S^2(\sigma\eta_1)]\} - \\ &- \sigma Q k^2 \{S(\sigma\eta) [\sin\eta + \varepsilon\varepsilon_1 C(\sigma\eta_1)] + C(\sigma\eta) [\cos\eta - \varepsilon\varepsilon_1 S(\sigma\eta_1)] - \\ &- \varepsilon\varepsilon_1 [\cos\eta_1\, C(\sigma\eta_1) + \sin\eta_1\, S(\sigma\eta_1)] - 2(\sqrt{|\eta|} - \varepsilon\varepsilon_1\sqrt{|\eta_1|})\} \\ \beta = &q_1 k [\cos\eta_1\, C(\eta_1, \eta) + \sigma \sin\eta_1\, S(\eta_1, \eta)] + \\ &+ p_1 k [\sin\eta_1\, C(\eta_1, \eta) - \sigma \cos\eta_1\, S(\eta_1, \eta)] + \\ &+ \tfrac{1}{2} Q k^2 \{C^2(\eta_1, \eta) + S^2(\eta_1, \eta) + (1 - \varepsilon\varepsilon_1) [C^2(\sigma\eta_1) + S^2(\sigma\eta_1)]\} + \\ &+ \sigma P k^2 \{S(\sigma\eta) [\sin\eta + \varepsilon\varepsilon_1 C(\sigma\eta_1)] + C(\sigma\eta) [\cos\eta - \varepsilon\varepsilon_1 S(\sigma\eta_1)] - \\ &- \varepsilon\varepsilon_1 [\cos\eta_1\, C(\sigma\eta_1) + \sin\eta_1\, S(\sigma\eta_1)] - 2(\sqrt{|\eta|} - \varepsilon\varepsilon_1\sqrt{|\eta_1|})\} \\ \text{and}\quad & \\ \gamma = &h^{-1}\zeta = \frac{\eta - \eta_1}{h} = \frac{R}{2} (t - t_1)(t + t_1 - 2t_0) \\ = &r_1 (t - t_1) + \tfrac{1}{2} R (t - t_1)^2\end{aligned}\right\} \tag{10.59}$$

Case 4: $h = 0$

In this case we have

$$\left.\begin{aligned}\alpha &= p_1 (t - t_1) + \tfrac{1}{2} P (t - t_1)^2 \\ \beta &= q_1 (t - t_1) + \tfrac{1}{2} Q (t - t_1)^2 \\ \gamma &= r_1 (t - t_1) + \tfrac{1}{2} R (t - t_1)^2\end{aligned}\right\} \tag{10.60}$$

Formulas (10.54), (10.55), (10.59) and (10.60) give the angles of rotation α, β, γ for each of the four cases considered.

The five components of the rotation affinor C, which, by means of formula (10.47), determines the instantaneous position of the body starting from its position $A_1 = A(t_1)$, are given by formulas (10.52). The EULER angles which describe the passage of the body from its initial position given by A_1 to its final position given by $A = A(t)$ are to be calculated from formulas (10.51). Hence we have described completely the motion of a symmetric rigid body subject to the action of forces, the torque of which with respect to the moving trihedral (coinciding with the principal axes of inertia of the body) has time-independent components.

10.5. Self-excited symmetric rigid body in a viscous medium. We consider here the motion of a symmetric rigid body about a fixed point in a viscous medium subject to a time-dependent self-excitement. This problem was considered by D. R. MERKIN [3] and R. LEE [1]. It will be shown that the components of $\vec{\omega}$ can be obtained by quadratures. In particular, when the self-excitement is constant and fixed in direction within the body, these components are expressible in terms of generalized sine and cosine integrals.

Sec. 10.5 ends with a discussion of the asymptotic motions of $\vec{\omega}$ relative to the body-fixed trihedral. These motions are further compared with the corresponding ones when friction is neglected. We shall see that the differences between the two are significant.

10.51. Equations of motion

In setting up the equations of motion of a body moving in a viscous medium, we need to consider the nature of the resisting force generated by the motion of the body. For the case of a force-free symmetric rigid body rotating about a fixed point F. KLEIN and A. SOMMERFELD [1] assumed that the drag force is equivalent to a torque vector with two components, one along the symmetry axis of the body and the other along the orthogonal projection of $\vec{\omega}$ on the equatorial plane. These components were assumed to be proportional to the components of $\vec{\omega}$ in these directions.

We shall follow this suggestion. Let the z-axis be the symmetry axis. Then we may denote the resistance components by $-\mu_0 r$ and $-\mu_1(p+q)$, where μ_0 and μ_1 are positive constants or, if friction is neglected, they are equal to zero.

Further, let

$$\vec{m} = m_1(t)\,\vec{i} + m_2(t)\,\vec{j} + m_3(t)\,\vec{k} \tag{10.61}$$

where $m_i(t)$ $(i = 1, 2, 3)$ are continuous functions of t, be the torque acting on the body. Then the equations of motion take the form

$$\left.\begin{aligned} A\dot{p} - (B - C)\,q\,r &= m_1(t) - \mu_1 p \\ B\dot{q} + (A - C)\,r\,p &= m_2(t) - \mu_1 q \\ C\dot{r} \qquad\qquad\quad &= m_3(t) - \mu_0 r \end{aligned}\right\} \tag{10.62}$$

Letting $(B - C)/A = (A - C)/B = h$, $\mu_1/A = \mu_1/B = \nu$, $\mu_0/C = \mu$, and $F_1(t) = m_1(t)/A$, $F_2(t) = m_2(t)/B$, $F_3(t) = m_3(t)/C$, we may write equations (10.62) in the form

$$\left.\begin{aligned} \dot{p} - h\,q\,r &= F_1(t) - \nu\,p \\ \dot{q} + h\,r\,p &= F_2(t) - \nu\,q \\ \dot{r} \qquad &= F_3(t) - \mu\,r \end{aligned}\right\} \tag{10.63}$$

If the torque vector is of fixed direction in the body, we may put

$$\left.\begin{aligned} F_1(t) &= (\alpha_1/A)\,F(t) = \lambda_1\,F(t) \\ F_2(t) &= (\alpha_2/B)\,F(t) = \lambda_2\,F(t) \\ F_3(t) &= (\alpha_3/C)\,F(t) = \lambda_3\,F(t) \end{aligned}\right\} \tag{10.64}$$

Here α_1, α_2, and α_3 denote the direction cosines of the fixed vector. In addition, if the acting moment is time-independent, $F(t)$ is a constant. Otherwise the torque vector will move in the body; its position at any time t is given by either (10.61) or (10.64).

Equations (10.63) together with the initial angular velocity $\vec{\omega}_0(p_0, q_0, r_0)$ at $t = 0$ determine uniquely the angular velocity of the body. We proceed now to the integration of these equations.

10.52. The angular velocity of a rigid body

The third equation of (10.63) can be integrated directly to give

$$r = r_0\,e^{-\mu t} + R(t) \tag{10.65}$$

where

$$R(t) = \int_0^t e^{-\mu(t-s)}\,F_3(s)\,ds$$

Multiplying the second equation of (10.63) by i and adding this to the first one, we obtain the equation

$$\dot{\pi} + \xi\,\pi = \Pi \tag{10.66}$$

where $\pi = p + i\,q$, $\Pi = F_1 + i\,F_2$, and $\xi = \nu + i\,h(r_0\,e^{-\mu t} + R)$ are complex-valued functions of t. The above linear equation has the general solution

$$\pi = \pi_0\,e^{-\zeta_1} + e^{-\zeta_1}\int_0^t e^{\zeta(s)}\,\Pi(s)\,ds \tag{10.67}$$

Here $\zeta_1 = \int_0^t \xi(s)\,ds$ and $\pi_0 = p_0 + i\,q_0$. By separating the real and imaginary parts of (10.67) we shall obtain explicit formulas for p and q.

We note first:

$$\pi_0\, e^{-\zeta_1} = e^{-\nu t}(p_0 \cos\zeta + q_0 \sin\zeta) + i\, e^{-\nu t}(-p_0 \sin\zeta + q_0 \cos\zeta)$$

$$\begin{aligned} e^{-\zeta_1} \int_0^t e^{\zeta_1(s)}\, \Pi(s)\, ds \\ = \int_0^t e^{-\nu(t-s)} \{F_1(s) \cos[\zeta(s) - \zeta(t)] - F_2(s) \sin[\zeta(s) - \zeta(t)]\}\, ds + \\ + i \int_0^t e^{-\nu(t-s)} \{F_1(s) \sin[\zeta(s) - \zeta(t)] + F_2(s) \cos[\zeta(s) - \zeta(t)]\}\, ds \end{aligned} \tag{10.68}$$

where

$$\zeta(t) = \int_0^t h[r_0\, e^{-\mu s} + R(s)]\, ds \tag{10.69}$$

Letting $\delta = \sqrt{p_0^2 + q_0^2}$ and $\sigma = \arctan(p_0/q_0)$, we have

$$\begin{aligned} p &= \delta\, e^{-\nu t} \sin(\zeta + \sigma) + \\ &+ \int_0^t e^{-\nu(t-s)} \{F_1(s) \cos[\zeta(s) - \zeta(t)] - F_2(s) \sin[\zeta(s) - \zeta(t)]\}\, ds \\ q &= \delta\, e^{-\nu t} \cos(\zeta + \sigma) + \\ &+ \int_0^t e^{-\nu(t-s)} \{F_1(s) \sin[\zeta(s) - \zeta(t)] + F_2(s) \cos[\zeta(s) - \zeta(t)]\}\, ds \end{aligned} \tag{10.70}$$

(i) *The torque vector fixed in direction within the body.* In the above analysis the direction of the self-excited torque vector is not restricted in any way. If we now assume that $\vec{m}$ is of fixed direction relative to the body so that (10.64) holds, then

$$\begin{aligned} p &= \delta e^{-\nu t} \sin(\zeta + \sigma) - \delta_1 \int_0^t e^{-\nu(t-s)}\, F(s) \sin[\zeta(s) - \zeta(t) - \sigma_1]\, ds \\ q &= \delta e^{-\nu t} \cos(\zeta + \sigma) + \delta_1 \int_0^t e^{-\nu(t-s)}\, F(s) \cos[\zeta(s) - \zeta(t) - \sigma_1]\, ds \end{aligned} \tag{10.71}$$

where $\delta_1 = \sqrt{\lambda_1^2 + \lambda_2^2}$ and $\sigma_1 = \arctan(\lambda_1/\lambda_2)$.

From (10.65) the expression for r becomes

$$r = r_0\, e^{-\mu t} + \lambda_3 \int_0^t e^{-\mu(t-s)}\, F(s)\, ds \tag{10.72}$$

(ii) *The resistance of the medium is negligible.* When a rigid body moves in a medium which we may assume frictionless, then $\mu_0 = \mu_1 = 0$, and equation (10.69) gives

$$\zeta(t) = \int_0^t h[r_0 + R(s)]\, ds \tag{10.73}$$

Further, formulas (10.70) and (10.65) give

$$\left.\begin{aligned} p &= \delta \sin(\zeta + \sigma) + \\ &\quad + \int_0^t F_1(s) \cos[\zeta(s) - \zeta(t)]\, ds - \int_0^t F_2(s) \sin[\zeta(s) - \zeta(t)]\, ds \\ q &= \delta \cos(\zeta + \sigma) + \\ &\quad + \int_0^t F_1(s) \sin[\zeta(s) - \zeta(t)]\, ds + \int_0^t F_2(s) \cos[\zeta(s) - \zeta(t)]\, ds \\ r &= r_0 + \int_0^t F_3(s)\, ds \end{aligned}\right\} \tag{10.74}$$

In addition, if the torque vector $\vec{m}$ is of fixed direction within the body, the above formulas can be further simplified. For now we have from (10.71) and (10.72)

$$\left.\begin{aligned} p &= \delta \sin(\zeta + \sigma) - \delta_1 \int_0^t F(s) \sin[\zeta(s) - \zeta(t) - \sigma_1]\, ds \\ q &= \delta \cos(\zeta + \sigma) + \delta_1 \int_0^t F(s) \cos[\zeta(s) - \zeta(t) - \sigma_1]\, ds \\ r &= r_0 + \lambda_3 \int_0^t F(s)\, ds \end{aligned}\right\} \tag{10.75}$$

These formulas then provide the angular velocity components of a rigid body under the action of a time-dependent self-excited torque vector fixed in direction within the body.

In all cases considered above, the solutions are given in terms of some integrals. The possibility of evaluating them by means of elementary or tabulated functions depends, of course, on the functions F_1, F_2, and F_3, or on the function F if the self-excited torque vector is fixed in direction. In what follows we shall discuss this possibility in detail for the case of a constant self-excitement of fixed direction within the body. Afterwards we shall discuss qualitatively the asymptotic motions of $\vec{\omega}$ subject to this type of torque.

10.53. Time-independent torque vector fixed in direction within the body. Denote by M the modulus of the torque vector. Let $m_i = \lambda_i M$ $(i = 1, 2, 3)$ and $m_{12} = \sqrt{m_1^2 + m_2^2}$. Then from formulas (10.71) and

(10.72) we have

$$\left.\begin{aligned} p &= \delta e^{-\nu t}\sin(\zeta+\sigma) - m_{12}\int_0^t e^{-\nu(t-s)}\sin[\zeta(s)-\zeta(t)-\sigma_1]\,ds \\ q &= \delta e^{-\nu t}\cos(\zeta+\sigma) + m_{12}\int_0^t e^{-\nu(t-s)}\cos[\zeta(s)-\zeta(t)-\sigma_1]\,ds \\ r &= (r_0 - m_3/\mu)\,e^{-\mu t} + m_3/\mu \end{aligned}\right\} \quad (10.76)$$

In case μ and ν are negligible, these expressions reduce to

$$\left.\begin{aligned} p &= \delta\sin(\zeta+\sigma) - m_{12}\int_0^t \sin[\zeta(s)-\zeta(t)-\sigma_1]\,ds \\ q &= \delta\cos(\zeta+\sigma) + m_{12}\int_0^t \cos[\zeta(s)-\zeta(t)-\sigma_1]\,ds \\ r &= r_0 + m_3 t \end{aligned}\right\} \quad (10.77)$$

For the two integrals appearing in (10.77), we have shown in Sec. 10.2 that they generally could be expressed in terms of the FRESNEL integrals.

For the corresponding integrals appearing in (10.76), we shall later show that in the general case they can be expressed in terms of integrals of the trigonometric type which includes the FRESNEL integrals.

As before we proceed now to discuss four cases; depending on the values of h, r_0 and m_3.

Case 1. $h \neq 0$, $r_0 = 0$, $m_3 = 0$. From equations (10.69) and (10.76) we obtain

$$\left.\begin{aligned} p &= (p_0 - m_1/\nu)\,e^{-\nu t} + m_1/\nu \\ q &= (q_0 - m_2/\nu)\,e^{-\nu t} + m_2/\nu \\ r &= 0 \end{aligned}\right\} \quad (10.78)$$

Case 2. $h \neq 0$, $r_0 \neq 0$, $m_3 = 0$. Under these assumptions we have

$$r = r_0\,e^{-\mu t} \quad (10.79)$$

and

$$\zeta(t) = h_1(1 - e^{-\mu t})$$

where $h_1 = h\,r_0/\mu$. Without any loss of generality we can assume in what follows that $h_1 > 0$ since the positive direction of the z-axis may be selected so that $h\,r_0 > 0$.

For the integral appearing in the first equation of (10.76), putting $\Phi = h_1\,e^{-\mu t} - \sigma_1$, we may write

$$\int_0^t e^{-\nu(t-s)}\sin[\zeta(s)-\zeta(t)-\sigma_1]\,ds$$

$$= e^{-\nu t}\sin\Phi\int_0^t e^{\nu s}\cos h_1\,e^{-\mu s}\,ds - e^{-\nu t}\cos\Phi\int_0^t \sin h_1\,e^{-\mu s}\,ds \quad (10.80)$$

The latter two integrals can be expressed in terms of the following two functions,

$$si(x, w) = -\int_x^\infty \frac{\sin u}{u^w}\, du$$
$$Ci(x, w) = -\int_x^\infty \frac{\cos u}{u^w}\, du \qquad (10.81)$$

which we may consider respectively as the generalized sine and cosine integrals. We shall allow w to take complex values. These integrals have been studied and their values partially tabulated by E. KREYSZIG [1, 2]. It is known that they converge for $\operatorname{Re} w > 0$. In our problem we shall later show that $\operatorname{Re} w = 1 + \nu/\mu$, where the ratio $\nu/\mu = (C/A)\,(\mu_1/\mu_0)$. If $\mu_1 = \mu_0$ or if they do not differ much, then for an elongated rigid body where A is much greater than C the ratio ν/μ is less than 1; accordingly, the Re w is between 1 and 2. For a flat-disked rigid body where now C is greater than A, the Re w is in general greater than 2. Consequently, we can use these integrals in expressing the components p and q.

If we make first a change of variables by means of the equation $u = h_1\, e^{-\mu s}$, the first and second integrals appearing on the right of (10.80) become respectively

$$\int_0^t e^{\nu s} \cos h_1\, e^{-\mu s}\, ds = h_2[Ci(h_1, \alpha_1) - Ci(h_1\, e^{-\mu t}, \alpha_1)]$$
$$\int_0^t e^{\nu s} \sin h_1\, e^{-\mu s}\, ds = h_2[si(h_1, \alpha_1) - si(h_1\, e^{-\mu t}, \alpha_1)] \qquad (10.82)$$

where $\alpha_1 = 1 + \nu/\mu$, $h_2 = h_1^{\nu/\mu}/\mu$. Thus we may write

$$p = \delta e^{-\nu t} \sin(\zeta + \sigma) + m_{12}\, e^{-\nu t}\, h_2[Ci(h_1, \alpha_1) - Ci(h_1\, e^{-\mu t}, \alpha_1)] \sin\Phi + {} + m_{12}\, e^{-\nu t}\, h_2[si(h_1, \alpha_1) - si(h_1\, e^{-\mu t}, \alpha_1)] \cos\Phi \qquad (10.83)$$

For the component q we find in a similar way that

$$q = \delta e^{-\nu t} \cos(\zeta + \sigma) + m_{12}\, e^{-\nu t}\, h_2[Ci(h_1, \alpha_1) - Ci(h_1\, e^{-\mu t}, \alpha_1)] \cos\Phi + {} + m_{12}\, e^{-\nu t}\, h_2[si(h_1, \alpha_1) - si(h_1\, e^{-\mu t}, \alpha_1)] \sin\Phi \qquad (10.84)$$

Case 3. $h \neq 0$, $m_3 \neq 0$. From formulas (10.69) and (10.76) we now have

$$r = (r_0 - m_3/\mu)\, e^{-\mu t} + m_3/\mu \qquad (10.85)$$

and $\zeta(t) = g(1 - e^{-\mu t}) + g_1 t$, where $g = h(r_0 - m_3/\mu)/\mu$, $g_1 = (m_3/\mu)\, h$.

Suppose first that $g \neq 0$. Putting

$$\psi = g\, e^{-\mu t} - g_1 t - \sigma_1$$
$$I_1 = \int_0^t e^{\nu s} \cos g_1 s \cos g\, e^{-\mu s}\, ds$$
$$I_2 = \int_0^t e^{\nu s} \sin g_1 s \cos g\, e^{-\mu s}\, ds$$
$$I_3 = \int_0^t e^{\nu s} \cos g_1 s \sin g\, e^{-\mu s}\, ds$$
$$I_4 = \int_0^t e^{\nu s} \sin g_1 s \sin g\, e^{-\mu s}\, ds$$

we observe that

$$\int_0^t e^{-\nu(t-s)} \sin[\zeta(s) - \zeta(t) - \sigma_1]\, ds = e^{-\nu t}(I_1 + I_4) \sin\psi + e^{-\nu t}(I_2 - I_3) \cos\psi$$

$$\int_0^t e^{-\nu(t-s)} \cos[\zeta(s) - \zeta(t) - \sigma_1]\, ds = e^{-\nu t}(I_1 + I_4) \cos\psi - e^{-\nu t}(I_2 - I_3) \sin\psi$$

Consequently, for the components p, q we have from (10.76) the expressions

$$\begin{aligned} p &= \delta e^{-\nu t} \sin(\zeta + \sigma) - m_{12}\, e^{-\nu t}[(I_1 + I_4) \sin\psi + (I_2 - I_3) \cos\psi] \\ q &= \delta e^{-\nu t} \cos(\zeta + \sigma) + m_{12}\, e^{-\nu t}[(I_1 + I_4) \cos\psi - (I_2 - I_3) \sin\psi] \end{aligned} \tag{10.86}$$

The integrals I_1 to I_4 can be expressed in terms of the generalized trigonometric integrals as follows.

Multiplying I_2 by i and adding this to I_1, we obtain

$$I_1 + i\, I_2 = \int_0^t e^{(\nu + i g_1)s} \cos \varepsilon\, g\, e^{-\mu s}\, ds$$

where $\varepsilon = \operatorname{sgn} g$ so that $\varepsilon g > 0$. If we make the substitution $u = \varepsilon g\, e^{-\mu s}$, then

$$I_1 + i\, I_2 = \frac{(\varepsilon g)^{(\nu + i g_1)/\mu}}{\mu} [Ci(\varepsilon g,\, w_1) - Ci(\varepsilon g\, e^{-\mu t},\, w_1)] \tag{10.87}$$

in which $w_1 = 1 + (\nu + i\, g_1)/\mu$, and the first factor on the right can be written as

$$\frac{(\varepsilon g)^{(\nu + i g_1)/\mu}}{\mu} = \frac{(\varepsilon g)^{\nu/\mu}}{\mu} \cos\left(\frac{g_1}{\mu} \ln \varepsilon g\right) + i \frac{(\varepsilon g)^{\nu/\mu}}{\mu} \sin\left(\frac{g_1}{\mu} \ln \varepsilon g\right)$$

Putting

$$g_2 = \frac{(\varepsilon g)^{\nu/\mu}}{\mu} \cos\left(\frac{g_1}{\mu} \ln \varepsilon g\right)$$

$$g_3 = \frac{(\varepsilon g)^{\nu/\mu}}{\mu} \sin\left(\frac{g_1}{\mu} \ln \varepsilon g\right)$$

$$Ci(\varepsilon g,\, w_1) = U_1 + i\, V_1$$

$$Ci(\varepsilon g\, e^{-\mu t},\, w_1) = U_2 + i\, V_2$$

and equating the real and imaginary parts of (10.87), we have

$$\begin{aligned} I_1 &= g_2 (U_1 - U_2) - g_3 (V_1 - V_2) \\ I_2 &= g_3 (U_1 - U_2) + g_2 (V_1 - V_2) \end{aligned} \tag{10.88}$$

In a similar way if we let

$$si(\varepsilon g,\, w_1) = U_3 + i\, V_3$$

$$si(\varepsilon g\, e^{-\mu t},\, w_1) = U_4 + i\, V_4$$

then

$$\begin{aligned} I_3 &= \varepsilon[g_2 (U_3 - U_4) - g_3 (V_3 - V_4)] \\ I_4 &= \varepsilon[g_3 (U_3 - U_4) + g_2 (V_3 - V_4)] \end{aligned} \tag{10.89}$$

Thus, formulas (10.86) together with (10.88) and (10.89) provide the components p and q when $g \neq 0$.

In case $g = 0$, these components can be expressed in terms of elementary functions. For now we have

$$\int_0^t e^{-\nu(t-s)} \sin[\zeta(s) - \zeta(t) - \sigma_1]\, ds = \frac{-1}{\delta_2} \sin(\sigma_1 + \sigma_2) + \frac{e^{-\nu t}}{\delta_2} \sin(g_1 t + \sigma_1 + \sigma_2)$$

$$\int_0^t e^{-\nu(t-s)} \cos[\zeta(s) - \zeta(t) - \sigma_1]\, ds = \frac{1}{\delta_2} \cos(\sigma_1 + \sigma_2) - \frac{e^{-\nu t}}{\delta_2} \cos(g_1 t + \sigma_1 + \sigma_2)$$

where $\delta_2 = \sqrt{\nu^2 + g_1^2}$, $\sigma_2 = \arctan(g_1/\nu)$.

It follows that

$$\left.\begin{aligned} p &= \delta e^{-\nu t} \sin(g_1 t + \sigma) - \frac{m_{12}}{\delta_2} e^{-\nu t} \sin(g_1 t + \sigma_1 + \sigma_2) + \frac{\nu m_1 + g_1 m_2}{\nu^2 + g_1^2} \\ q &= \delta e^{-\nu t} \cos(g_1 t + \sigma) - \frac{m_{12}}{\delta_2} e^{-\nu t} \cos(g_1 t + \sigma_1 + \sigma_2) + \frac{\nu m_2 - g_1 m_1}{\nu^2 + g_1^2} \end{aligned}\right. \tag{10.90}$$

Case 4. $h = 0$. This is the case of dynamic symmetry. Directly from equations (10.63) we obtain

$$\left.\begin{aligned} p &= m_1/\nu + (p_0 - m_1/\nu)\, e^{-\nu t} \\ q &= m_2/\nu + (q_0 - m_2/\nu)\, e^{-\nu t} \\ r &= m_3/\mu + (r_0 - m_3/\mu)\, e^{-\mu t} \end{aligned}\right\} \tag{10.91}$$

10.54. The asymptotic motion of the spin vector. When a rigid body rotates in a viscous medium, its angular velocity will in general be limited by friction. This statement is substantiated by the following boundedness

Theorem: The angular velocity of a self-excited symmetric rigid body rotating about a fixed point in a viscous medium is bounded, if the modulus of the self-excited torque vector is bounded.

The truth of this theorem follows easily from the expressions for p, q, and r. For from equations (10.70) we have

$$|p| \leq |\delta e^{-\nu t} \sin(\zeta + \sigma)| + \left| \int_0^t e^{-\nu(t-s)} F_1(s) \cos[\zeta(s) - \zeta(t)]\, ds \right| + \left| \int_0^t e^{-\nu(t-s)} F_2(s) \sin[\zeta(s) - \zeta(t)]\, ds \right| \leq$$

$$\leq \delta e^{-\nu t} + e^{-\nu t} M_1' \int_0^t e^{\nu s}\, ds + e^{-\nu t} M_2' \int_0^t e^{\nu s}\, ds$$

where $|F_1(s)| \leq M_1'$, $|F_2(s)| \leq M_2'$.

Integrating the expressions on the right, we find that

$$|p| \leq \delta e^{-\nu t} + \frac{M_1'}{\nu} (1 - e^{-\nu t}) + \frac{M_2'}{\nu} (1 - e^{-\nu t})$$

which in turn implies that

$$|p| \leq \delta + (M_1' + M_2')/\nu \qquad (\text{for all } t \geq 0)$$

Similarly, we obtain

$$|q| \leq \delta + (M_1' + M_2')/\nu \qquad (\text{for all } t \geq 0)$$

From equation (10.65), we have

$$|r| \leqq |r_0 e^{-\mu t}| + |e^{-\mu t} \int_0^t F_3(s)\, e^{\mu s} a s|$$

If $|F_3(t)| \leqq M_3'$, then

$$|r| \leqq |r_0| + M_3'/\mu \qquad \text{(for all } t \geqq 0)$$

It is easily seen that this theorem need not hold if friction is neglected. Moreover, even though the speed of rotation is bounded, the spin vector may still wander in space without approaching any limit. Nevertheless, if the self-excitement is time-independent and fixed in direction relative to the body, we shall show in what follows that $\vec{\omega}$ approaches asymptotically a constant vector in the body-fixed trihedral. The rigid body itself then assumes asymptotically a uniform rotation about an axis fixed in the body as well as in space.

We shall consider this in four cases as in Sec. 10.53, and in doing so we shall also compare these motions with the corresponding ones in which friction is neglected. There are essential differences.

Case 1. $h \neq 0$, $r_0 = 0$, $m_3 = 0$. Since $r = 0$, the motion of the endpoint of $\vec{\omega}$ takes place wholly in the equatorial plane. The system

$$\left.\begin{aligned} \dot{p} - h\,q\,r &= m_1 - \nu\,p \\ \dot{q} + h\,r\,p &= m_2 - \nu\,q \\ \dot{r} \qquad &= m_3 - \mu\,r \end{aligned}\right\} \qquad (10.92)$$

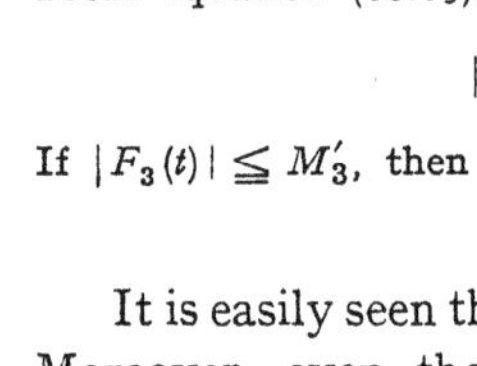

Fig. 12

with the assumption $m_3 = 0$ possesses a unique singular point $(m_1/\nu, m_2/\nu, 0)$ in the $p\,q$-plane. According to (10.78) every trajectory in this plane is rectilinear and approaches the singular point as illustrated in Fig. 12. This point represents a state of permanent rotation of the body, and this rotation is stable, as we shall see in the discussion of Case 2.

When friction is negligible, r remains identically zero, and the endpoint of $\vec{\omega}$, as before, always lies in the $p\,q$-plane. But the phase picture is essentially

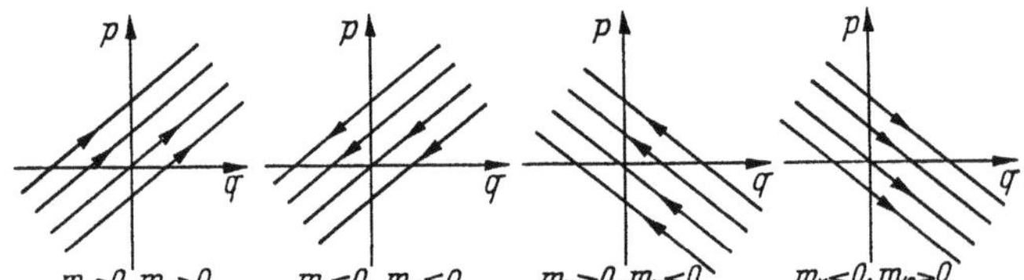

Fig. 13

different. For now the system has no singular point in the $p\,q$-plane, unless $m_1 = m_2 = 0$. From (10.78) we have

$$p = p_0 + m_1 t$$
$$q = q_0 + m_2 t$$

implying that in the $p\,q$-plane the trajectories are parallel lines. These are shown in Fig. 13.

This example verifies our previous statement that the angular velocity need not be bounded if friction is neglected, even though the modulus of the self-excited torque vector is bounded.

Case 2. $h \neq 0$, $r_0 \neq 0$, $m_3 = 0$. Turning to the second case, we consider first the frictionless situation since the motion is then simpler to describe. With $\mu = \nu = 0$ and $r_0 \neq 0$, equation (10.79) gives $r = r_0$. Thus the endpoint of $\vec{\omega}$, initially lying in the plane $z = r_0$, will remain forever in this plane. The following system, obtained from (10.92) by putting ν, μ, and m_3 equal to zero,

$$\left.\begin{aligned} \dot{p} - h\,q\,r &= m_1 \\ \dot{q} + h\,r\,p &= m_2 \\ \dot{r} \qquad\quad &= 0 \end{aligned}\right\} \tag{10.93}$$

has in the phase space (p, q, r) a set of singular points consisting of the space curve defined as the intersection of the two hyperbolic cylinders

$$q\,r = -m_1/h$$

$$p\,r = m_2/h$$

Furthermore, since

$$p/q = -m_2/m_1 \tag{10.94}$$

this curve lies wholly on the plane defined by (10.94). The nature of this curve in the plane can be seen if we rotate the $p\,q$-plane about the r-axis by an angle $\Theta = \arctan(-m_2/m_1)$. In the rotated coordinate system the equations of the curve are

$$\begin{aligned} q' \cos\Theta - p' \sin\Theta &= -m_1/h\,r' \\ q' \sin\Theta + p' \cos\Theta &= m_2/h\,r' \end{aligned} \tag{10.95}$$

Here p', q', and r' denote the new coordinates. Equations (10.95) are equivalent to

$$q'\,r' = k' = \text{const.}$$

$$p' = 0$$

and hence this curve is a hyperbola as illustrated in Fig. 14, in which the p'-axis is perpendicular to the plane of the paper.

This curve intersects the plane $r = r_0\,(\neq 0)$ at the point $(m_2/h\,r_0,\ -m_1/h\,r_0,\ r_0)$. The trajectories in this plane form a continuum of concentric circles with the center at the singular point. For from equations (10.77), we have

$$p = \delta \sin(h\,r_0\,t + \sigma) + \frac{1}{h\,r_0}\{m_1 \sin(h\,r_0\,t) + m_2[1 - \cos(h\,r_0\,t)]\}$$

If we put

$$\Omega_1 = [(p_0 - m_2/h\,r_0)^2 + (q + m_1/h\,r_0)^2]^{1/2}$$

$$\Delta_1 = \arctan(p_0 - m_2/h\,r_0)/(q_0 + m_1/h\,r_0)$$

then

$$p - m_2/h\,r_0 = \Omega_1 \sin(h\,r_0\,t + \Delta_1)$$

In the same way, we have

$$q + m_1/h\,r_0 = \Omega_1 \cos(h\,r_0\,t + \Delta_1)$$

These last two equations define parametrically a circle of radius Ω_1. Consequently in the plane $z = r_0$, the endpoint of $\vec{\omega}$ moves in a circle. The phase trajectories in the plane $r = r_0$ are shown in Fig. 15.

Moreover, since r is constant, these circles represent a continuous family of periodic motions about a stable permanent rotation. The period of each of these motions is clearly

$$T = 2\pi/|h\, r_0|$$

which varies inversely with r_0 for a given mass distribution, represented by the constant h. From the results of Cases 1 and 2 we have the following

Theorem: For a self-excited symmetric rigid body rotating about a fixed point in a frictionless medium, if the time-independent torque vector with constant direction is fixed in the equatorial plane, the motion of the endpoint of $\vec{\omega}$ takes place wholly in the initial plane $z = r_0$. In the phase space (p, q, r) the trajectories in the plane $r = r_0$

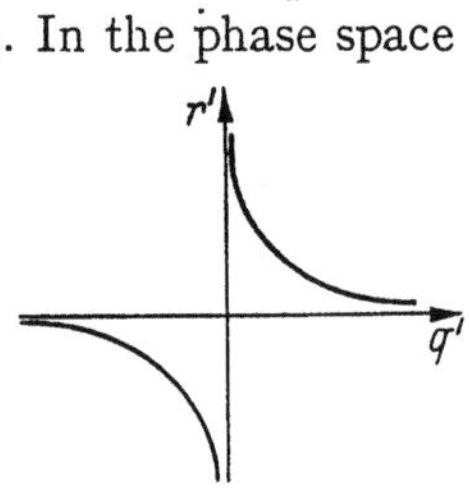

Fig. 14

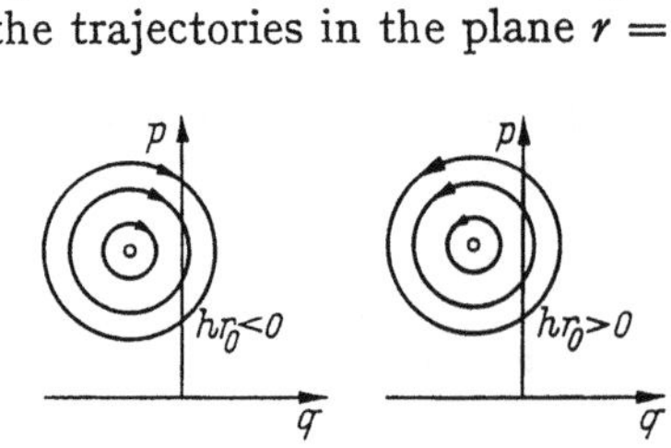

Fig. 15

form a continuum of concentric circles with center at the singular point $(m_2/h\, r_0,\ -m_1/h\, r_0,\ r_0)$, representing a state of stable permanent rotation. If $r_0 = 0$, the center being considered at infinity, these circles degenerate into parallel lines.

Consider now the situation when friction is not negligible. The motion of the endpoint of $\vec{\omega}$ becomes more complicated. For now, since $r = r_0\, e^{-\mu t}$, the motion no longer takes place in a plane perpendicular to the z-axis. Instead, the endpoint approaches asymptotically the $x\,y$-plane.

As mentioned in Case 1, the system (10.92) with $m_3 = 0$ has only one singular point $(m_1/\nu,\ m_2/\nu,\ 0)$. On the basis of the Lemma mentioned on page 57 and formulas (10.76), we observe that as $t \to \infty$ the components p, q, and r approach respectively the values m_1/ν, m_2/ν, and 0. Thus, every trajectory in the phase space approaches the singular point.

Case 3. $h \neq 0$, $m_3 \neq 0$. If friction is neglected, system (10.92) has no singular point. Every trajectory in the phase space (p, q, r) tends to infinity. The components of $\vec{\omega}$ are now given by formulas (10.77). As the two integrals on the right of (10.77) can be expressed in terms of the Fresnel integrals and the trigonometric functions, and since these two kinds of functions are bounded for $t \geqq 0$, the magnitudes of p and q are thus bounded also. On the other hand, $|r|$ increases indefinitely when $t \to \infty$, and as a result the vector $\vec{\omega}$ approaches asymptotically the positive or negative z-axis, according as the algebraic sign of m_3 is positive or negative.

However, if the body is further acted upon by a torque due to friction, system (10.92) will then possess a unique singular point which can be obtained by solving the system of equations

$$\left.\begin{aligned} \nu p - h q r - m_1 &= 0 \\ \nu q + h p r - m_2 &= 0 \\ \mu r \qquad\quad - m_3 &= 0 \end{aligned}\right\} \qquad (10.96)$$

the solution of which is

$$p = (m_1 \nu + m_2 g_1)/(\nu^2 + g_1^2), \qquad \text{where} \quad g_1 = h\, m_3/\mu$$
$$q = (m_2 \nu - m_1 g_1)/(\nu^2 + g_1^2)$$
$$r = m_3/\mu$$

The following lemma will show that all other trajectories tend asymptotically to this point.

Lemma: For $\zeta(t) = g_1 t - g\, e^{-\mu t} + g$, $\mu > 0$, $\nu > 0$, as $t \to \infty$

$$J_1 = \int_0^t e^{-\nu(t-s)} \sin[\zeta(s) - \zeta(t) - \sigma_1]\, ds \to \frac{-1}{\nu^2 + g_1^2} (\nu \sin\sigma_1 + g_1 \cos\sigma_1) \tag{i}$$

$$J_2 = \int_0^t e^{-\nu(t-s)} \cos[\zeta(s) - \zeta(t) - \sigma_1]\, ds \to \frac{1}{\nu^2 + g_1^2} (\nu \cos\sigma_1 - g_1 \sin\sigma_1) \tag{ii}$$

Proof: Integration by parts yields

$$J_1 = -\frac{\sin\sigma_1}{\nu} - \frac{e^{-\nu t}}{\nu} \sin[\zeta(0) - \zeta(t) - \sigma_1]$$
$$- \frac{\mu}{\nu} g\, e^{-\nu t} \int_0^t e^{-(\mu-\nu)s} \cos[\zeta(s) - \zeta(t) - \sigma_1]\, ds$$
$$- \frac{g_1}{\nu} \int_0^t e^{-\nu(t-s)} \cos[\zeta(s) - \zeta(t) - \sigma_1]\, ds$$

The third term on the right approaches zero as $t \to \infty$. For we have

$$\left| e^{-\nu t} \int_0^t e^{-(\mu-\nu)s} \cos[\zeta(s) - \zeta(t) - \sigma_1]\, ds \right| \leq e^{-\nu t} \int_0^t e^{-(\mu-\nu)s}\, ds$$
$$= \frac{e^{-\nu t}}{\nu - \mu} [e^{-(\mu-\nu)t} - 1]$$

Since the right hand side tends to zero when $t \to \infty$, the left hand side does also. Consequently, we may write

$$J_1 + \frac{g_1}{\nu} J_2 = \frac{-\sin\sigma_1}{\nu} + \varepsilon_1, \qquad \text{where} \quad \varepsilon_1 \to 0 \quad \text{as} \quad t \to \infty \tag{10.97}$$

The same procedure yields

$$J_2 = \frac{\cos\sigma_1}{\nu} - \frac{e^{-\nu t}}{\nu} \cos[\zeta(0) - \zeta(t) - \sigma_1]$$
$$+ \frac{\mu}{\nu} g\, e^{-\nu t} \int_0^t e^{-(\mu-\nu)s} \sin[\zeta(s) - \zeta(t) - \sigma_1]\, ds$$
$$+ \frac{g_1}{\nu} \int_0^t e^{-\nu(t-s)} \sin[\zeta(s) - \zeta(t) - \sigma_1]\, ds$$

Because the second and third terms tend to zero as $t \to \infty$, we may write

$$J_2 - \frac{g_1}{\nu} J_1 = \frac{\cos\sigma_1}{\nu} + \varepsilon_2, \quad \text{where} \quad \varepsilon_2 \to 0 \quad \text{as} \quad t \to \infty \tag{10.98}$$

Solving (10.97) and (10.98) for J_1 and J_2, we obtain

$$J_1 = \frac{-1}{\nu^2 + g_1^2} (\nu \sin\sigma_1 + g_1 \cos\sigma_1) + \varepsilon_3, \quad \text{where} \quad \varepsilon_3 \to 0 \quad \text{as} \quad t \to \infty$$

and

$$J_2 = \frac{1}{\nu^2 + g_1^2} (\nu \cos\sigma_1 - g_1 \sin\sigma_1) + \varepsilon_4, \quad \text{where} \quad \varepsilon_4 \to 0 \quad \text{as} \quad t \to \infty.$$

Thus from formulas (10.76) and the above Lemma, we have as $t \to \infty$

$$\begin{aligned} p &= (\nu m_1 + g_1 m_2)/(\nu^2 + g_1^2) \\ q &= (\nu m_2 - g_1 m_1)/(\nu^2 + g_1^2) \\ r &= m_3/\mu \end{aligned}$$

Since all other trajectories tend to the singular point, the state of permanent rotation represented by this point is clearly stable.

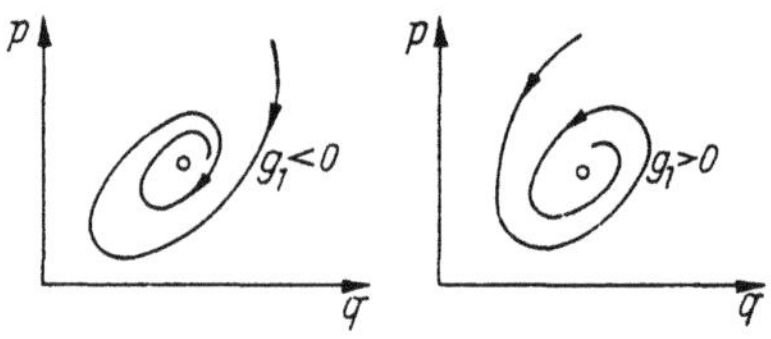

Fig. 16

In the case where the initial position of the endpoint of $\vec{\omega}$ lies in the plane $z = m_3/\mu$, equation (10.85) shows that it will remain forever in this plane. In the phase space (p, q, r) this means that every trajectory originated at any point in the plane $r = m_3/\mu$, which contains the singular point of the system, will remain forever in this plane. Furthermore, multiplying the first equation of (10.90) by i and adding this to the second one, we have

$$\begin{aligned} &[q - (\nu m_2 - g_1 m_1)/(\nu^2 + g_1^2)] + i[p - (\nu m_1 + g_1 m_2)/(\nu^2 + g_1^2)] \\ &\qquad = \delta e^{-\nu t} e^{i(g_1 t + \sigma)} - \frac{m_{12}}{\delta_2} e^{-\nu t} e^{i(g_1 t + \sigma_1 + \sigma_2)} \\ &\qquad = e^{-\nu t} e^{i g_1 t} \left[\delta e^{i\sigma} - \frac{m_{12}}{\delta_2} e^{i(\sigma_1 + \sigma_2)}\right] \end{aligned}$$

If we let δ_3 and σ_3 denote respectively the modulus and argument of the bracket expression, then

$$\left(q - \frac{\nu m_2 - g_1 m_1}{\nu^2 + g_1^2}\right) + i\left(p - \frac{\nu m_1 + g_1 m_2}{\nu^2 + g_1^2}\right) = \delta_3 e^{-\nu t} e^{i(g_1 t + \sigma_3)}$$

This equation shows that the trajectories in the plane $r = m_3/\mu$ spiral toward the singular point, as illustrated in Fig. 16.

Case 4. $h = 0$. In the case of dynamic symmetry and negligible friction, we have from (10.63)

$$\begin{aligned} p &= p_0 + m_1 t \\ q &= q_0 + m_2 t \\ r &= r_0 + m_3 t \end{aligned}$$

These expressions show that the angular velocity increases indefinitely at a uniform rate.

Again this increase will be restricted when friction is considered. For equations (10.91) show that all trajectories in the phase space are rectilinear, and tend asymptotically to the only singular point $(m_1/\nu, m_2/\nu, m_3/\mu)$.

The qualitative results obtained in this Section concerning the motion of a symmetric rigid body subject to a time-independent self-excitement with fixed direction in the body can be conveniently summarized into the following

Theorem: For a self-excited symmetric rigid body rotating about a fixed point in a viscous medium, the motion of the body approaches asymptotically a uniform rotation about an axis fixed in the body as well as in space. The angular velocity components along the x, y, z axes of the body-fixed trihedral are respectively

$$(\nu m_1 + g_1 m_2)/(\nu^2 + g_1^2), \ (\nu m_2 - g_1 m_1)/(\nu^2 + g_1^2), \ m_3/\mu$$

If friction is neglected, the angular velocity in general increases indefinitely, except in the case when the torque vector lies in the equatorial plane with the initial position of $\vec{\omega}$ outside this plane and the body is not dynamically symmetric. In such a case, either $\vec{\omega}$ assumes a stable periodic motion with period equal to $T = |2\pi/h r_0|$, or the body assumes a stable permanent rotation with $m_2/h r_0$, $-m_1/h r_0$, r_0 as the angular velocity components in the direction of the x, y, z axes respectively.

§ 11. Self-excited asymmetric rigid body

11.1. Torque vector fixed along the axis of either the largest or the smallest principal moment of inertia. Assume that the torque vector $\vec{\mathfrak{M}}$ fixed along one of the above axes is time-dependent and in magnitude equal to a constant plus a term which is proportional to the square of the modulus of the spin vector $\vec{\omega}$, i.e. let

$$|\vec{\mathfrak{M}}| = M + \mu_0 |\vec{\omega}|^2 \tag{11.1}$$

where M and μ_0 are constants. In such a case it will be possible to obtain explicit expressions for the components p, q and r of $\vec{\omega}$ in terms of an auxiliary variable φ. An additional quadrature, followed by an inversion of a relation between φ and t, will give us the components p, q and r as functions of the time t.

In addition, we shall discuss qualitatively, following R. LEE [1], the types of motion $\vec{\omega}$ can have with respect to the moving trihedral fixed in the rigid body. This can be achieved by a phase-plane analysis of a related dynamical system which, as we shall see, is an extension of a simple pendulum.

11.11. *Torque vector fixed along the largest principal axis*

Let us consider first the case where the torque vector $\vec{\mathfrak{M}}$ acts along the axis of the largest principal moment of inertia, say, the x-axis. In such a case EULER's equations assume the form

$$\left.\begin{aligned} \dot{p} - a\,q\,r &= m + \mu\,|\vec{\omega}|^2 \\ \dot{q} + b\,r\,p &= 0 \\ \dot{r} - c\,p\,q &= 0 \end{aligned}\right\} \tag{11.2}$$

where

$$a = \frac{B-C}{A}\,, \qquad b = \frac{A-C}{B}\,, \qquad c = \frac{A-B}{C}$$

and $m = M/A$, $\mu = \mu_0/A$ are constants. Without loss of generality we may assume that $A > B > C$ so that the inertia numbers a, b, c are all positive.

Next, introduce an auxiliary variable φ by means of the relation

$$d\varphi = p\,dt$$

Then the second and third equations of system (11.2) become

$$\begin{aligned} \frac{dq}{d\varphi} + b\,r &= 0 \\ \frac{dr}{d\varphi} - c\,q &= 0 \end{aligned} \tag{11.3}$$

This is a system of two linear equations in q and r, the solution of which is

$$\begin{aligned} q &= \alpha_1 \sqrt{b} \cos \sqrt{b\,c}\,\varphi - \alpha_2 \sqrt{b}\,\sin \sqrt{b\,c}\,\varphi \\ r &= \alpha_1 \sqrt{c}\,\sin \sqrt{b\,c}\,\varphi + \alpha_2 \sqrt{c} \cos \sqrt{b\,c}\,\varphi \end{aligned} \tag{11.4}$$

The constants α_1 and α_2 are determined by the initial values of q and r, say q_0 and r_0, for $t = 0$.

If both constants α_1 and α_2 are zero, then $q = r = 0$. The rigid body then rotates permanently about the x-axis, fixed in the body as well as in space, with angular velocity p given by the equation

$$\dot{p} = m + \mu\,p^2 \tag{11.4'}$$

Depending on the relative magnitudes of m and μ, we obtain from the last equation the following expressions for p (in each of these expressions p_0 is the value of p for $t = 0$):

(i) If $m > 0$ and $\mu > 0$, then

$$p = \sqrt{m/\mu} \tan\left[\sqrt{m\,\mu}\,t + \operatorname{arc\,tan} p_0 \sqrt{\mu/m}\right]$$

(ii) If $m < 0$ and $\mu < 0$, then

$$p = \sqrt{m/\mu} \cot\left[\sqrt{m\,\mu}\,t + \operatorname{arc\,cot} p_0 \sqrt{\mu/m}\right]$$

(iii) If $m > 0$ and $\mu < 0$, then

$$p = \sqrt{-m/\mu}\tanh[\sqrt{-m\,\mu}\,t + \operatorname{ar\,tanh} p_0\sqrt{\mu/-m}] \quad \text{for} \quad |\mu|\, p_0^2 < m$$
$$p = \sqrt{-m/\mu}\coth[\sqrt{-m\,\mu}\,t + \operatorname{ar\,coth} p_0\sqrt{\mu/-m}] \quad \text{for} \quad |\mu|\, p_0^2 > m$$

(iv) If $m < 0$ and $\mu > 0$, then

$$p = -\sqrt{-m/\mu}\tanh[\sqrt{-m\,\mu}\,t - \operatorname{ar\,tanh} p_0\sqrt{\mu/-m}] \quad \text{for} \quad \mu\, p_0^2 < |m|$$
$$p = -\sqrt{-m/\mu}\coth[\sqrt{-m\,\mu}\,t - \operatorname{ar\,coth} p_0\sqrt{\mu/-m}] \quad \text{for} \quad \mu\, p_0^2 > |m|$$

(v) If $\mu = 0$, then $p = m\,t + p_0$

(vi) If $m = 0$, $\mu \neq 0$, then $p^{-1} = p_0^{-1} - \mu\,t$ for $p_0 \neq 0$

$p = 0$ for $p_0 = 0$

In what follows we shall assume that at least one of the constants α_1 and α_2 is different from zero. According to system (11.4) this implies that at least one of the initial values q_0 or r_0 is different from zero. Under such an assumption we can write system (11.4) in the form

$$\begin{aligned} q &= \delta\sqrt{b}\cos(\sqrt{b\,c}\,\varphi + \sigma_0) \\ r &= \delta\sqrt{c}\sin(\sqrt{b\,c}\,\varphi + \sigma_0) \end{aligned} \tag{11.4a}$$

where $\delta = \sqrt{\alpha_1^2 + \alpha_2^2}$ and $\sigma_0 = \operatorname{arc\,tan}(\alpha_2/\alpha_1)$ are constants determined by the initial values q_0, r_0. We proceed to express p as a function of φ.

If we substitute expressions (11.4a) into the first equation of system (11.2) and make use of the fact that

$$|\vec{\omega}|^2 = p^2 + q^2 + r^2 \quad \text{and} \quad p = d\varphi/dt$$

then we obtain the following second order nonlinear differential equation for the variable φ:

$$\frac{d^2\varphi}{dt^2} - \mu\left(\frac{d\varphi}{dt}\right)^2 = m + \frac{\delta^2 a}{2}\sqrt{b\,c}\sin 2(\sqrt{b\,c}\,\varphi + \sigma_0) + $$
$$+ \mu\,\delta^2[b\cos^2(\sqrt{b\,c}\,\varphi + \sigma_0) + c\sin^2(\sqrt{b\,c}\,\varphi + \sigma_0)] \tag{11.5}$$

To solve this equation we shall apply a sequence of transformations of the dependent as well as of the independent variables to reduce equation (11.5) to a system of two first order differential equations. For this purpose put

$$\varphi_1 = \sqrt{b\,c}\,\varphi + \sigma_0$$

Then equation (11.5) assumes the form

$$\frac{d^2\varphi_1}{dt^2} - \frac{\mu}{\sqrt{b\,c}}\left(\frac{d\varphi_1}{dt}\right)^2 = m\sqrt{b\,c} + \frac{\delta^2 a}{2}\,b\,c\sin 2\varphi_1 + $$
$$+ \mu\,\delta^2\sqrt{b\,c}\,(b\cos^2\varphi_1 + c\sin^2\varphi_1) \tag{11.6}$$

If we make use of the trigonometric identities

$$2\cos^2\varphi_1 = 1 + \cos 2\varphi_1, \qquad 2\sin^2\varphi_1 = 1 - \cos 2\varphi_1$$

equation (11.6) becomes

$$\frac{d^2\varphi_1}{dt^2} = \left[m + \frac{\mu\,\delta^2}{2}(b + c)\right]\sqrt{b\,c} + \frac{\mu}{\sqrt{b\,c}}\left(\frac{d\varphi_1}{dt}\right)^2 + d\sin 2(\varphi_1 + \sigma_1) \quad (11.7)$$

where

$$d^2 = \left(\frac{\delta^2 a\,b\,c}{2}\right)^2 + \left[\mu\frac{\delta^2}{2}\sqrt{b\,c}(b - c)\right]^2, \qquad \sigma_1 = \frac{1}{2}\arctan\frac{\mu(b-c)}{a\sqrt{b\,c}}$$

Putting

$$\varphi_2 = 2(\varphi_1 + \sigma_1)$$

we obtain

$$\frac{d^2\varphi_2}{dt^2} = [2m + \mu\,\delta^2(b + c)]\sqrt{b\,c} + \frac{\mu}{2\sqrt{b\,c}}\left(\frac{d\varphi_2}{dt}\right)^2 + 2d\sin\varphi_2 \quad (11.8)$$

Further, introducing a new independent variable τ by means of the equation

$$t = \tau/\sqrt{2d}$$

we can write (11.8) in the form

$$\frac{d^2\varphi_2}{d\tau^2} = \frac{1}{2d}[2m + \mu\,\delta^2(b + c)]\sqrt{b\,c} + \frac{\mu}{2\sqrt{b\,c}}\left(\frac{d\varphi_2}{d\tau}\right)^2 + \sin\varphi_2] \quad (11.9)$$

Finally, if we introduce the notations

$$\left.\begin{aligned} m_1 &= \frac{1}{2d}[2m + \mu\,\delta^2(b + c)]\sqrt{b\,c} \\ \mu_1 &= \frac{\mu}{2\sqrt{b\,c}} \\ p_1 &= \frac{d\varphi_2}{d\tau} \end{aligned}\right\} \quad (11.10)$$

equation (11.9) is equivalent to the following system of two first order equations

$$\begin{aligned} \frac{dp_1}{d\tau} &= m_1 + \sin\varphi_2 + \mu_1 p_1^2 \\ \frac{d\varphi_2}{d\tau} &= p_1 \end{aligned} \quad (11.11)$$

Further, we have the following relations

$$\begin{aligned} \varphi_2 &= 2(\sqrt{b\,c}\,\varphi + \sigma_0 + \sigma_1) \\ p_1 &= \sqrt{2b\,c/d}\,p \end{aligned} \quad (11.12)$$

System (11.11) can be easily integrated. First from (11.11) we obtain the equation

$$\frac{d p_1^2}{d \varphi_2} - 2\mu_1 p_1^2 = 2m_1 + 2\sin\varphi_2 \tag{11.13}$$

which is linear in p_1^2.

Integrating equation (11.13) we obtain

$$p_1^2 = C_1 e^{2\mu_1\varphi_2} - \frac{2}{1+4\mu_1^2}(2\mu_1 \sin\varphi_2 + \cos\varphi_2) - \frac{m_1}{\mu_1} \tag{11.14}$$

where C_1 is an arbitrary constant of integration which is to be calculated from the given initial conditions. For example, if we put $\varphi = \varphi_0$, $p = p_0$ for $t = 0$, and if we denote

$$\varphi_2(0) = 2\left(\sqrt{b\,c}\,\varphi + \sigma_0 + \sigma_1\right) = \varphi_{20}$$

$$p_1(0) = \sqrt{2b\,c/d}\,p_0 = p_{10}$$

then

$$C_1 = e^{-2\mu_1\varphi_{20}}\left[p_{10}^2 + \frac{2}{1+4\mu_1^2}(2\mu_1 \sin\varphi_{20} + \cos\varphi_{20}) + \frac{m_1}{\mu_1}\right] \tag{11.15}$$

From the second relation of system (11.12) we obtain the following expression for the square of the x-component of $\vec{\omega}$:

$$p^2 = (d/2b\,c)\left[C_1 e^{2\mu_1\varphi_2} - \frac{2}{1+4\mu_1^2}(2\mu_1 \sin\varphi_2 + \cos\varphi_2) - \frac{m_1}{\mu_1}\right] \tag{11.16}$$

If $U(\varphi)$ denotes the right-hand side of equation (11.16), with φ_2 replaced by $2(\sqrt{b\,c}\,\varphi + \sigma_0 + \sigma_1)$, then

$$p = \sqrt{U(\varphi)} \tag{11.17}$$

Thus formula (11.17) together with formulas (11.4) gives us p, q, and r as functions of φ. Since

$$d\varphi = p\,dt = \sqrt{U(\varphi)}\,dt$$

we have that

$$t = \int_0^{\varphi} \frac{d\varphi}{\sqrt{U(\varphi)}} \tag{11.18}$$

By inversion of relation (11.18) we shall obtain φ as a function of t. Let this result be denoted by

$$\varphi = \varphi(t) \tag{11.19}$$

If in equations (11.4) and (11.16) we replace φ by t as given by (11.19), then we shall obtain p, q and r as functions of t.

However, a complete description of the motion of a rigid body in space calls for the coordinates of the body as functions of the time t.

This is achieved if we know the EULER angles Θ, ψ, and Φ as functions of t. Having obtained p, q, and r as functions of the auxiliary variable φ, we can make use of the formulas in Sec. 10.3 and express the EULER angles in terms of expressions which contain integrals of functions of φ. By means of (11.19) we may then calculate Θ, ψ, and Φ for any value of t, and thus obtain the position of the body in space.

11.111 A qualitative discussion of the motion

From system (11.4) it follows that

$$\frac{q^2}{\delta^2 b} + \frac{r^2}{\delta^2 c} = 1$$

i.e. the projection of the endpoint of the spin vector $\vec{\omega}$ onto the $y\,z$-plane lies on an ellipse with center at the origin and with the major and minor semi-axes along the two principal axes in that plane. The endpoint of $\vec{\omega}$ thus moves with respect to the moving trihedral which is fixed in the body on the surface of an elliptic cylinder defined by the equation

$$\frac{y^2}{\delta^2 b} + \frac{z^2}{\delta^2 c} = 1 \tag{11.20}$$

At the same time the projection p of $\vec{\omega}$ onto the x-axis varies according to equation (11.5) or its equivalent (11.11) which involves p_1. Since p_1 is a scalar multiple of p, the qualitative behavior of p_1 must also be that of p. In order to see the possible variation of p_1, and thus of p, we shall give a phase-plane analysis of the dynamical system described by system (11.11). From such an analysis we shall then be able to deduce the possible types of motion that the endpoint of $\vec{\omega}$ can have with respect to the body-fixed trihedral under the torque vector $\vec{\mathfrak{M}}$. For brevity, we shall sometimes simply say the motion of $\vec{\omega}$. We shall distinguish two cases, according as μ_0 is zero or different from zero. These will be called respectively the unperturbed and perturbed cases. It is interesting to note that system (11.11) is an extension of the dynamical system describing the motion of a simple pendulum.

11.112. The motion of the spin vector in the unperturbed case

This means that the rigid body is acted upon by a constant torque along the largest principal axis. This problem was considered first by R. GRAMMEL [7]. He showed by a method different from that of R. LEE [1] that $\vec{\omega}$ can have essentially three types of motion; two are asymptotic and the other is periodic. We shall give here a complete discussion of this case by the method of phase-plane analysis. It will be seen that R. GRAMMEL's results follow very easily from the topology of the phase-plane defined by system (11.11) for $\mu_1 = 0$. Furthermore, using this method we shall be able to study the types of motion of $\vec{\omega}$ for different values of the parameters of this system. For a particular set of values of these parameters we can also indicate the regions of stable and unstable motions. In addition, we can establish conditions for the existence and non-existence of periodic motion.

For $\mu_1 = 0$ system (11.11) becomes

$$\frac{dp_1}{d\tau} = m_1 + \sin\varphi_2, \qquad \frac{d\varphi_2}{d\tau} = p_1 \tag{11.21}$$

which represents a conservative dynamical system. The topology of the phase-plane is defined by the equation

$$\frac{dp_1}{d\varphi_2} = \frac{m_1 + \sin\varphi_2}{p_1} \tag{11.22}$$

It is known that for such a dynamical system an energy integral exists which can be obtained from equation (11.22). In fact, by separation of variables and integration we obtain

$$\tfrac{1}{2}\, p_1^2 - m_1\, \varphi_2 + \cos\varphi_2 = E$$

where E denotes the energy constant.

The equilibrium positions of this system occur at the points where the numerator and the denominator of the expression on the right hand side of (11.22) vanish simultaneously, i.e. they occur at the points along the φ_2-axis determined by the equation

$$m_1 + \sin\varphi_2 = 0$$

It is clear that if $|m_1| > 1$, the system has no equilibrium points since $|\sin\varphi_2| \leqq 1$. Since the topological structure of the phase-plane depends on the values of m_1, we shall consider separately the following cases.

Case 1. $0 < |m_1| < 1$. The system (11.21) has then an infinite number of singular points which occur in pairs corresponding to the points of maximum and minimum potential energy. These points are spaced equally apart at a distance of 2π along the φ_2-axis. Since the

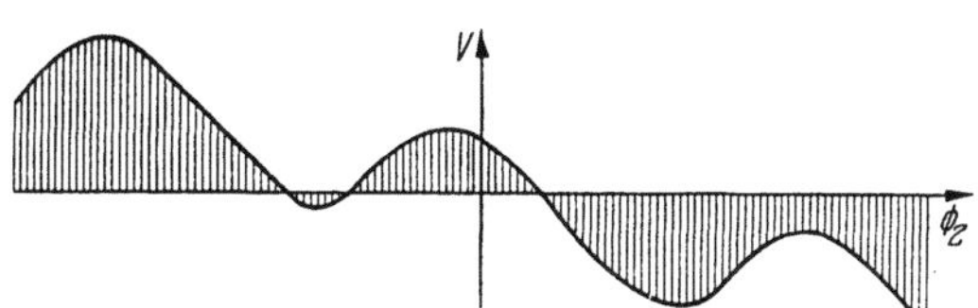

Fig. 17

dynamical system is conservative, the singular points corresponding to points of minimum potential energy are centers, while those corresponding to points of maximum potential energy are saddle points. Fig. 17 shows the graph of the potential energy V versus φ_2 for

$0 < m_1 < 1$. Fig. 18 shows the phase-plane trajectories for the same values of m_1. These trajectories are curves of constant energy. They are symmetric about the φ_2-axis, and except at the singular points, they cross the φ_2-axis at right angles. The equations of the separatrices (heavy loops in Fig. 18) can be obtained from the condition that these

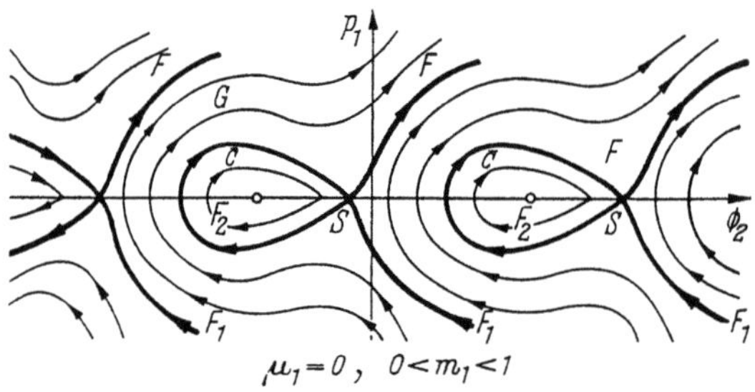

Fig. 18

trajectories pass through the saddle points. If, for example, φ_{2s} denotes the value of φ_2 at the saddle point S, then the equation of the separatrix which passes through S is

$$p_1^2 = 2[m_1(\varphi_2 - \varphi_{2s}) - \cos\varphi_2 + \cos\varphi_{2s}]$$

Along any trajectory G in the regions formed by two consecutive separatrices p_1 eventually goes to infinity. Since $p = \sqrt{d/2bc}\, p_1$, clearly p is also unbounded. This means that the spin vector $\vec{\omega}$ approaches the positive x-axis, for the angle between $\vec{\omega}$ and the x-axis eventually approaches zero. On the branch of each separatrix which eventually goes to infinity, say F, $\vec{\omega}$ approaches the x-axis for the same reason. Along the other two branches F_1 and F_2 which approach a saddle point S as $\tau \to +\infty$, $p \to 0$ as $t \to +\infty$, and φ_2 approaches a finite value, say φ_{2s}. Thus as $t \to +\infty$ the spin vector $\vec{\omega}$ approaches the vector

$$\delta\sqrt{b}\cos(\varphi_{2s}/2)\,\vec{j} + \delta\sqrt{c}\sin(\varphi_{2s}/2)\,\vec{k}$$

lying in the yz-plane of the moving trihedral.

Inside the closed branches of the separatrices we have regions of periodic motion since p, q and r return to the same values after the period

$$T = (1/\sqrt{2d})\int \frac{d\varphi_2}{p_1}$$

where the line integral is taken along a closed trajectory $\bar{C}$. It should be noted, however, that periodicity of $\vec{\omega}$ does not necessarily imply that the corresponding motion of the rigid body is periodic in space.

It is clear that any periodic motion inside the closed branch of each separatrix is stable.

Motions about the singular points correspond to permanent rotations of the rigid body since p, q and r become constant and consequently the derivative $d\vec{\omega}/dt$ reduces to zero. Therefore the rigid body will rotate with a constant angular velocity about an axis fixed in the body as well as in space. Moreover, at the saddle points we have unstable permanent rotations, while at the centers we have stable permanent rotations.

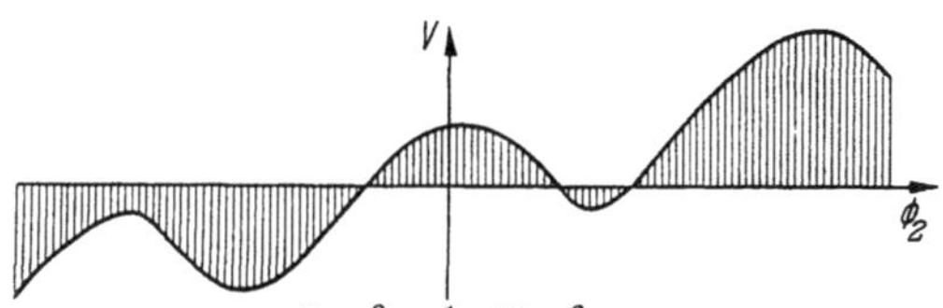

Fig. 19

These are the only types of permanent rotations a rigid body subject to a torque specified above may have.

R. Grammel [7] calls asymptotic motions of the first kind all motions for which $\vec{\omega}$ approaches the x-axis with its modulus tending to infinity. It is seen from Fig. 18 that these motions occur in the regions bounded by two consecutive separatrices and on the branches of separatrices which eventually go to infinity. R. Grammel calls asymptotic motions of the second kind all motions for which $\vec{\omega}$ approaches a finite vector fixed in the yz-plane. We see that such motions occur along the branches F_1 and F_2 of separatrices which approach the saddle points. Finally, periodic motions of $\vec{\omega}$ noted by R. Grammel are motions about the centers.

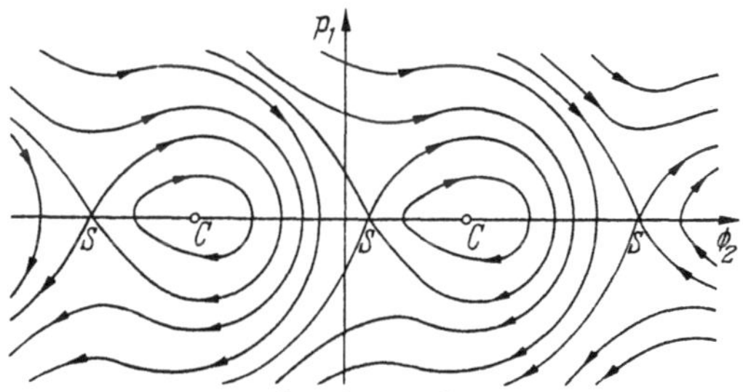

Fig. 20

Figs. 19 and 20 show the graph of the potential energy V versus φ_2 and the phase-plane trajectories for $-1 < m_1 < 0$. As seen from

Fig. 20, the motions of $\vec{\omega}$ are essentially the same as before, except for the asymptotic motions of the first kind where now the spin vector approaches the negative x-axis instead of the positive x-axis.

Case 2: $|m_1| = 1$. In this case the singular points in each pair coalesce into one singular point which is neither a saddle point nor a

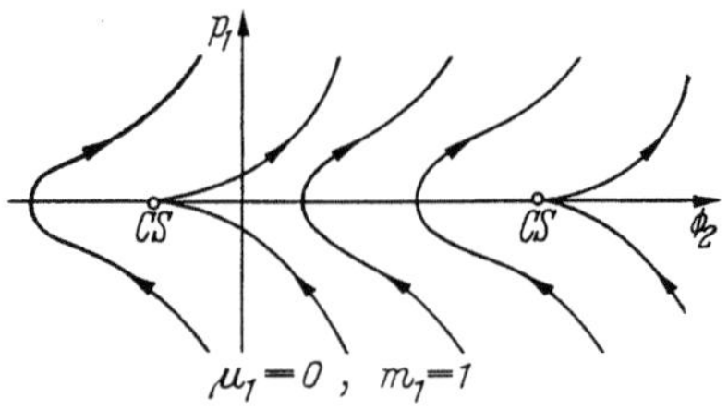

Fig. 21

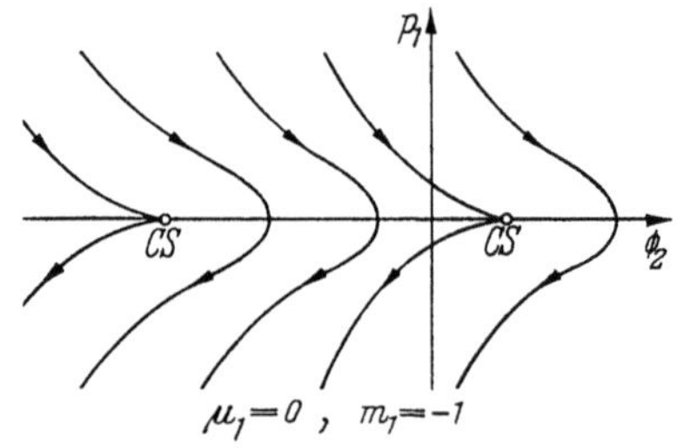

Fig. 22

center. They are equally spaced at intervals of 2π along the φ_2-axis. The potential energy curve has neither maximum nor minimum points. Fig. 21 shows the phase plane for $m_1 = 1$, and Fig. 22 for $m_1 = -1$. At the singular points CS the direction of the field according to equation (11.22) is indeterminate. However, by application of l'Hospital's rule we obtain that

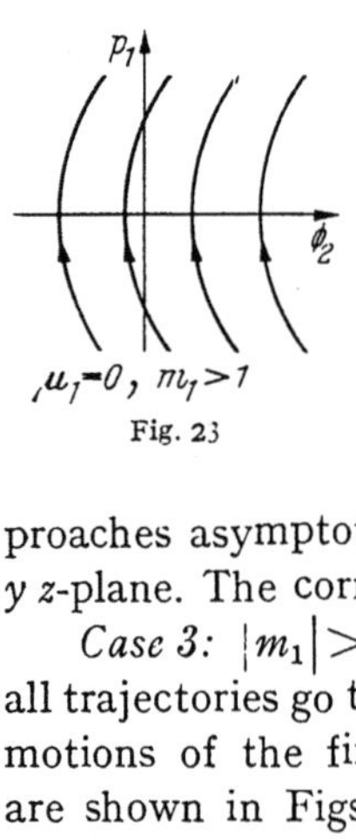

Fig. 23

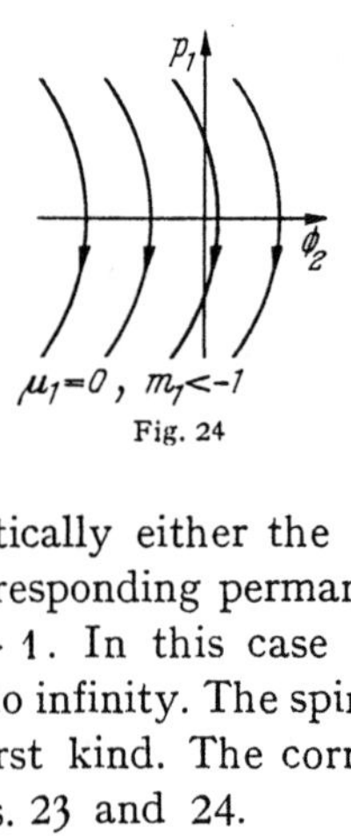

Fig. 24

$$\left(\frac{dp}{d\varphi_2}\right)^2 = 0$$

Thus the separatrices approach these points with zero slope. Since there are no closed trajectories, no periodic motions can exist; $\vec{\omega}$ approaches asymptotically either the x-axis or a constant vector in the $y\,z$-plane. The corresponding permanent rotations are clearly unstable.

Case 3: $|m_1| > 1$. In this case there are no singular points, and all trajectories go to infinity. The spin vector $\vec{\omega}$ assumes only asymptotic motions of the first kind. The corresponding phase-plane trajectories are shown in Figs. 23 and 24.

Case 4: $m_1 = 0$. For completeness of the discussion the phase-plane trajectories for $m_1 = m = 0$ are also given. In this case equations (11.21) reduce to the equations of a simple pendulum. The corresponding motions of $\vec{\omega}$ are those of a force-free asymmetric rigid body, described by the polhodes on the Poinsot ellipsoid. Note, however, that we can also describe these motions from Fig. 25. This completes the discussion of the topological structure of the phase-plane defined by system (11.21). The critical values of m_1 are $1, 0, -1$ since for these values the topological structure of the phase-plane is radically changed.

In the perturbed case ($\mu \neq 0$), using the same method, R. LEE [1] established, besides the motions mentioned above, a second type of periodic motion. For motions of this type, the spin vector with its endpoint on the surface of the elliptic cylinder (11.20) rotates about the axis of this cylinder and returns to its initial position after one

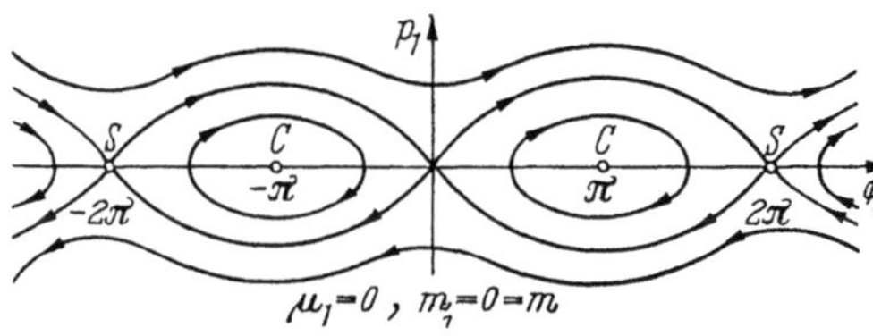

Fig. 25

complete revolution. The components p, q, and r as functions of the time t can be expressed in terms of the JACOBIAN elliptic functions, and the period of rotation in terms of the complete elliptic integral of the first kind. He showed also that the perturbation term $\mu_0 |\vec{\omega}|^2$ can significantly affect the motions of $\vec{\omega}$. We refer the reader to his work for a detailed analysis of the perturbed case.

11.12. Torque vector fixed along the smallest principal axis

If the torque vector $\vec{\mathfrak{M}}$ acts along the z-axis, being the axis of the smallest principal moment of inertia, then due to the particular form of EULER's equations the solutions are obtained from the preceding solutions by interchanging appropriate quantities which characterize such a motion.

11.2. Torque vector fixed along the middle principal axis. However, if $\vec{\mathfrak{M}}$ acts along the middle principal axis, i.e. along the y-axis, such a method will not give us the solutions since in this case new functions are involved.

11.21. Equations of motion and their integration

If the torque vector of modulus $|\vec{\mathfrak{M}}| = M + \mu_0 |\vec{\omega}|^2$ acts along the middle principal axis, then EULER's dynamical equations assume the form

$$\left.\begin{aligned} \dot{p} - a\,q\,r &= 0 \\ \dot{q} + b\,r\,p &= m + u|\vec{\omega}|^2 \\ \dot{r} - c\,p\,q &= 0 \end{aligned}\right\} \qquad (11.23)$$

where

$$m = M/B, \qquad \mu = \mu_0/B$$

Again, introducing the auxiliary variable φ by means of the differential relation

$$d\varphi = q\,dt \tag{11.24}$$

we obtain for p and r the system of equations

$$\begin{aligned} \frac{dp}{d\varphi} - a\,r &= 0 \\ \frac{dr}{d\varphi} - c\,p &= 0 \end{aligned} \tag{11.25}$$

The general solution of system (11.25), being a system of linear equations with constant coefficients, is

$$\begin{aligned} p &= \alpha_1 \sqrt{a} \cosh \sqrt{a\,c}\,\varphi + \alpha_2 \sqrt{a} \sinh \sqrt{a\,c}\,\varphi \\ r &= \alpha_1 \sqrt{c}\, \sinh \sqrt{a\,c}\,\varphi + \alpha_2 \sqrt{c} \cosh \sqrt{a\,c}\,\varphi \end{aligned} \tag{11.26}$$

where α_1 and α_2 are arbitrary constants determined by the initial values of p and r. For example, if at $t = 0$ we have $\varphi = \varphi_0$, $p = p_0$, $r = r_0$, then α_1 and α_2 are obtained by solving the algebraic equations

$$\begin{aligned} p_0 &= \alpha_1 \sqrt{a} \cosh \sqrt{a\,c}\,\varphi_0 + \alpha_2 \sqrt{a} \sinh \sqrt{a\,c}\,\varphi_0 \\ r_0 &= \alpha_1 \sqrt{c}\, \sinh \sqrt{a\,c}\,\varphi_0 + \alpha_2 \sqrt{c} \cosh \sqrt{a\,c}\,\varphi_0 \end{aligned}$$

Since the determinant of the coefficients of the α's is different from zero, $\alpha_1 = \alpha_2 = 0$ if and only if $p_0 = r_0 = 0$.

When both α_1 and α_2 are zero, the second equation of system (11.23) gives

$$\dot{q} - \mu\,q^2 = m \tag{11.27}$$

In such a case the rigid body rotates permanently about the middle axis with angular velocity q. The solution of equation (11.27) can be obtained from that of equation (11.4') by replacing p with q since both variables satisfy an identical equation.

In what follows we shall assume that at least one of the α's is different from zero. In other words, at least one of the initial values p_0 or r_0 is different from zero. The component q of $\vec{\omega}$ then satisfies the equation

$$\dot{q} = m - b\,p\,r + \mu(p^2 + q^2 + r^2) \tag{11.28}$$

while the other two components p and r are given by formulas (11.26). Substituting into equation (11.28) the expressions (11.26) and making use of the formulas

$$\sinh^2 \sqrt{a\,c}\,\varphi = (1/2)\left(\cosh 2\sqrt{a\,c}\,\varphi - 1\right)$$
$$\cosh^2 \sqrt{a\,c}\,\varphi = (1/2)\left(\cosh 2\sqrt{a\,c}\,\varphi + 1\right)$$

we can write (11.28) in the form

$$\dot q - \mu\, q^2 = m_0 + k_1 \sinh 2\sqrt{a\,c}\,\varphi + k_2 \cosh 2\sqrt{a\,c}\,\varphi \tag{11.29}$$

where

$$\left.\begin{aligned} m_0 &= m + (\mu/2)\,(a-c)\,(\alpha_1^2 - \alpha_2^2)\\ k_1 &= \mu\,\alpha_1\,\alpha_2(a+c) - (1/2)\,b\sqrt{a\,c}\,(\alpha_1^2+\alpha_2^2)\\ k_2 &= (\mu/2)\,(a+c)\,(\alpha_1^2+\alpha_2^2) - b\,\alpha_1\,\alpha_2\sqrt{a\,c}\end{aligned}\right\} \tag{11.30}$$

Since $d\varphi = q\,dt$, we obtain from equation (11.29) the following equation for φ:

$$\frac{d^2\varphi}{dt^2} - \mu\left(\frac{d\varphi}{dt}\right)^2 = m_0 + k_1 \sinh 2\sqrt{a\,c}\,\varphi + k_2 \cosh 2\sqrt{a\,c}\,\varphi \tag{11.31}$$

In order to integrate equation (11.31) and to facilitate our later discussions we shall reduce it to an equivalent system of first order equations, changing the dependent as well as the independent variables. For this purpose we shall consider three cases depending on the relative values of k_1 and k_2.

Case (i): $|k_1| > |k_2|$. Under this assumption we may write

$$k_1 \sinh 2\sqrt{a\,c}\,\varphi + k_2 \cosh 2\sqrt{a\,c}\,\varphi = \varepsilon_1\, d_1 \sinh\left(2\sqrt{a\,c}\,\varphi + \sigma_1\right)$$

where

$$d_1 = \sqrt{k_1^2 - k_2^2}, \qquad \sigma_1 = \operatorname{ar}\tanh(k_2/k_1), \qquad \varepsilon_1 = \operatorname{sgn} k_1$$

Equation (11.31) then assumes the form

$$\frac{d^2\varphi}{dt^2} - \mu\left(\frac{d\varphi}{dt}\right)^2 = m_0 + \varepsilon_1\, d_1 \sinh\left(2\sqrt{a\,c}\,\varphi + \sigma_1\right) \tag{11.32}$$

Introduce the new variables $\varphi_2 = 2\sqrt{a\,c}\,\varphi + \sigma_1$ and $t = \tau/\sqrt{2 d_1 \sqrt{a\,c}}$. Then equation (11.32) becomes

$$\frac{d^2\varphi_2}{d\tau^2} - \frac{\mu}{2\sqrt{a\,c}}\left(\frac{d\varphi_2}{d\tau}\right)^2 = \frac{m_0}{d_1} + \varepsilon_1 \sinh\varphi_2 \tag{11.33}$$

Further, putting

$$\mu_1 = \frac{\mu}{2\sqrt{a\,c}}, \qquad m_1 = \frac{m_0}{d_1}, \qquad \frac{d\varphi_2}{d\tau} = q_1 \tag{11.34}$$

we obtain the following system of first order equations

$$\left.\begin{aligned} \frac{dq_1}{d\tau} &= m_1 + \varepsilon_1 \sinh\varphi_2 + \mu_1\, q_1^2\\ \frac{d\varphi_2}{d\tau} &= q_1 \end{aligned}\right\} \tag{11.35}$$

Obviously, we have

$$\varphi = \frac{1}{2\sqrt{a\,c}}\,(\varphi_2 - \sigma_1) \quad \text{and} \quad q = \sqrt{\frac{d_1}{2\sqrt{a\,c}}}\,q_1$$

From equations (11.35) we obtain the following first order linear equation in q_1^2:

$$\frac{d q_1^2}{d\varphi_2} - 2\mu_1\,q_1^2 = 2m_1 + 2\varepsilon_1 \sinh\varphi_2 \tag{11.36}$$

Integrating this equation we obtain

$$q_1^2 = C_1\,e^{2\mu_1\varphi_2} + \frac{2\varepsilon_1}{1 - 4\mu_1^2}\,(2\mu_1 \sinh\varphi_2 + \cosh\varphi_2) - \frac{m_1}{\mu_1} \tag{11.37}$$

where the constant C_1 is to be determined from the initial condition. Integrating equation (11.36) we assume that $|\mu_1| \neq \frac{1}{2}$. A solution obtained disregarding this assumption may be considered as a limiting case of solution (11.37).

Case (ii): $|k_1| < |k_2|$. Under this assumption we can write

$$k_1 \sinh 2\sqrt{a\,c}\,\varphi + k_2 \cosh 2\sqrt{a\,c}\,\varphi = \varepsilon_2\,d_2 \cosh\left(2\sqrt{a\,c} + \sigma_2\right)$$

where

$$d_2 = \sqrt{k_2^2 - k_1^2}\,, \qquad \sigma_2 = \operatorname{ar\,tanh}(k_1/k_2)\,, \qquad \varepsilon_2 = \operatorname{sgn} k_2$$

As before, with

$$\varphi_2 = 2\sqrt{a\,c}\,\varphi + \sigma_2 \quad \text{and} \quad t = \tau/\sqrt{2 d_2 \sqrt{a\,c}}$$

equation (11.31) takes the form

$$\frac{d^2\varphi_2}{d\tau^2} - \frac{\mu}{2\sqrt{a\,c}}\left(\frac{d\varphi_2}{d\tau}\right)^2 = \frac{m_0}{d_2} + \varepsilon_2 \cosh\varphi_2 \tag{11.38}$$

Corresponding to equations (11.35) in the preceding case, we have now the system of first order equations

$$\begin{aligned} \frac{dq_1}{d\tau} &= m_2 + \varepsilon_2 \cosh\varphi_2 + \mu_1\,q_1^2 \\ \frac{d\varphi_2}{d\tau} &= q_1 \end{aligned} \tag{11.39}$$

where

$$\mu_1 = \mu/2\sqrt{a\,c}\,, \qquad m_2 = m_0/d_2$$

Moreover, we have

$$\varphi = \left(1/2\sqrt{a\,c}\right)(\varphi_2 - \sigma_2)\,, \qquad q = \sqrt{d_2/2\sqrt{a\,c}}\,q_1$$

Equations (11.39) lead to the following first order linear equation in q_1^2:

$$\frac{d q_1^2}{d\varphi_2} - 2\mu_1\,q_1^2 = 2m_2 + 2\varepsilon_2 \cosh\varphi_2 \tag{11.40}$$

the general solution of which is

$$q_1^2 = C_1 e^{2\mu_1 \varphi_2} + \frac{2\varepsilon_2}{1-4\mu_1^2}(2\mu_1 \cosh \varphi_2 + \sinh \varphi_2) - \frac{m_2}{\mu_1} \tag{11.41}$$

C_1 being a constant of integration.

Case (iii): $|k_1| = |k_2| \neq 0$. In this case

$$\begin{aligned} k_1 \sinh 2\sqrt{a\,c}\,\varphi + k_2 \cosh 2\sqrt{a\,c}\,\varphi &= (\operatorname{sgn} k_1)\,|k_1|\,e^{2\sqrt{a c}\,\varphi} \quad \text{if } k_1 = k_2 \\ &= -(\operatorname{sgn} k_1)\,|k_1|\,e^{-2\sqrt{a c}\,\varphi} \text{ if } k_1 = -k_2 \end{aligned}$$

Equation (11.31) then becomes

$$\frac{d^2\varphi}{dt^2} - \mu\left(\frac{d\varphi}{dt}\right)^2 = m_0 + (\operatorname{sgn} k_1)\,|k_1|\,e^{2\sqrt{a c}\,\varphi} \tag{11.42a}$$

and

$$\frac{d^2\varphi}{dt^2} - \mu\left(\frac{d\varphi}{dt}\right)^2 = m_0 - (\operatorname{sgn} k_1)\,|k_1|\,e^{-2\sqrt{a c}\,\varphi} \tag{11.42b}$$

respectively. If we let

$$\varphi_2 = 2\sqrt{a\,c}\,\varphi, \qquad t = \tau/\sqrt{2\,|k_1|\,\sqrt{a\,c}}$$

then equations (11.42a) and (11.42b) become

$$\frac{d^2\varphi_2}{d\tau^2} - \frac{\mu}{2\sqrt{a\,c}}\left(\frac{d\varphi_2}{d\tau}\right)^2 = \frac{m_0}{|k_1|} + (\operatorname{sgn} k_1)\,e^{\varphi_2} \tag{11.43a}$$

and

$$\frac{d^2\varphi_2}{d\tau^2} - \frac{\mu}{2\sqrt{a\,c}}\left(\frac{d\varphi_2}{d\tau}\right)^2 = \frac{m_0}{|k_1|} - (\operatorname{sgn} k_1)\,e^{-\varphi_2} \tag{11.43b}$$

Further, putting $\mu_1 = \frac{\mu}{2\sqrt{a\,c}}$, $m_3 = \frac{m_0}{|k_1|}$, $\frac{d\varphi_2}{d\tau} = q_1$, we obtain the following systems of first order equations which are equivalent to equations (11.43a) and (11.43b) respectively,

$$\left.\begin{aligned} \frac{dq_1}{d\tau} &= m_3 + (\operatorname{sgn} k_1)\,e^{\varphi_2} + \mu_1 q_1^2 \\ \frac{d\varphi_2}{d\tau} &= q_1 \end{aligned}\right\} \tag{11.44a}$$

$$\left.\begin{aligned} \frac{dq_1}{d\tau} &= m_3 - (\operatorname{sgn} k_1)\,e^{-\varphi_2} + \mu_1 q_1^2 \\ \frac{d\varphi_2}{d\tau} &= q_1 \end{aligned}\right\} \tag{11.44b}$$

Furthermore, in each case we have the relations

$$\varphi = (1/2\sqrt{a\,c})\,\varphi_2, \qquad q = \sqrt{|k_1|/2\sqrt{a\,c}}\,q_1$$

System (11.44a) gives the first order equation

$$\frac{dq_1^2}{d\varphi_2} - 2\mu_1 q_1^2 = 2m_3 + (\operatorname{sgn} k_1)\,2e^{\varphi_2} \tag{11.45}$$

the general solution of which is

$$q_1^2 = C_1 e^{2\mu_1 \varphi_2} + (\operatorname{sgn} k_1) \frac{2}{1 - 2\mu_1} e^{\varphi_2} - \frac{m_3}{\mu_1} \tag{11.46}$$

where C_1 is an arbitrary constant determined by the initial values q_{10} and φ_{20}. Similarly, if $k_1 = -k_2$, we obtain from equation (11.44b) that

$$q_1^2 = C_1 e^{2\mu_1 \varphi_2} + (\operatorname{sgn} k_1) \frac{2}{1 + 2\mu_1} e^{-\varphi_2} - \frac{m_3}{\mu_1} \tag{11.47}$$

where the arbitrary constant is to be determined by means of the initial values q_{10} and φ_{20}.

Accordingly, from equations (11.37), (11.41), (11.46) or (11.47) we can express q as a function of φ since in each case q is a scalar multiple of q_1 and φ_2 is a linear function of φ. Let this relation be denoted symbolically by

$$q = V(\varphi) \tag{11.48}$$

Then the inversion of the relation

$$t = \int_{\varphi_0}^{\varphi} \frac{d\varphi}{V(\varphi)} \tag{11.49}$$

gives us φ as a function of t, say

$$\varphi = \varphi(t) \tag{11.50}$$

Equation (11.50) then provides the solution of equation (11.31), and in conjunction with equation (11.48) it gives q as a function of t. If in system (11.26) we replace φ by $\varphi(t)$, then we obtain p and r in terms of the time t.

The EULER angles for the position of the body in space can be computed as mentioned in Section 11.11.

11.22. A qualitative discussion of the motion of the spin vector $\vec{\omega}$ with respect to the moving trihedral in the unperturbed case

Formulas (11.26) and (11.48) give the x, y and z-components of the spin vector $\vec{\omega}$ as functions of φ. The description of the motion of $\vec{\omega}$ with respect to the moving trihedral in terms of the time t requires the evaluation and inversion of integral (11.49). In general these calculations must be performed by some kind of numerical procedures since the integral is rather complicated and it is not tabulated except in special cases when the parameters assume certain particular values and the integral becomes an elliptic integral. In what follows we shall give a qualitative description of the motion of $\vec{\omega}$ with respect to the moving trihedral.

First, from formulas (11.26) it follows that

$$\frac{p^2}{a} - \frac{r^2}{c} = \alpha_1^2 - \alpha_2^2 \tag{11.51}$$

or

$$\frac{r^2}{c} - \frac{p^2}{a} = \alpha_2^2 - \alpha_1^2 \tag{11.52}$$

This means that the projection of the endpoint of $\vec{\omega}$ onto the xz-plane always lies along one of the conjugate hyperbolas

$$\frac{x^2}{a} - \frac{z^2}{c} = \alpha_1^2 - \alpha_2^2 \quad \text{if} \quad |\alpha_1| > |\alpha_2| \tag{11.53}$$

and

$$\frac{z^2}{c} - \frac{x^2}{a} = \alpha_2^2 - \alpha_1^2 \quad \text{if} \quad |\alpha_2| > |\alpha_1| \tag{11.54}$$

These hyperbolas are graphically illustrated in Fig. 26, where $\vec{\omega}'$ is the orthogonal projection of $\vec{\omega}$ onto the xz-plane. Then the motion of the spin vector $\vec{\omega}$ is such that its endpoint always lies on the surface of one of the hyperbolic cylinders defined by equations (11.53) and (11.54).

If $|\alpha_1| = |\alpha_2|$ equation (11.51) or (11.52) degenerates into equations

$$\frac{p}{\sqrt{a}} + \frac{r}{\sqrt{c}} = 0 \tag{11.55}$$

and

$$\frac{p}{\sqrt{a}} - \frac{r}{\sqrt{c}} = 0 \tag{11.56}$$

and the above conjugate hyperbolas degenerate into the straight lines

$$z = \sqrt{\frac{c}{a}}\, x \tag{11.57}$$

and

$$z = -\sqrt{\frac{c}{a}}\, x \tag{11.58}$$

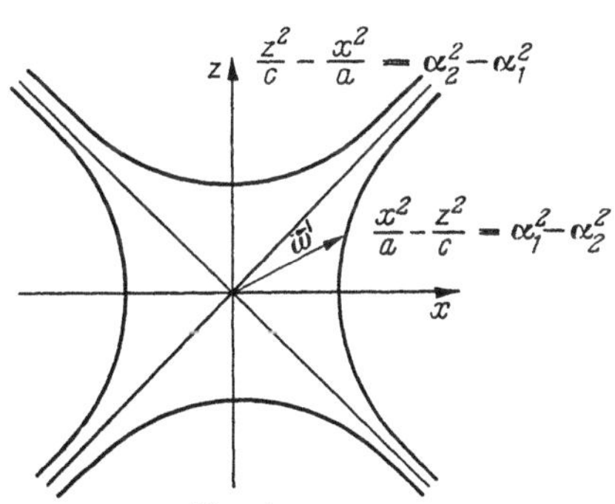

Fig. 26

The vector $\vec{\omega}'$ will lie always along one of these two lines while $\vec{\omega}$ itself will move in such a way that its endpoint will remain for all t in one of the two planes, say I and II, defined by equation (11.57) and (11.58) respectively. These two planes are the planes of the separating polhodes of the force-free asymmetric gyroscope (Fig. 27).

Furthermore, by virtue of the fact that formulas (11.26) now assume the form

$$p = \alpha_1 \sqrt{a}\, e^{\sqrt{ac}\,\varphi}, \qquad r = \alpha_1 \sqrt{c}\, e^{\sqrt{ac}\,\varphi} \qquad \text{if} \quad \alpha_1 = \alpha_2 \tag{11.59}$$

and

$$p = \alpha_1 \sqrt{a}\, e^{-\sqrt{ac}\,\varphi}, \qquad r = -\alpha_1 \sqrt{c}\, e^{-\sqrt{ac}\,\varphi} \quad \text{if} \quad \alpha_1 = -\alpha_2 \tag{11.60}$$

the spin vector $\vec{\omega}$ will, according to the algebraic sign of α_1, remain in either the upper or the lower half of plane I if $\alpha_1 = \alpha_2$ and in either the upper or lower-half of plane II if $\alpha_1 = -\alpha_2$.

The locus of the endpoint of $\vec{\omega}$ on these surfaces or planes is determined by differential equation (11.31). For the unperturbed case ($\mu = 0$) we shall show that its locus is a closed curve on the surfaces of the cylinders, and with one exception is also a closed curve on the two planes of the separating polhodes.

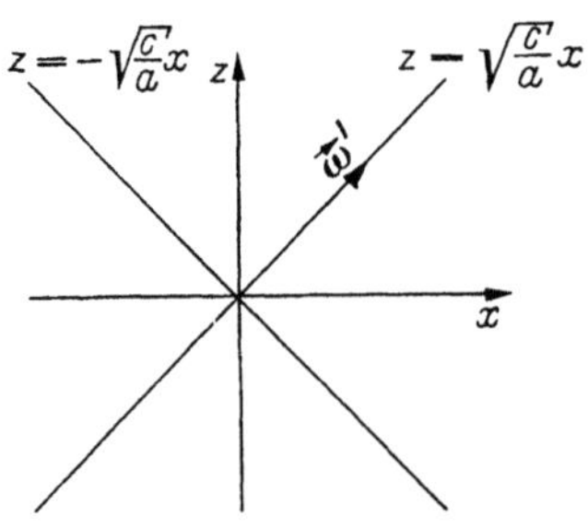

Fig. 27

For $\mu = 0$, the expressions (11.30) for m_0, k_1, k_2 becomes

$$\left.\begin{aligned} m_0 &= m \\ k_1 &= -(\tfrac{1}{2})\, b \sqrt{a\,c}\, (\alpha_1^2 + \alpha_2^2) \\ k_2 &= -b\, \alpha_1 \alpha_2 \sqrt{a\,c} \end{aligned}\right\} \qquad (11.61)$$

so that if $|\alpha_1| \neq |\alpha_2|$, then $|k_1| > |k_2|$ and sgn k_1 is negative. Equation (11.31) is equivalent to system (11.35) which now assumes the form

$$\begin{aligned} d q_1/d\tau &= m_1 - \sinh \varphi_2 \\ d \varphi_2/d\tau &= q_1 \end{aligned} \qquad (11.62)$$

where $\varphi = (1/2\sqrt{a\,c})\,(\varphi_2 - \sigma_1)$ and $q = \sqrt{d_1/2\sqrt{a\,c}}\, q_1$. Since a, c and d_1 are constants, the qualitative behavior of q and φ will follow from that of q_1 and φ_2. The singular points of system (11.62) occur on the φ_2-axis at the points where

$$m_1 - \sinh \varphi_2 = 0 \qquad (11.63)$$

i.e. for

$$\varphi_2 = \operatorname{ar\,sinh} m_1$$

Since equations (11.62) represent a conservative dynamical system, an energy integral

$$\tfrac{1}{2}\, q_1^2 + (-m_1\, \varphi_2 + \cosh \varphi_2) = E \qquad (11.64)$$

exists. Here E denotes the energy constant and

$$V = -(m_1\, \varphi_2 - \cosh \varphi_2) \qquad (11.65)$$

is the potential energy of the system. Since

$$\begin{aligned} dV/d\varphi_2 &= -m_1 + \sinh \varphi_2 \\ d^2V/d\varphi_2^2 &= \cosh \varphi_2 > 0 \quad \text{for all} \quad \varphi_2 \end{aligned}$$

the potential energy curve is concave upward, and the singular points correspond to minimum potential energy. The phase-plane trajectories for such a system are shown in Fig. 28. Consequently the locus of the endpoint of the spin vector $\vec{\omega}$ on the corresponding surface of the cylinder is a closed curve. In other words, $\vec{\omega}$ assumes a periodic motion.

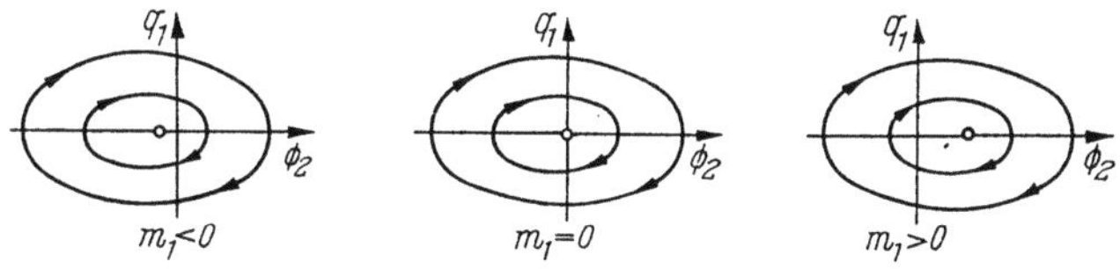

Fig. 28

If $|\alpha_1| = |\alpha_2|$ so that $\vec{\omega}$ moves in one of the two planes of the separating polhodes, equation (11.31) can be reduced to the system

$$\begin{aligned} dq_1/d\tau &= m_3 - e^{\varphi_2} \\ d\varphi_2/d\tau &= q_1 \qquad \text{if} \quad \alpha_1 = \alpha_2 \end{aligned} \tag{11.66}$$

or

$$\begin{aligned} dq_1/d\tau &= m_3 + e^{-\varphi_2} \\ d\varphi_2/d\tau &= q_1 \qquad \text{if} \quad \alpha_1 = -\alpha_2 \end{aligned} \tag{11.66a}$$

The singular points of the above systems occur respectively at the points along the φ_2-axis at which

$$m_3 - e^{\varphi_2} = 0 \tag{11.67}$$

or

$$m_3 + e^{-\varphi_2} = 0 \tag{11.67a}$$

It is obvious that no equilibrium positions can occur for system (11.66) if $m_3 \leqq 0$, and for system (11.66a) if $m_3 \geqq 0$. The two systems have respectively the energy integrals

$$\tfrac{1}{2} q_1^2 - m_3 \varphi_2 + e^{\varphi_2} = E \tag{11.68}$$

or

$$\tfrac{1}{2} q_1^2 - m_3 \varphi_2 + e^{-\varphi_2} = E \tag{11.68a}$$

In Figs. 29 and 30 are given respectively the phase-plane trajectories. These can be established with reference to the corresponding potential energy curve of the system under consideration, or with reference to equations (11.67) to (11.68a). In the subcases (d) and (e) of each Figure are given the phase-plane trajectories when the corresponding systems have no singular points. Naturally all trajectories then tend to infinity. However, in the subcase (e) both $|q_1|$ and $|\varphi_2| \to \infty$ so that $|q| \to \infty$, and according to (11.59) and (11.60) both p and $r \to 0$. Consequently $\vec{\omega}$ tends asymptotically toward the negative or the positive y-axis with infinite modulus. On the other hand in the subcase (d) the spin vector $\vec{\omega}$ tends toward the y-axis but its modulus remains finite since the trajectories approach asymptotically a horizontal line in the phase-plane. In the latter case the motion of $\vec{\omega}$ is also well-

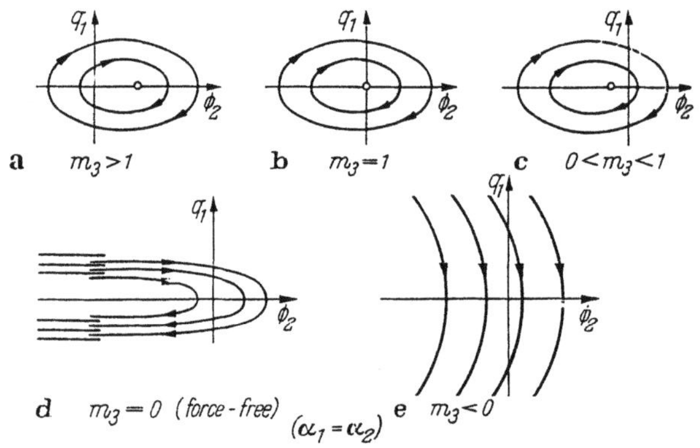

Fig. 29

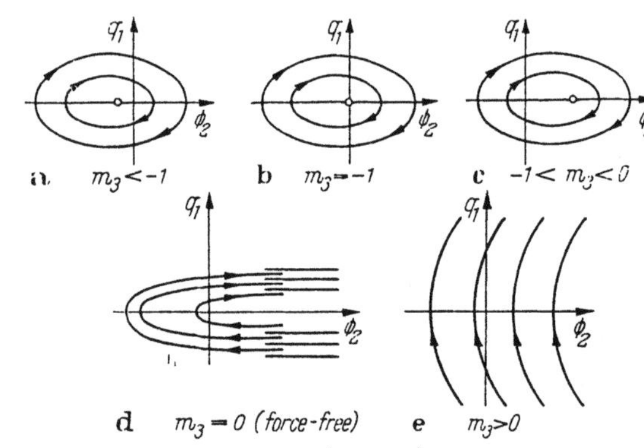

Fig. 30

known from the consideration of the Poinsot ellipsoid. Thus the exception to closed curves occurs when systems (11.66) and (11.66a) have no singular points.

We again refer the reader to R. Lee [1] for a detailed analysis of the perturbed case.

§ 12. Approximate solutions

12.1. Periodic solutions. In the preceding sections we established that under suitable conditions the spin vector $\vec{\omega}$ of an asymmetric rigid body, subject to a time-dependent torque vector fixed along one of the principal axes, can assume periodic motions in a sufficiently small neighborhood of a stable permanent rotation, represented by a center in the corresponding phase plane. In the following three sections we are concerned with the construction of these periodic solutions, using the method of a small parameter due to POINCARÉ.

For this purpose let us consider first the differential equation

$$\frac{d^2\Phi}{d\tau^2} + k^2\,\Phi = \beta_2 \frac{\Phi^2}{2!} + \beta_3 \frac{\Phi^3}{3!} + \beta_4 \frac{\Phi^4}{4!} \tag{12.1}$$

to which all of our later equations can be reduced. Here $k > 0$ and β_i $(i = 2, 3, 4)$ are arbitrary constants. This equation, when reduced to a system of two first order equations, belongs to a general class of differential systems known as LYAPUNOV-systems. It is known that in the $(\Phi, d\Phi/d\tau)$ phase plane, in a sufficiently small neighborhood of the origin, equation (12.1) possesses periodic solutions which are analytic functions of the initial value η of Φ, and reduce to the trivial solution $\Phi = d\Phi/d\tau = 0$ for $\eta = 0$. Furthermore, the period of each of these solutions is also an analytic function of η, reducing to $2\pi/k$ when $\eta = 0$. We shall proceed to the construction of these periodic solutions for $\eta \neq 0$.

Since (12.1) is autonomous, there is no loss of generality in assuming that the initial value of $d\Phi/d\tau$ is zero. Denote by $\mathcal{T}$ the period of the periodic solution Φ corresponding to the initial value η. We may write

$$\mathcal{T} = (2\pi/k)\,(1 + h_1\,\eta + h_2\,\eta^2 + h_3\,\eta^3 + \cdots) \tag{12.2}$$

where h_i $(i = 1, 2, \ldots)$ are constant coefficients to be determined later.

Following I. G. MALKIN [1, 2], we shall first make a change of the time scale by letting

$$\tau = (s/k)\,(1 + h_1\,\eta + h_2\,\eta^2 + h_3\,\eta^3 + \cdots) \tag{12.3}$$

Then equation (12.1) assumes the form

$$\begin{aligned}\frac{d^2\Phi}{ds^2} &+ (1 + h_1\,\eta + h_2\,\eta^2 + \cdots)^2\,\Phi \\ &= (1 + h_1\,\eta + h_2\,\eta^2 + \cdots)^2 \left(\frac{\beta_2}{k^2}\,\frac{\Phi^2}{2!} + \frac{\beta_3}{k^2}\,\frac{\Phi^3}{3!} + \frac{\beta_4}{k^2}\,\frac{\Phi^4}{4!}\right)\end{aligned} \tag{12.4}$$

Clearly, the periodic solution Φ as a function of s is analytic and of period 2π. Thus we shall seek the solution Φ of (12.4) in the form

$$\Phi = \eta\cos s + \eta^2\,\Phi_2(s) + \eta^3\,\Phi_3(s) + \cdots \tag{12.5}$$

where $\Phi_i(s)$ are periodic functions of period 2π which have the initial values $\Phi_i(0) = d\Phi_i(0)/ds = 0$, $i = 2, 3, \ldots$

Substituting the expression (12.5) in (12.4) and equating the coefficients of like powers of η, we obtain for Φ_2 the equation

$$\frac{d^2\Phi_2}{ds^2} + \Phi_2 = \frac{\beta_2}{4k^2} + 2h_1 \cos s + \frac{\beta_2}{4k^2}\cos 2s$$

For this equation to have a solution of period 2π, the coefficient of $\cos s$ must be zero. We obtain then

$$h_1 = 0$$

$$\Phi_2(s) = \frac{\beta_2}{4k^2} - \frac{\beta_2}{6k^2}\cos s - \frac{\beta_2}{12k^2}\cos 2s \tag{12.6}$$

For the function $\Phi_3(s)$ we have

$$\frac{d^2\Phi_3}{ds^2} + \Phi_3 = -\frac{\beta_2^2}{12k^4} + \left(\frac{5\beta_2^2}{24k^4} + \frac{3\beta_3}{24k^2} - 2h_2\right)\cos s - \frac{\beta_2^2}{12k^4}\cos 2s + \\ + \left(-\frac{\beta_2^2}{24k^4} + \frac{\beta_3}{24k^2}\right)\cos 3s$$

whence we find as before

$$h_2 = \frac{5\beta_2^2}{48k^4} + \frac{\beta_3}{16k^2}$$

$$\Phi_3(s) = -\frac{\beta_2^2}{12k^4} + \left(\frac{29\beta_2^2}{576k^4} + \frac{\beta_3}{192k^2}\right)\cos s + \frac{\beta_2^2}{36k^4}\cos 2s + \\ + \left(\frac{\beta_2^2}{192k^4} - \frac{\beta_3}{192k^2}\right)\cos 3s \tag{12.7}$$

The constant h_3 is obtained by equating to zero the coefficient of $\cos s$ in the differential equation for $\Phi_4(s)$. This then yields

$$h_3 = -\frac{5\beta_2^3}{144k^6} - \frac{\beta_2\beta_3}{48k^4} \tag{12.8}$$

We can proceed in the same way as above to obtain higher order approximations; however, the computations soon become long and complicated. Thus we shall go no further, and consider that the required periodic solution Φ and its period $\mathcal{T}$ are expressed approximately by formulas (12.2), (12.3), and (12.5) to (12.8).

12.11. Periodic solutions in the case of a time-dependent torque vector fixed along the largest principal axis

It was shown that the components of $\vec{\omega}$ can be expressed in terms of the variables φ or φ_2 since the two are related by the first equation of (11.12). Further, φ_2 satisfies the differential equation (for $\mu_1 = 0$)

$$\frac{d^2\varphi_2}{d\tau^2} = m_1 + \sin\varphi_2 \tag{12.9}$$

When $|m_1| < 1$, it was established that the singular points of (12.9) occurred on the φ_2-axis, and some of these singular points were centers surrounded by closed trajectories which represent periodic solutions of (12.9). Our purpose here is to construct these periodic solutions, using the theory described in Sec. 12.1. Putting these solutions in the expressions for p, q, and r, and replacing τ by $t\sqrt{2d}$, as defined in Sec. 11.11, we shall obtain the components p, q, and r as periodic functions of t. Such procedures avoid evaluating and inverting the integral in (11.18).

Denote by φ_{2c} the value of φ_2 at a singular point of the center type, and let $\varphi_2 = \varphi_{2c} + \Phi$. Then (12.9) becomes

$$\frac{d^2\Phi}{d\tau^2} + \sqrt{1-m_1^2}\sin\Phi = m_1 - m_1\cos\Phi \tag{12.10}$$

Here we have used the fact that at the center $\cos\varphi_{2c}$ is equal to $-\sqrt{1-m_1^2}$. If we replace $\sin\Phi$ and $\cos\Phi$ by

$$\sin\Phi = \Phi - \frac{\Phi^3}{3!}$$

$$\cos\Phi = 1 - \frac{\Phi^2}{2!} + \frac{\Phi^4}{4!}$$

then equation (12.10) assumes the form

$$\frac{d^2\Phi}{d\tau^2} + \sqrt{1-m_1^2}\,\Phi = m_1\frac{\Phi^2}{2!} + \sqrt{1-m_1^2}\,\frac{\Phi^3}{3!} - m_1\frac{\Phi^4}{4!} \tag{12.11}$$

Comparing this equation with (12.1), we can immediately write down, from formulas (12.2), (12.3) and (12.5) to (12.8), the periodic solution Φ corresponding to the initial value η of Φ. Hence, the required period solution φ_2, occurring in a sufficiently small neighborhood of the center, and its period $\mathcal{T}$ are determined by the following formulas:

$$\left.\begin{aligned}
s &= (1-m_1^2)^{1/4}(1 + h_1\eta + h_2\eta^2 + h_3\eta^3 + \cdots)^{-1}\tau \\
\varphi_2 &= \varphi_{2c} + \eta\cos s + \eta^2\Phi_2(s) + \eta^3\Phi_3(s) + \cdots \\
\Phi_2(s) &= \frac{m_1}{4\sqrt{1-m_1^2}} - \frac{m_1}{6\sqrt{1-m_1^2}}\cos s - \frac{m_1}{12\sqrt{1-m_1^2}}\cos 2s \\
\Phi_3(s) &= -\frac{m_1^2}{12(1-m_1^2)} + \left[\frac{29m_1^2}{576(1-m_1^2)} + \frac{1}{192}\right]\cos s + \\
&\quad + \frac{m_1^2}{36(1-m_1^2)}\cos 2s + \left[\frac{m_1^2}{192(1-m_1^2)} - \frac{1}{192}\right]\cos 3s \\
h_1 &= 0 \\
h_2 &= \frac{5m_1^2}{48(1-m_1^2)} + \frac{1}{16} \\
h_3 &= -\frac{5m_1^3}{144(1-m_1^2)^{3/2}} - \frac{m_1}{48\sqrt{1-m_1^2}} \\
\mathcal{T} &= \frac{2\pi}{(1-m_1^2)^{1/4}}(1 + h_1\eta + h_2\eta^2 + h_3\eta^3 + \cdots)
\end{aligned}\right\} \tag{12.12}$$

If the torque vector is fixed along the smallest principal axis, the corresponding periodic solutions can be obtained by simply interchanging the appropriate quantities as mentioned in Section 11.12.

12.12. Periodic solutions in the case of a time-dependent torque vector fixed along the middle principal axis

Here we need to consider three subcases, because the behavior of $\vec{\omega}$ depends on the initial position of $\vec{\omega}$ relative to the two planes of the separating polhodes. Although the differential equations satisfied by the variable φ_2 are slightly different in form, the steps taken to obtain the corresponding periodic solutions are identical with the ones in Sec. 12.11. Hence, in what follows we shall simply summarize the results.

1. *The initial position of $\vec{\omega}$ not in planes I and II* [see equation (11.62)]

$$\left.\begin{aligned}
s &= (1+m_1^2)^{1/4}\,(1+h_1\eta+h_2\eta^2+h_3\eta^3+\cdots)\\
\varphi_2 &= \varphi_{2e}+\eta\cos s+\eta^2\,\Phi_2(s)+\eta^3\,\Phi_3(s)+\cdots\\
\varphi_{2e} &= \operatorname{ar\,sinh} m_1\\
\Phi_2(s) &= -\frac{m_1}{4\sqrt{1+m_1^2}}+\frac{m_1}{6\sqrt{1+m_1^2}}\cos s+\frac{m_1}{12\sqrt{1+m_1^2}}\cos 2s\\
\Phi_3(s) &= -\frac{m_1^2}{12(1+m_1^2)}+\left[\frac{29m_1^2}{576(1+m_1^2)}-\frac{1}{192}\right]\cos s+\\
&\quad+\frac{m_1^2}{36(1+m_1^2)}\cos 2s+\left[\frac{m_1^2}{192(1+m_1^2)}+\frac{1}{192}\right]\cos 3s\\
h_1 &= 0\\
h_2 &= \frac{5m_1^2}{48(1+m_1^2)}-\frac{1}{16}\\
h_3 &= \frac{5m_1^3}{144(1+m_1^2)^{3/2}}-\frac{m_1}{48\sqrt{1+m_1^2}}\\
\mathcal{T} &= \frac{2\pi}{(1+m_1^2)^{1/4}}\,(1+h_1\eta+h_2\eta^2+h_3\eta^3+\cdots)
\end{aligned}\right\}\qquad(12.13)$$

2. *The initial position of $\vec{\omega}$ in plane I* [see equations (11.66)]

$$\left.\begin{aligned}
s &= \sqrt{m_3}\,(1+h_1\eta+h_2\eta^2+h_3\eta^3+\cdots)^{-1}\,\tau,\qquad m_3>0\\
\varphi_2 &= \varphi_{2e}+\eta\cos s+\eta^2\,\Phi_2(s)+\eta^3\,\Phi_3(s)+\cdots\\
\varphi_{2e} &= \ln m_3\\
\Phi_2(s) &= -\frac{1}{4}+\frac{1}{6}\cos s+\frac{1}{12}\cos 2s\\
\Phi_3(s) &= -\frac{1}{12}+\frac{26}{576}\cos s+\frac{1}{36}\cos 2s+\frac{1}{96}\cos 3s\\
h_1 &= 0,\quad h_2=\frac{1}{24},\quad h_3=\frac{1}{72}\\
\mathcal{T} &= \frac{2\pi}{\sqrt{m_3}}\,(1+h_1\eta+h_2\eta^2+h_3\eta^3+\cdots)
\end{aligned}\right\}\qquad(12.14)$$

3. *The initial position of $\vec{\omega}$ in plane II* [see equations (11.66a)]

$$\left.\begin{aligned}
s &= \sqrt{-m_3}\,(1 + h_1\eta + h_2\eta^2 + h_3\eta^3 + \cdots)^{-1}\,\tau, \qquad m_3 < 0\\
\varphi_2 &= \varphi_{2e} + \eta\cos s + \eta^2\,\Phi_2(s) + \eta^3\,\Phi_3(s) + \cdots\\
\varphi_{2e} &= \ln\left(-\frac{1}{m_3}\right)\\
\Phi_2(s) &= \frac{1}{4} - \frac{1}{6}\cos s - \frac{1}{12}\cos 2s\\
\Phi_3(s) &= -\frac{1}{12} + \frac{26}{576}\cos s + \frac{1}{36}\cos 2s + \frac{1}{96}\cos 3s\\
h_1 &= 0, \quad h_2 = \frac{1}{24}, \quad h_3 = -\frac{1}{72}\\
\mathcal{T} &= \frac{2\pi}{\sqrt{-m_3}}\,(1 + h_1\eta + h_2\eta^2 + h_3\eta^3 + \cdots)
\end{aligned}\right\} \qquad (12.15)$$

12.2. Iterative solutions. In the case of a symmetric rigid body subject to a time-dependent or time-independent self-excitement with variable or fixed axis, we were able, as indicated in Sec. 10, to obtain explicit expressions for the components of the angular velocity vector in terms of some integrals. However, for an asymmetric rigid body subject to the same type of self-excitement, we are far from the above situation. Even for the case of a time-independent torque vector fixed in the body, the general solution has not been obtained, and the corresponding solution in the case of a time-dependent torque vector is not even in sight. The difficulty in obtaining a general solution can be foreseen when we consider the fact that already in the classical case of EULER and POINSOT ($\vec{\mathfrak{M}} = 0$) the general solution is given in terms of elliptic functions.

Faced with this fact, R. GRAMMEL [7] attempted to construct approximate solutions of EULER's equations. He first considered the case where the torque vector $\vec{\mathfrak{M}}$ acts along the smallest or the largest principal axis and the initial angular velocity vector $\vec{\omega}_0$ has a small component perpendicular to the chosen axis. Using iterative methods, he constructed successive approximations for p, q, and r. Next he considered the case where the torque vector $\vec{\mathfrak{M}}$ deviates little from one of the above mentioned axes and the initial angular velocity vector $\vec{\omega}_0$ lies on the chosen axis. A proper combination of these two cases gives an approximate solution of EULER's equation for time-dependent or time-independent self-excitements with fixed as well as variable axes in the vicinity of the largest or smallest principal axis. However, the convergence of these successive approximations was not proved. We refer the reader to his paper for a complete discussion of these solutions.

§ 13. Regulation of rotations about fixed axes by self-excitements with fixed axes

13.1. Time-independent rotations caused by time-independent self-excitement. The motion of a rigid body with one fixed point in space as well as in the body is governed by the EULER dynamical equations

$$\left.\begin{aligned} A\dot{p} + (C-B)\,q\,r &= M_x \\ B\dot{q} + (A-C)\,r\,p &= M_y \\ C\dot{r} + (B-A)\,p\,q &= M_z \end{aligned}\right\} \tag{13.1}$$

where M_x, M_y, M_z are the constant components of the torque $\vec{\mathfrak{M}}$ in a body-fixed coordinate trihedral. The time-independent rotations and their stability were investigated first by R. GRAMMEL [6], who showed that for $\vec{\mathfrak{M}} \neq 0$ in the case of a symmetric body $(B = C)$ there are ∞^3 time-independent (stationary) rotations which are "almost all" stable while in the case of an asymmetric body there are ∞^3 stationary rotations which are "almost all" unstable.

Symmetric rigid body. Let $B = C$. Then from the first equation of system (13.1) it follows that for $\dot{p} = \dot{q} = \dot{r} = 0$, of necessity $M_x = 0$. If, further, the body-fixed torque vector is directed along the y-axis, then $M_z = 0$. Also, system (13.1) shows that for $p = a = \text{const.}$, $r = c = \text{const.}$, q must vanish, and that the constants a and c are related to $M_y = |\vec{\mathfrak{M}}|$ by the relation

$$M_y = (A - B)\,a\,c \tag{13.2}$$

Hence the stationary motion is a uniform rotation of the body about an axis g through O in a meridian plane, perpendicular to which is directed the torque vector $\vec{\mathfrak{M}}$ of magnitude (13.2), the velocity of rotation being $\omega = \sqrt{a^2 + c^2}$. The direction of rotation about the axis g is not determined since only the sign of the product $a\,c$ is known. The motion is a regular precession without proper rotation. Permanent rotations about the symmetry axis (for which $c = 0$) or about axes lying in the equatorial plane $(a = 0)$ are not taken into consideration since for them $M_y = 0$.

In order to investigate the stability of the stationary motions subject to moment (13.2), consider the perturbed motion

$$p = a + \varepsilon_1 e^{\varrho t}, \qquad q = \varepsilon_2 e^{\varrho t}, \qquad r = c + \varepsilon_3 e^{\varrho t} \tag{13.3}$$

where ε_1, ε_2, ε_3 are small quantities. If we substitute (13.3) into system (13.1), put $B = C$, $M_x = M_z = 0$, and neglect the products of the small quantities ε_i, then we obtain the system

$$\left.\begin{aligned} &A\,\varrho\,\varepsilon_1 &&= 0 \\ &B\,\varrho\,\varepsilon_2 + (A - B)\,(c\,\varepsilon_1 + a\,\varepsilon_3) &&= 0 \\ &B\,\varrho\,\varepsilon_3 - (A - B)\,a\,\varepsilon_2 &&= 0 \end{aligned}\right\} \tag{13.4}$$

In order that there exists a nontrivial solution for ε_1, ε_2, ε_3, the determinant of the coefficients of system (13.4) must vanish. Hence we obtain the equation

$$\begin{vmatrix} A\,\varrho & 0 & 0 \\ (A - B)\,c & B\,\varrho & (A - B)\,a \\ 0 & -(A - B)\,a & B\,\varrho \end{vmatrix} = 0 \tag{13.5}$$

the roots of which are

$$\varrho_1 = 0, \qquad \varrho_{2,3} = \pm \frac{A - B}{B}\,a\,i \tag{13.6}$$

Since for $a \neq 0$ there are no roots with positive real parts, permanent rotations of a self-excited symmetric rigid body, as long as its component $a \neq 0$, are all stable or, more exactly expressed, parastable (see R. Grammel [4_1]). The fact that one root of equation (13.5) is zero indicates that the stability is not complete in the sense that any arbitrary perturbation may not result in oscillations about the unperturbed motion.

Permanent rotations about the symmetry axis ($c = 0$) are stable while those about axes lying in the equatorial plane ($a = 0$) are unstable. This follows from (13.6) by virtue of the fact that for $a = 0$ all three roots ϱ_1, ϱ_2, ϱ_3 are equal to zero. As a consequence there are solutions of the form $\varepsilon + \varepsilon' t + \varepsilon'' t^2$.

Asymmetric rigid body. Without any loss of generality assume that

$$A > B > C \tag{13.7}$$

Then for fixed values of p, q, r, say,

$$p = a, \qquad q = b, \qquad r = c \tag{13.8}$$

it follows from system (13.1) that

$$\left.\begin{aligned} (B - C)\,b\,c &= -M_x \\ (A - C)\,c\,a &= M_y \\ (A - B)\,a\,b &= -M_z \end{aligned}\right\} \tag{13.9}$$

System (13.9) shows that none of the components M_x, M_y, M_z of $\vec{\mathfrak{M}}$ can vanish. Hence torque vectors which lie in one of the three principal planes of inertia at O (the principal axes of inertia being excluded), cannot maintain any permanent rotation.

On the other hand there exist permanent rotations if two components of the torque vector $\vec{\mathfrak{M}}$ vanish, i.e. torque vectors fixed along one of the principal axes may maintain a permanent rotation, the axis of which lies in that principal plane which is perpendicular to the principal axis under consideration.

The components of rotation b and c, c and a, a and b, and the corresponding torque components M_x, M_y and M_z are related by the first, second or third equation of system (13.9).

Finally, if all three components of $\vec{\mathfrak{M}}$ are different from zero, system (13.9) gives the following expressions for the squares of the components a, b, c of the permanent rotation (13.8),

$$\left.\begin{aligned} a^2 &= \frac{B-C}{(A-B)(A-C)}\,\frac{M_y M_z}{M_x} \\ b^2 &= \frac{A-C}{(A-B)(B-C)}\,\frac{M_z M_x}{M_y} \\ c^2 &= \frac{A-B}{(A-C)(B-C)}\,\frac{M_x M_y}{M_z} \end{aligned}\right\} \tag{13.10}$$

The signs of a, b, c are determined by system (13.9), which gives us the following correspondence between the signs of M_x, M_y, M_z and those of a, b, c:

M_x	M_y	M_z	a	b	c
+	+	+	+	−	+
			−	+	−
+	−	−	+	+	−
			−	−	+
−	+	−	−	−	−
			+	+	+
−	−	+	−	+	+
			+	−	−
+	+	−	no real values for a, b, c		
+	−	+			
−	+	+			
−	−	−			

Hence to every given permanent rotation vector $\vec{\omega}(a, b, c)$ there corresponds a body-fixed torque vector $\vec{\mathfrak{M}}(M_x, M_y, M_z)$ which may maintain this rotation. However, not any given torque vector $\vec{\mathfrak{M}}$ can maintain a real permanent rotation $\vec{\omega}$ but only those torques $\vec{\mathfrak{M}}$ which lie in one of the four octants made up by the directed axes of the largest, middle and smallest principal moments of inertia and constitute a right-hand system. The corresponding rotation vector $\vec{\omega}$ lies in the octant derived from that of $\vec{\mathfrak{M}}$ by a quarter rotation about the directed axis of the largest principal moment of inertia A in the sense of a right-hand screw, or in the octant which is symmetric to the above obtained with respect to the origin O (to any permissible torque $\vec{\mathfrak{M}}$ there are two permanent rotations of opposite directions but equal magnitudes).

If equations (13.9) are multiplied by $-a$, $+b$, $-c$ respectively and added, then the relation

$$\vec{\omega} \cdot \vec{\mathfrak{M}} = a\,M_x + b\,M_y + c\,M_z = 0$$

results. This relation shows that the body-fixed torque $\vec{\mathfrak{M}}$ is always perpendicular to the velocity vector $\vec{\omega}$ of the permanent rotation maintained by $\vec{\mathfrak{M}}$.

Since in the rotation of the body the vector $\vec{\mathfrak{M}}$ does no work, it follows that such a body-fixed torque can only maintain, but not generate, the rotation $\vec{\omega}$. In order to set up a rotation $\vec{\omega}$ an initial torque with components $A\,a$, $B\,b$, $C\,c$ is required.

13.11. Stability of time-independent rotations

Consider the perturbed motion

$$p = a + \varepsilon_1\, e^{\varrho t}, \qquad q = b + \varepsilon_2\, e^{\varrho t}, \qquad r = c + \varepsilon_3\, e^{\varrho t} \qquad (13.11)$$

where a, b, c characterize the unperturbed motion (13.8). Substituting (13.11) into system (13.1), neglecting the products of the small quantities ε_i $(i = 1, 2, 3)$ and observing system (13.9), we obtain a system of linear equations in ε_i. In order that this system be consistent, its determinant must vanish, i.e.

$$\begin{vmatrix} A\,\varrho & -(B-C)\,c & -(B-C)\,b \\ (A-C)\,c & B\varrho & (A-C)\,a \\ -(A-B)\,b & -(A-B)\,a & C\,\varrho \end{vmatrix} = 0$$

or

$$\alpha_0\,\varrho^3 + \alpha_1\,\varrho^2 + \alpha_2\,\varrho + \alpha_3 = 0 \qquad (13.12)$$

where

$$\left.\begin{aligned} \alpha_0 &= A\,B\,C \\ \alpha_1 &= 0 \\ \alpha_2 &= A(A-B)(A-C)\,a^2 - B(A-B)(B-C)\,b^2 + \\ &\qquad + C(A-C)(B-C)\,c^2 \\ \alpha_3 &= 2(A-B)(A-C)(B-C)\,a\,b\,c \end{aligned}\right\} \qquad (13.13)$$

Consider first the special case $a = 0$ (the axis of rotation lies in the $y\,z$-plane). Then, besides the root $\varrho_1 = 0$, equation (13.12) has

$$\left.\begin{array}{c}\text{two equal real roots with}\\ \text{opposite signs}\\ \text{a pair of imaginary roots}\end{array}\right\}\ \varrho_2,\ \varrho_3 \text{ provided that } \frac{c^2}{b^2} \lessgtr \frac{B(A-B)}{C(A-C)} \tag{13.14}$$

The presence of a positive real root leads to instability of rotation even for the limiting case when the equality sign in (13.14) holds. The presence of a zero root and a pair of imaginary roots leads to parastability.

In the case $b = 0$ (the axis of rotation lies in the $x\,z$-plane) we have besides the root $\varrho_1 = 0$ a pair of imaginary roots. This again means parastability of the rotation.

Finally for $c = 0$ (the axis of rotation lies in the $x\,y$-plane) we have $\varrho_1 = 0$ and

$$\left.\begin{array}{c}\text{two equal real roots with}\\ \text{opposite signs}\\ \text{a pair of imaginary roots}\end{array}\right\}\ \varrho_2,\ \varrho_3 \text{ provided that } \frac{a^2}{b^2} \lessgtr \frac{B(B-C)}{A(A-C)} \tag{13.15}$$

Hence in the first case the rotation is unstable while in the second case it is parastable.

The above results can be summarized as follows.

Except the stable permanent rotations about the axis of the largest moment of inertia and that of the smallest moment of inertia ($\vec{\mathfrak{M}} = 0$) all permanent rotations of a self-excited asymmetric rigid body about axes lying in the principal plane determined by the largest and smallest moments of inertia are parastable. Also parastable are all rotations about axes lying in the principal plane determined by the smallest and middle moments of inertia, and forming with the middle principal axis an angle

$$\varphi > \varphi_1 = \operatorname{arc\,tan} \sqrt{\frac{B(A-B)}{C(A-C)}} \tag{13.16}$$

and similarly about axes lying in the principal plane determined by the largest and middle moments of inertia, and forming with the middle principal axis an angle

$$\varphi > \varphi_2 = \operatorname{arc\,tan} \sqrt{\frac{B(B-C)}{A(A-C)}} \tag{13.17}$$

Permanent rotations about other axes through O lying in the plane determined by the smallest and middle as well as by the largest and middle moments of inertia ($\varphi \leqq \varphi_1$ and $\varphi \leqq \varphi_2$) are unstable.

Next let us prove that permanent rotations about all other axes through O which do not lie in any of the three principal planes are unstable.

In fact, according to the HERMITE-HURWITZ criterion the conditions for equation (13.12) to have roots with no positive real parts are

$$\alpha_0 > 0, \quad \alpha_1 \geqq 0, \quad \alpha_3 \geqq 0, \quad \alpha_1 \alpha_2 - \alpha_0 \alpha_3 > 0 \qquad (13.18)$$

In our case we have $\alpha_0 > 0$, $\alpha_1 = 0$ and $\alpha_3 \neq 0$ for $a \neq 0$, $b \neq 0$, $c \neq 0$. Hence the third and fourth conditions of (13.18) contradict each other. Consequently there is at least one root with a positive real part, and as a consequence all permanent rotations about axes for which $a \neq 0$, $b \neq 0$, $c \neq 0$ are unstable.

This result of R. GRAMMEL, which states that permanent rotations of an asymmetric rigid body maintained by a body-fixed torque $\vec{\mathfrak{M}}$ are "almost all" unstable, contradicts an earlier result on the stability of permanent rotations. According to Sec. 8.3 in the case of an asymmetric rigid body which moves subject to gravity and the mass center of which differs from the fixed point O, there are infinitely many axes through O which, when put in the vertical position, may serve as the axes of permanent rotations of the body. The mass center cone, which passes through the three principal axes of inertia, contains also an infinity of axes which do not lie along any of the principal axes. From the results obtained in Sec. 8.3 on the stability of permanent rotations it follows that the mass center cone contains besides an infinity of unstable axes of permanent rotations also an infinity of axes that are stable. The apparent discrepancy between the earlier results on stability of permanent rotations and the above result by R. GRAMMEL can be explained as follows. In the case of the STAUDE-MLODZEEVSKII permanent rotations the torque vector $\vec{\mathfrak{M}}_g$, fixed within the body, is the moment of the weight of the body about the fixed point O. Such a torque, however, can also be generated by a body-fixed internal reaction, the moment $\vec{\mathfrak{M}}$ of which is equal to $\vec{\mathfrak{M}}_g$, instead of by gravity. The apparent change of an axis of permanent rotation subject to $\vec{\mathfrak{M}}_g$ from stable to unstable has its explanation in the fact that in the case of STAUDE-MLODZEEVSKII permanent rotations the moment of gravity $\vec{\mathfrak{M}}_g$ is body-fixed, while in the case of a perturbed permanent rotation the vector $\vec{\mathfrak{M}}_g$ becomes of variable direction within the body (as a consequence of a displacement of the mass center G with respect to O). However, a torque vector which by its nature is of fixed direction within the body, in general, also remains fixed after the disturbance.

13.12. *Stabilization of unstable time-independent rotations*

Let us note that the stability of the STAUDE-MLODZEEVSKII axes of permanent rotations was achieved by virtue of the fact that the moment $\vec{\mathfrak{M}}_g$ due to gravity acted as a stabilizing factor. Hence, in order to stabilize the unstable permanent rotations of a self-excited asymmetric rigid body, the moment $\vec{\mathfrak{M}}$ must be correspondingly regulated. Let us also note that if such a regulation is possible, then the moment $\vec{\mathfrak{M}}$ cannot be any longer a constant vector relative to the body-fixed coordinate trihedral.

Since the rotation of a rigid body can be detected by devices belonging to the body, the components M_x, M_y, M_z of a regulating moment $\vec{\mathfrak{M}}$ must depend upon the components p, q, r of $\vec{\omega}$. The simplest dependence we can think of is

$$\left.\begin{aligned} M_x &= M_x^0 - m_x\, p \\ M_y &= M_y^0 - m_y\, q \\ M_z &= M_z^0 - m_z\, r \end{aligned}\right\} \tag{13.19}$$

where M_x^0, M_y^0, M_z^0; m_x, m_y, m_z are constants. Then by virtue of (13.1) and (13.19) the components a, b, c of the corresponding permanent rotation vector $\vec{\omega}$ must satisfy the equations

$$\left.\begin{aligned} M_x^0 &= m_x\, a - (B - C)\, b\, c \\ M_y^0 &= m_y\, b + (A - C)\, c\, a \\ M_z^0 &= m_z\, c - (A - B)\, a\, b \end{aligned}\right\} \tag{13.20}$$

We cannot solve these equations for a, b, c; however, given a, b, c and a set of regulating constants m_x, m_y, m_z, we can calculate the "principal components" M_x^0, M_y^0, M_z^0 of $\vec{\mathfrak{M}}$.

Substituting the perturbed motion (13.11) into system (13.1), where M_x, M_y, M_z are given by (13.19), observing (13.20) and neglecting the products of the small quantities ε_i $(i = 1, 2, 3)$, we obtain for the consistency of the linear system so obtained an equation of the form (13.12), the coefficients of which have the expressions

$$\left.\begin{aligned} \alpha_0 &= A\,B\,C \\ \alpha_1 &= B\,C\,m_x + C\,A\,m_y + A\,B\,m_z \\ \alpha_2 &= A\,(A - B)\,(A - C)\,a^2 - B\,(A - B)\,(B - C)\,b^2 + \\ &\quad + C\,(A - C)\,(B - C)\,c^2 + A\,m_y\,m_z + B\,m_z\,m_x + \\ &\quad + C\,m_x\,m_y \\ \alpha_3 &= 2\,(A - B)\,(A - C)\,(B - C)\,a\,b\,c + m_x\,m_y\,m_z + \\ &\quad + (A - B)\,(A - C)\,a^2\,m_x - \\ &\quad - (A - B)\,(B - C)\,b^2\,m_y + (A - C)\,(B - C)\,c^2\,m_z \end{aligned}\right\} \tag{13.21}$$

Since by the HERMITE-HURWITZ stability criterion the coefficients α_0, α_1, α_3 must be positive, let us take for the regulating constants m_x, m_z

some positive values and for m_y the value zero. Then

$$\begin{aligned}\alpha_1\alpha_2-\alpha_0\alpha_3 &= B(A-C)\times\\ &\times[(A-B)A^2a^2m_z-2(A-B)(B-C)A\,C\,a\,b\,c+(B-C)C^2c^2m_x]+\\ &+B^2[m_x m_z-(A-B)(B-C)b^2](C\,m_x+A\,m_z)\end{aligned} \tag{13.22}$$

For the stability of rotations for which $b \neq 0$ and at least one of the components a and c is also different from zero, it is sufficient to select the regulating components m_x and m_z as follows

$$m_x=(A-B)\,|b|,\qquad m_y=0,\qquad m_z=(B-C)\,|b| \tag{13.23}$$

In fact, for $b \gtrless 0$ we have

$$\left.\begin{aligned}\alpha_0 &= A\,B\,C>0\\ \alpha_1 &= B^2(A-C)\,|b|>0\\ \alpha_3 &= (A-C)\,[(A-B)\,a\pm(B-C)\,c]^2\,|b|>0\\ \alpha_1\alpha_2-\alpha_0\alpha_3 &= B(A-B)(A-C)(B-C)(A\,a\mp C\,c)^2\,|b|>0\end{aligned}\right\} \tag{13.24}$$

Hence the stability conditions (13.18) are satisfied and therefore all axes of rotations for which $a \neq 0$, $b \neq 0$, $c \neq 0$ as well as those for which one of the components a or c vanishes are now stable; stable also are those axes which lie in the three principal planes and previously were found to be parastable.

Since the axis of the largest moment of inertia ($b = c = 0$) and that of the smallest moment of inertia ($a = b = 0$) are stable, we are left only with the problem of stabilizing the axis of the middle moment of inertia ($a = c = 0$, $b \neq 0$) for which $M_x^0 = M_z^0 = 0$ and $M_y^0 = b\,m_y$. This is so since we have the case of a permanent rotation with $\vec{\mathfrak{M}} = 0$. For $a = c = 0$ it follows from (13.21) that

$$\begin{aligned}\alpha_3 &= [m_x m_z-(A-B)(B-C)b^2]\,m_y\\ \alpha_1\alpha_2-\alpha_0\alpha_3 &= B^2[m_x m_z-(A-B)(B-C)b^2](C\,m_x+A\,m_z)+\\ &\quad+2A\,B\,C\,m_x m_y m_z+A\,C\,m_y^2(C\,m_x+A\,m_z)+\\ &\quad+B\,m_y(C^2m_x^2+A^2m_z^2)\end{aligned} \tag{13.25}$$

which for

$$m_x>(A-B)\,|b|,\qquad m_y>0,\qquad m_z=(B-C)\,|b| \tag{13.26}$$

satisfy the stability conditions (13.18). Hence the middle principal inertia axis also has become stable.

It is easy to verify that in all earlier cases under conditions (13.26) the stability is as well secured as under conditions (13.23). Hence if the regulating constants m_x, m_y, m_z satisfy conditions (13.26), then all permanent rotations of a self-excited rigid body can be made stable.

Finally let us note that stabilization can also be achieved under a much broader assumption than (13.19). Namely, assuming that

$$M_x = F_1(p), \quad M_y = F_2(q), \quad M_z = F_3(r)$$

where F_i $(i = 1, 2, 3)$ are given functions which are differentiable for $p = a$, $q = b$, $r = c$ respectively, and that a, b, c satisfy a system analogous to (13.20), the regulating constants become $-F_1'(a)$, $-F_2'(b)$, $-F_3'(c)$.

13.21. *Time-dependent rotations about fixed axes caused by self-excitements with fixed axes*

In situations where it is necessary to regulate the rotation of a rigid body, rotations generated by a self-excitement with a constant direction in the body are of importance. In the preceding section we have considered the case of time-independent rotations. Let us now turn to the more general case of time-dependent rotations.

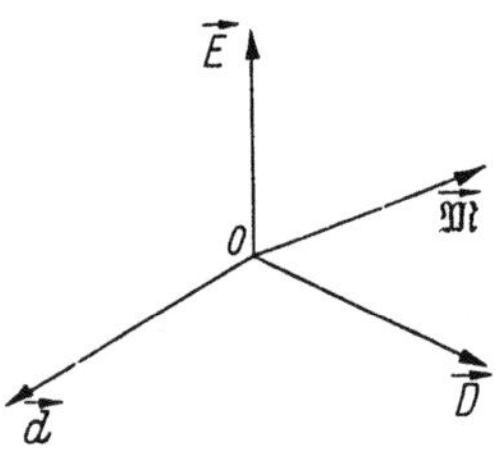

Fig. 31

Each time-dependent rotation $\vec{\omega}$ about an axis fixed in the body and in space can be written in the form

$$p = \alpha_1 g(t), \quad q = \alpha_2 g(t), \quad r = \alpha_3 g(t) \qquad (13.27)$$

where $g(t)$ is an arbitrary continuously differentiable function not identically zero, and α_1, α_2, α_3 are constants, at least one of which is not zero. Introduce the two vectors

$$\vec{d} = \vec{d}(\alpha_1, \alpha_2, \alpha_3), \quad \vec{D} = \vec{D}(A\,\alpha_1, B\,\alpha_2, C\,\alpha_3) \qquad (13.28)$$

which are fixed in the moving trihedral, and let $\vec{E} = \vec{d} \times \vec{D}$ (see Fig. 31). Then the angular momentum $\vec{I}$ of the body is given by

$$\vec{I} = A\,\alpha_1\, g(t)\,\vec{i} + B\,\alpha_2\, g(t)\,\vec{j} + C\,\alpha_3\, g(t)\,\vec{k}$$

and the Euler equations assume the vector form

$$\dot{g}(t)\,\vec{D} + g^2(t)\,\vec{E} = \vec{\mathfrak{M}} \qquad (13.29)$$

Since $\vec{E}$ is perpendicular to $\vec{D}$, it follows that, if $\vec{\mathfrak{M}}$ is time-independent and fixed in direction within the body, the above vector equation can be satisfied if and only if $\vec{E} = 0$, or equivalently the conditions

$$\alpha_1 : \alpha_2 : \alpha_3 = A\,\alpha_1 : B\,\alpha_2 : C\,\alpha_3 \tag{13.30}$$

must hold.

For a symmetric rigid body the above conditions can easily be met. However, for an asymmetric body, this is possible if an only if two of the constants α_1, α_2, α_3 are zero. In other words, the vector $\vec{d}$ must lie along a principal axis. Hence we have proved the following theorem due to R. GRAMMEL [7].

Theorem: There does not exist a time-independent torque vector, fixed along an axis in the body other than one of the principal axes, which can generate or annihilate a time-dependent rotation of a rigid body about an axis fixed in the body as well as in space, or which can even influence the velocity of rotation.

Assume now that $\vec{\mathfrak{M}}$ is time-dependent with constant direction and the body is not symmetric. Further, suppose that under such conditions a time-dependent rotation about a fixed axis other than a principal axis exists. Then it follows from (13.29) that

$$\dot{g}(t) = \varrho\, g^2(t) \quad \left[\text{if} \quad \vec{\mathfrak{M}} = \vec{M}_0\, g^2(t),\ \vec{M}_0 = \text{const.}\right] \tag{13.31}$$

where ϱ is a constant determined by the direction numbers of $\vec{\mathfrak{M}}$ and the fixed axis. We may assume $\varrho \neq 0$; otherwise $g(t) = \text{constant}$. The solution of (13.31) is given by

$$\begin{aligned} g^{-1}(t) &= g_0^{-1} - \varrho\, t \quad &\text{if} \quad g_0 \neq 0 \\ g(t) &= 0 \quad &\text{if} \quad g_0 = 0 \end{aligned}$$

where g_0 is a constant. Hence the angular velocity components of the time-dependent rotation must be

$$p = \alpha_1 (g_0^{-1} - \varrho\, t)^{-1}, \quad q = \alpha_2 (g_0^{-1} - \varrho\, t)^{-1}, \quad r = \alpha_3 (g_0^{-1} - \varrho\, t)^{-1} \tag{13.32}$$

and the corresponding components of the time-dependent torque vector $\vec{\mathfrak{M}}$ are

$$\left.\begin{aligned} M_x &= (g_0^{-1} - \varrho\, t)^{-2}\,[A\,\varrho\,\alpha_1 - (B - C)\,\alpha_2\,\alpha_3] \\ M_y &= (g_0^{-1} - \varrho\, t)^{-2}\,[B\,\varrho\,\alpha_2 - (C - A)\,\alpha_3\,\alpha_1] \\ M_z &= (g_0^{-1} - \varrho\, t)^{-2}\,[C\,\varrho\,\alpha_3 - (A - B)\,\alpha_1\,\alpha_2] \end{aligned}\right\} \tag{13.33}$$

On the other hand, for any values of ϱ and g_0 (13.32) is a solution of EULER's equations with M_x, M_y, and M_z given by (13.33). Thus we see that equations (13.32) and (13.33) characterize the time-dependent rotations about fixed axes in an asymmetric rigid body subject to time-dependent self-excitements with constant directions in the body.

We may note that the moments (13.33) can never put a rigid body into motion starting from rest or bring an existing motion to rest since the rest position as seen from (13.32) corresponds to $t = \pm\infty$.

13.22. Time-dependent rotations about fixed axes caused by self-excitements with variable axes

In view of the negative results obtained in the preceding section, we see that if we disregard the cases of rotations about one of the principal axes, a rotation about an axis fixed in the body as well as in space can be generated or annihilated or in general regulated differently than that given by (13.32) only by a self-excitement with a variable axis. According to (13.29) the torque vector $\vec{\mathfrak{M}}$ does not move arbitrarily in the body but is restricted to move in the plane which is fixed in the body and determined by the vectors $\vec{d}$ and $\vec{D}$ given by (13.28).

The component $\dot{g}(t)\,\vec{D}$ of $\vec{\mathfrak{M}}$ has to regulate the velocity of rotation, while the other component $g^2(t)\,\vec{E}$ has to compensate the moment of the flight force generated by the accelerated rotation of the body, thus keeping the axis of rotation fixed in its position.

If in particular the rotation has to start from rest and become uniformly accelerated, i.e. if $g(t) = t$, then the component $\dot{g}(t)\,\vec{D}$ becomes the constant vector $\vec{D}$, while the component $g^2(t)\,\vec{E} = t^2\,\vec{E}$ increases with constant acceleration. The torque vector $\vec{\mathfrak{M}}$ then approaches asymptotically the vector $\vec{E}$, which is orthogonal to the axis of rotation.

Chapter III

Externally excited rigid body

§ 14. Symmetric rigid body subject to a periodic torque

14. 1. Statement of the problem. W. Braunbek [1] considers the motion of a symmetric rigid body suspended at the center of mass G and subject to a torque

$$\vec{\mathfrak{M}} = \mu\left[\vec{a} \times \left(\vec{H}_0 + \vec{H}_1\right)\right]$$

generated as follows. A magnetic bar with magnetic moment μ is situated along the axis of symmetry of the body, subject to a homogeneous magnetic field $\vec{H} = \vec{H}_0 + \vec{H}_1$, where $\vec{H}_0$ is a constant field and $\vec{H}_1$ is an alternating field which varies periodically with respect to the time t. The unit vector $\vec{a}$ lies in the direction of the symmetry axis of the body. Subject to the constant field $\vec{H}_0$ alone, the motion of the body is identical to that of a body under the influence of gravity. BRAUNBEK studies two cases: (i) the alternating field $\vec{H}_1$ is parallel to the constant field $\vec{H}_0$, and (ii) $\vec{H}_1$ is orthogonal to $\vec{H}_0$.

14.11. Alternating field parallel to the constant field

If the direction of the Z-axis is chosen parallel to that of $\vec{H}_0$ and $\vec{H}_1$, then

$$H_z = H_0 + H_1 \cos\omega t = H_0(1 + \alpha \cos\omega t)$$

where $\alpha = H_1/H_0$ and ω is the frequency of the field $\vec{H}_1$ (Fig. 32). The symmetry axis of the body is subject to the torque (positive if it tends to increase Θ)

$$M = -\mu H_0(1 + \alpha \cos\omega t) \sin\Theta$$

perpendicular to the Z, a-plane, and the EULER angles satisfy the nonlinear differential equations

$$\ddot{\Phi} + \ddot{\psi} \cos\Theta - \dot{\psi}\,\dot{\Theta} \sin\Theta = 0 \tag{14.1}$$

$$A\,\ddot{\psi} \sin\Theta - C\,\dot{\Phi}\,\dot{\Theta} - (C - 2A)\,\dot{\psi}\,\dot{\Theta} \cos\Theta = 0 \tag{14.2}$$

$$A\,\ddot{\Theta} + C\,\dot{\Phi}\,\dot{\psi} \sin\Theta + (C - A)\,\dot{\psi}^2 \sin\Theta \cos\Theta = -\mu H_0(1 + \alpha \cos\omega t) \sin\Theta \tag{14.3}$$

which for $\alpha = 0$ reduce to the equations of motion of a heavy symmetric body. The general method known to represent the EULER angles in terms of elliptic functions, however, does not apply in the above case because of the presence of the term in α. On the other hand several methods known from the theory of the motion of a heavy rigid body can be extended to our case. Such methods are: (i) the method of superimposing small oscillations over the regular precession in the field $\vec{H}_0$, (ii) the method of setting up a single differential equation for Θ alone, (iii) the method of small angles, i.e. discussion of motions for which the axis of symmetry of the body devi-

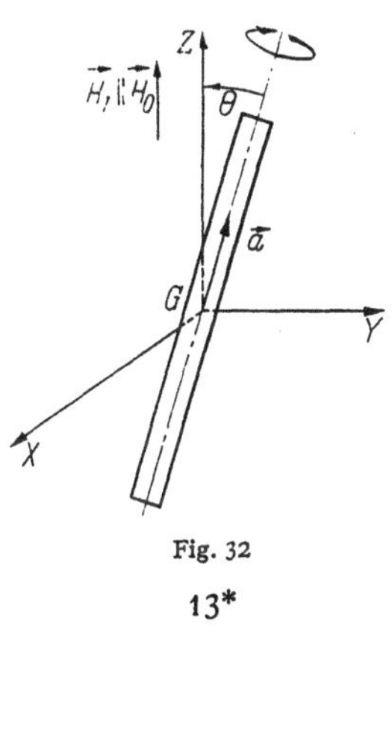

Fig. 32

ates slightly from the field axis, (iv) the method of reducing to the MATHIEU differential equation. We shall consider only the first method, i.e.

Superposition of small oscillations over the regular precession in $\vec{H}_0$. For $\alpha = 0$ a particular solution representing the regular precession in the field $\vec{H}_0$ is given by the equations

$$\Phi = \omega_1 t, \quad \psi = \omega_2 t, \quad \Theta = \Theta_0 = \text{const.} \tag{14.4}$$

where

$$C\,\omega_1\,\omega_2 + (C - A)\,\omega_2^2 \cos\Theta_0 = -\mu\,H_0 \tag{14.5}$$

ω_1 is the frequency of rotation of the body about its axis of symmetry, and ω_2 is the frequency of the free, regular precession in the field $\vec{H}_0$.

For $\alpha \neq 0$ but $\alpha \ll 1$ let us look for a neighbouring solution of (14.4) in the form

$$\left.\begin{aligned} \Phi &= \omega_1 t + a\,e^{i\omega_3 t} + a'\,e^{i\omega t} \\ \psi &= \omega_2 t + b\,e^{i\omega_3 t} + b'\,e^{i\omega t} \\ \Theta &= \Theta_0 + c\,e^{i\omega_3 t} + c'\,e^{i\omega t} \end{aligned}\right\} \tag{14.6}$$

where the second terms on the right represent "free nutation" with frequency ω_3 and the third terms "forced nutation" with frequency ω. In addition, assume that

$$\begin{aligned} \omega_3\,|a| &\ll \omega_1, & \omega\,|a'| &\ll \omega_1 \\ \omega_3\,|b| &\ll \omega_2, & \omega\,|b'| &\ll \omega_2 \\ c &\ll \Theta_0, & c' &\ll \Theta_0 \end{aligned}$$

Substituting (14.6) into equations (14.1) to (14.3) and omitting quadratic and higher order terms with respect to the small quantities, we obtain three homogeneous linear equations for a, b, c, and two homogeneous and one nonhomogeneous linear equations for a', b', c'. From the condition of consistency of the system of homogeneous equations we obtain for the frequency of the free nutation the expression

$$\omega_3 = \frac{1}{A}\sqrt{[C\,\omega_1 + (C - 2A)\,\omega_2 \cos\Theta_0]^2 + A^2\,\omega_2^2 \sin^2\Theta_0} \tag{14.7}$$

Further, the solution of the homogeneous system is given by

$$\begin{aligned} \frac{b}{c} &= -i\,\frac{C\,\omega_1 + (C - 2A)\,\omega_2 \cos\Theta_0}{A\,\omega_3 \sin\Theta_0} \\ \frac{a}{c} &= i\,\frac{C\,\omega_1 \cos\Theta_0 + C\,\omega_2 \cos^2\Theta_0 - A\,\omega_2(1 + \cos^2\Theta_0)}{A\,\omega_3 \sin\Theta_0} \end{aligned} \tag{14.8}$$

and that of the nonhomogeneous system is

$$\left.\begin{aligned} a' &= i\frac{\mu H_1}{A^2\omega(\omega^2-\omega_3^2)}[C\,\omega_1\cos\Theta_0 + C\,\omega_2\cos^2\Theta_0 - \\ &\quad - A\,\omega_2(1+\cos^2\Theta_0)] \\ b' &= -i\frac{\mu H_1}{A^2\omega(\omega^2-\omega_3^2)}[C\,\omega_1 + (C-2A)\,\omega_2\cos\Theta_0] \\ c' &= \frac{\mu H_1 \sin\Theta_0}{A(\omega^2-\omega_3^2)} \end{aligned}\right\} \quad (14.9)$$

Hence the motion is represented by the superposition of the regular precession, free nutation and forced nutation, the free nutation being known from the theory of the motion of a heavy symmetric rigid body. As $\omega \to \omega_3$ the amplitude of the forced nutation approaches infinity, i.e. for $\omega = \omega_3$ resonance occurs.

Starting from the "slow precession", i.e. $\omega_2 \ll \omega_1$ and approaching the limit case of the "fast rotating body" we obtain

$$\left.\begin{aligned} &\omega_3 \approx \frac{C}{A}\omega_1 \\ &\frac{a}{c} \approx i\cot\Theta_0, \qquad \frac{a'}{c'} \approx i\frac{\omega_3}{\omega}\cot\Theta_0 \\ &\frac{b}{c} \approx -\frac{i}{\sin\Theta_0}, \qquad \frac{b'}{c'} \approx -i\frac{\omega_3}{\omega}\frac{1}{\sin\Theta_0} \end{aligned}\right\} \quad (14.10)$$

The formulas obtained differ from those for the forced nutation of the axis of the Earth for two reasons. First, the Earth axis is subject to two moment vectors $\vec{M}_\Theta$ and $\vec{M}_\psi$ which are both of the same order of magnitude. Second, due to the fact that in the case of Earth's nutation we are far away from the resonance case ($\omega_3 \approx 2\pi$ days, $\omega = 2\pi \cdot 19$ years), the calculations are different from the beginning. As a consequence the coefficients b' and c' in the case of Earth's nutation can be of the same order of magnitude, contrary to formulas (14.10).

14.12. Alternating field orthogonal to the constant field

Consider the case of a "fast rotating body" for which $\dot{\psi} \ll \dot{\Phi}$, and hence the axis of the angular momentum $\vec{I}$ and that of the body almost coincide. For $\omega_2 \ll \omega_1$ the modulus of the angular momentum $\vec{I}$ assumes the form

$$I \approx C\,\omega_1$$

and by equation (14.5) we have that

$$\omega_2 \approx -\frac{\mu H_0}{I}$$

Hence the moment $\vec{\mathfrak{M}} = \mu(\vec{a} \times \vec{H})$ of the field $\vec{H}$ about the center of mass G can be written in the form

$$\vec{\mathfrak{M}} = \frac{\mu}{I}\left(\vec{I} \times \vec{H}\right) = -\frac{\omega_2}{H_0}\left(\vec{I} \times \vec{H}\right)$$

Thus the basic equation of the motion of the body $\dot{\vec{I}} = \vec{\mathfrak{M}}$ reduces to the linear homogeneous vector differential equation for $\vec{I}$,

$$\dot{\vec{I}} = -\frac{\omega_2}{H_0}\left(\vec{I} \times \vec{H}\right) \tag{14.11}$$

Since $\dot{\vec{I}}$ is orthogonal to $\vec{I}$, the magnitude of the angular momentum $\vec{I}$ remains constant.

In the general case the endpoint of the vector $\vec{H}_1$, the latter being a harmonic function of frequency ω, describes an ellipse (elliptically polarized alternating field). If the X- and Y-axes are chosen along the axes of this ellipse (Z-axis being parallel to $\vec{H}_0$), then the field $\vec{H}$ can be represented in the form

$$\begin{aligned} H_X &= H_1 \cos\omega t \qquad & H_1 &= \alpha H_0 \\ H_Y &= H_1' \sin\omega t \qquad & H_1' &= \alpha' H_0 \\ H_Z &= H_0 \end{aligned}$$

In such a case the vector differential equation (14.11) is equivalent to the following three scalar equations

$$\left.\begin{aligned} \dot{I}_X &= \omega_2(-I_Y + \alpha' I_Z \sin\omega t) \\ \dot{I}_Y &= \omega_2(I_X - \alpha I_Z \cos\omega t) \\ \dot{I}_Z &= \omega_2(-\alpha' I_X \sin\omega t + \alpha I_Y \cos\omega t) \end{aligned}\right\} \tag{14.12}$$

Since I is constant, then

$$I^2 = I_X^2 + I_Y^2 + I_Z^2 = \text{const.} \tag{14.12'}$$

Formally, this equation is obtained by multiplying the above equations by I_X, I_Y, I_Z respectively and adding.

(i) *Circularly polarized alternating field:* $\alpha' = \alpha$. This is the simplest case in which system (14.12) can be completely solved in terms of three arbitrary constants. For this purpose introduce a coordinate trihedral $G\,x\,y\,z$ which rotates with angular velocity ω about the Z-axis and which for $t = 0$ coincides with the fixed coordinate trihedral $G\,X\,Y\,Z$. Then system (14.12) reduces to the system with constant coefficients,

$$\dot{I}_x = (\omega - \omega_2)\, I_y, \quad \dot{I}_y = -(\omega - \omega_2)\, I_x - \alpha\,\omega_2\, I_z, \quad \dot{I}_z = \alpha\,\omega_2\, I_y \tag{14.13}$$

the general solution of which is

$$\left.\begin{aligned} I_x &= C_1 \frac{\omega-\omega_2}{\lambda}\sin\lambda t - C_2 \frac{\omega-\omega_2}{\lambda}\cos\lambda t + C_3 \frac{\alpha\,\omega_2}{\lambda} \\ I_y &= C_1\cos\lambda t + C_2 \sin\lambda t \\ I_z &= C_1 \frac{\alpha\,\omega_2}{\lambda}\sin\lambda t - C_2 \frac{\alpha\,\omega_2}{\lambda}\cos\lambda t - C_3 \frac{\omega-\omega_2}{\lambda} \end{aligned}\right\} \tag{14.14}$$

where

$$\lambda^2 = (\omega-\omega_2)^2 + \alpha^2\,\omega_2^2 \tag{14.14'}$$

The constants of integration C_1, C_2, C_3 can be expressed in terms of the initial values I_{X0}, I_{Y0}, I_{Z0}.

Relative to the fixed coordinate trihedral the solution is

$$\left.\begin{aligned} I_X = {} & C_1\left(\frac{\omega-\omega_2}{\lambda}\cos\omega t\sin\lambda t - \sin\omega t\cos\lambda t\right) - \\ & - C_2\left(\frac{\omega-\omega_2}{\lambda}\cos\omega t\cos\lambda t + \sin\omega t\sin\lambda t\right) + \\ & + C_3 \frac{\alpha\,\omega_2}{\lambda}\cos\omega t \\ I_Y = {} & C_1\left(\cos\omega t\cos\lambda t + \frac{\omega-\omega_2}{\lambda}\sin\omega t\sin\lambda t\right) + \\ & + C_2\left(\cos\omega t\sin\lambda t - \frac{\omega-\omega_2}{\lambda}\sin\omega t\cos\lambda t\right) + \\ & + C_3\frac{\alpha\,\omega_2}{\lambda}\sin\omega t \\ I_Z = {} & C_1\frac{\alpha\,\omega_2}{\lambda}\sin\lambda t - C_2\frac{\alpha\,\omega_2}{\lambda}\cos\lambda t - C_3\frac{\omega-\omega_2}{\lambda} \end{aligned}\right\} \tag{14.15}$$

From the above expressions the following particular cases result.

(a) If $\alpha = 0$ and thus $\lambda = \omega - \omega_2$, we have free regular precession in the field $\vec{H}_0$, given by the equations

$$\left.\begin{aligned} I_X &= -C_1\sin\omega_2 t - C_2\cos\omega_2 t \\ I_Y &= C_1\cos\omega_2 t - C_2\sin\omega_2 t \\ I_Z &= -C_3 \end{aligned}\right\} \tag{14.16}$$

(b) If $\alpha \ll 1$ (quadratic and higher order terms in α being neglected) we have a weak alternating field $\vec{H}_1$:

$$\left.\begin{aligned} I_X &= -C_1\sin\omega_2 t - C_2\cos\omega_2 t + \alpha\frac{\omega_2}{\omega-\omega_2}C_3\cos\omega t \\ I_Y &= C_1\cos\omega_2 t - C_2\sin\omega_2 t + \alpha\frac{\omega_2}{\omega-\omega_2}C_3\sin\omega t \\ I_Z &= -C_3 + \alpha\frac{\omega_2}{\omega-\omega_2}\times \\ &\quad\times[-C_2\cos(\omega-\omega_2)t + C_1\sin(\omega-\omega_2)t] \end{aligned}\right\} \tag{14.17}$$

In the case $C_1 = C_2 = 0$ the resulting motion can be interpreted as forced precession, given by the equations

$$I_X = \alpha \frac{\omega_2}{\omega - \omega_2} C_3 \cos\omega t, \qquad I_Y = \alpha \frac{\omega_2}{\omega - \omega_2} C_3 \sin\omega t, \qquad I_Z = -C_3 \tag{14.18}$$

It is easy to verify that (14.18) is a solution of system (14.12) for $\alpha = \alpha'$. The constant of integration C_3 can be calculated from equation (14.12′) in terms of the magnitude I of the angular momentum.

The solution (14.17) for I_X and I_Y can be interpreted as the result of superposition of free regular precession (14.16) with frequency ω_2 and forced precession (14.18) with frequency ω. For $\omega = \omega_2$ the amplitude of the forced precession becomes infinite (the approximation procedure used fails), and resonance with the frequency of the free precession occurs. This phenomenon has been widely studied in the theory of a nucleus in a high-frequency magnetic field.

In order to visualize the general solution (14.14) geometrically, let us introduce the complex quantity $L = C_2 + i C_1$. Then (14.14) assumes the form

$$\left.\begin{aligned} I_x &= -\frac{\omega - \omega_2}{\lambda} \operatorname{Re}(L e^{i\lambda t}) + C_3 \frac{\alpha \omega_2}{\lambda} \\ I_y &= \operatorname{Re}(-i L e^{i\lambda t}) \\ I_z &= -\frac{\alpha \omega_2}{\lambda} \operatorname{Re}(L e^{i\lambda t}) - C_3 \frac{\omega - \omega_2}{\lambda} \end{aligned}\right\} \tag{14.19}$$

or, in vector form,

$$\vec{I} = \vec{I}_2 + \vec{I}_1$$

where

$$\vec{I}_2 = \vec{I}_2 \left[-\frac{\omega - \omega_2}{\lambda} \operatorname{Re}(L e^{i\lambda t}),\ \operatorname{Re}(-i L e^{i\lambda t}),\ -\frac{\alpha \omega_2}{\lambda} \operatorname{Re}(L e^{i\lambda t}) \right]$$

and

$$\vec{I}_1 = \vec{I}_1 \left(C_3 \frac{\alpha \omega_2}{\lambda},\ 0,\ -C_3 \frac{\omega - \omega_2}{\lambda} \right)$$

System (14.19) shows that $\vec{I}_2$ is orthogonal to $\vec{I}_1$, and its magnitude by (14.14′) is constant and equal to $|L|$. The endpoint of $\vec{I}$ moves in a plane perpendicular to the fixed vector $\vec{I}_1$, describing in this plane a circle of radius $|L|$ about the endpoint of $\vec{I}_1$ as the center. The vector $\vec{I}$ itself, being the axis of symmetry of the body in our approximation, describes a corresponding circular cone (Fig. 33).

In order to obtain the motion in the space-fixed coordinate trihedral $G X Y Z$, the above system must be subjected to a rotation about the z-axis with angular velocity ω.

The angle Θ, made by the vector $\vec{I}$ with the Z-axis, varies between two limits Θ_1 and Θ_2 which depend upon the position of the $\vec{I}$-cone relative to the Z-axis. The resonance case $\omega = \omega_2$ is characterized by the fact that in the rotating coordinate trihedral $G x y z$ the fixed vector $\vec{I}_1$ coincides with the x-axis while the vector $\vec{I}_2$ rotates in a plane

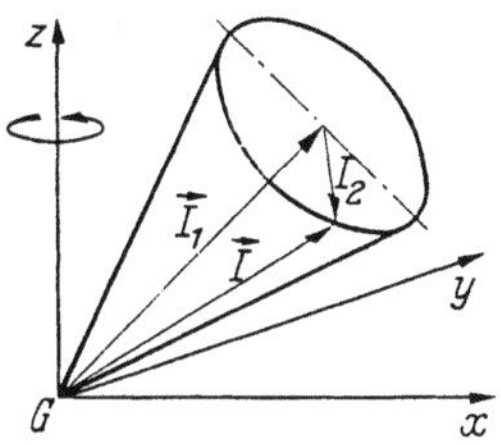

Fig. 33

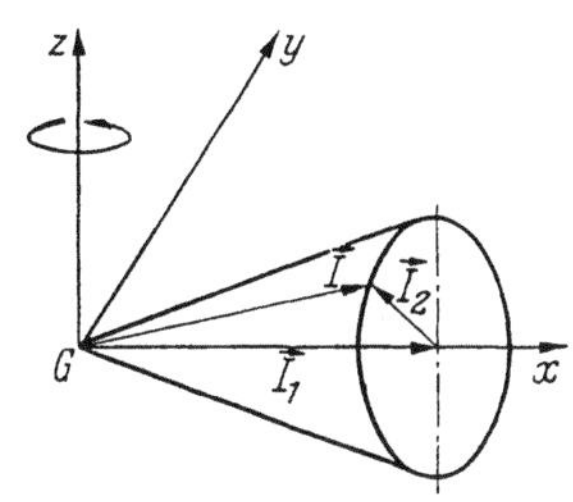

Fig. 34

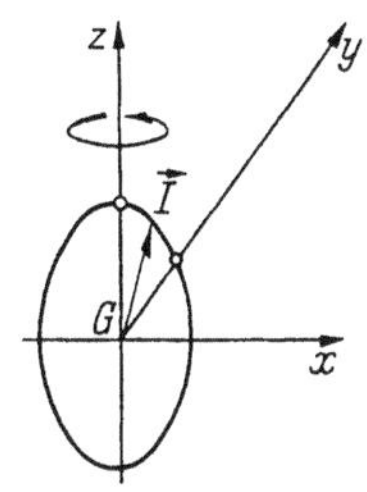

Fig. 35

parallel to the $y z$-plane, and hence $\Theta_2 = \pi - \Theta_1$ (Fig. 34). If in the resonance case the vector $\vec{I}$, being the axis of symmetry of the body, at a certain instant, say $t = 0$, lies in the $y z$-plane, then $C_3 = 0$, and the I-cone degenerates into the $y z$-plane (Fig. 35). This is the only case in which the angle Θ assumes all the values between 0 and π, and the endpoint of the vector $\vec{I}$, relative to the space-fixed system $G X Y Z$, describes a spiral which extends over the whole unit sphere.

The simplest case results when in system (14.19) the complex amplitude L is equal to zero, and hence $\vec{I}_2 = 0$. Then, relative to the $G x y z$-trihedral, the direction of $\vec{H}_0$ (Z-axis), the axis of symmetry of the body ($\vec{I} = \vec{I}_1$) and the direction of $\vec{H}_1$ (x-axis) all lie in the same plane ($x z$-plane), which relative to space rotates with angular velocity ω about the Z-axis. This case corresponds to F. Bloch's [1] solution of system (14.12) for $\alpha' = \alpha$ and represents forced precession of the rigid body.

(ii) *Linearly and elliptically polarized alternating fields.* The case of linearly polarized alternating field ($\alpha' = 0$) and that of an elliptically polarized field ($\alpha \neq \alpha' \neq 0$) are much more difficult than the circularly polarized alternating field ($\alpha' = \alpha$). Braunbek solved these two cases (for $\alpha \ll 1$ and $\alpha \ll 1$, $\alpha' \ll 1$) by the method of successive approximations, expanding I_X, I_Y, I_Z for small values of the parameters α

and α, α' respectively in powers of α and α, α'. Convergence of the series thus obtained was not investigated.

14.2. Symmetric rigid body subject to gravitation and an orthogonal sinusoidal periodic force. Consider a symmetric rigid body which rotates with constant angular velocity about its axis and is subject to a constant gravitation field in the direction of the negative

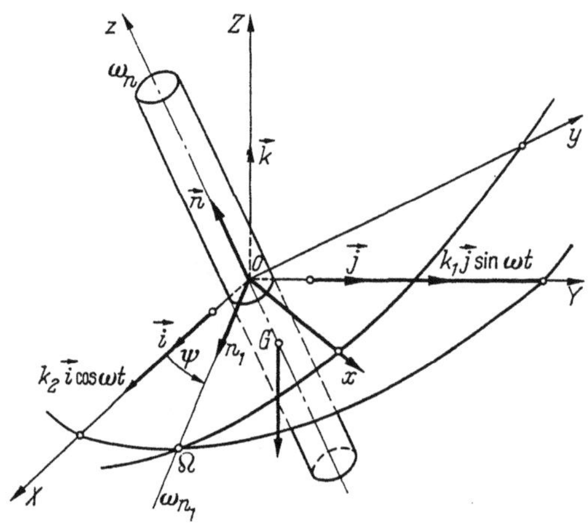

Fig. 36

Z-axis, and to a force having the components $k_2 \cos\omega\, t$, $k_1 \sin\omega\, t$, 0 in the X, Y, Z-directions respectively.

In terms of the principal moments of inertia, I_n, I_{n_1} along the axis of symmetry of the body and the perpendicular to the $Z\,z$-plane respectively, the angular momentum vector $\vec{I}$ can be expressed in the form

$$\vec{I} = \omega_n I_n \vec{n} + \omega_{n_1} I_{n_1} \vec{n}_1$$

where ω_n, ω_{n_1} and $\vec{n}$, $\vec{n}_1$ are the corresponding angular velocity components and unit vectors (Fig. 36).

Since

$$\frac{d\vec{n}}{dt} = \vec{\Omega} \times \vec{n}$$

$$\vec{\Omega} = \omega_n \vec{n} + \omega_{n_1} \vec{n}_1$$

then

$$\vec{n} \times \frac{d\vec{n}}{dt} = \vec{n} \times \left(\vec{\Omega} \times \vec{n}\right) = \vec{\Omega} - \omega_n \vec{n} = \omega_{n_1} \vec{n}_1$$

Hence

$$\vec{I} = \omega_n I_n \vec{n} + I_{n_1} \left(\vec{n} \times \frac{d\vec{n}}{dt}\right) \tag{14.20}$$

In addition to gravity, $-m g \vec{k}$, the rigid body is subject to a sinusoidal force of the form

$$k_1 \vec{j} \sin\omega t \pm k_2 \vec{i} \cos\omega t$$

where the double sign $(\pm)$ indicates the direction of rotation of the force vector about the axis of symmetry of the body. We shall distinguish the following particular cases: (i) linear force (for example, $k_2 = 0$), (ii) circular force $(k_1 = k_2)$, and (iii) elliptical force $(k_1, k_2$ arbitrary).

Let the radius vectors of the points of application of the above forces to the symmetry axis of the body be $-s\vec{n}$, $s_1 \vec{n}$ and $s_2 \vec{n}$ respectively, ($|s|$ being the distance to the mass center G; if G lies below O, s is positive provided that $\vec{n}$ is directed upward.) Then, neglecting friction, by application of the basic equation of dynamics

$$\frac{d\vec{I}}{dt} = \vec{\mathfrak{M}}$$

we obtain that

$$\omega_n I_n \frac{d\vec{n}}{dt} + I_{n_1}\left(\vec{n} \times \frac{d^2\vec{n}}{dt^2}\right) = \pm k_2 s_2 \left(\vec{n} \times \vec{i}\right) \cos\omega t + \\ + k_1 s_1 \left(\vec{n} \times \vec{j}\right) \sin\omega t + m g s \left(\vec{n} \times \vec{k}\right) \qquad (14.21)$$

Dividing both sides of equation (14.21) by I_{n_1} and projecting it on the axes of the coordinate trihedral $O X Y Z$, we obtain for the components n_X, n_Y, n_Z of $\vec{n}$ the following nonlinear system of differential equations

$$\left.\begin{aligned}
\frac{N}{I_{n_1}} \dot{n}_X - n_Z \ddot{n}_Y + n_Y \ddot{n}_Z - \Omega_X^2 n_Y &= -\frac{k_1 s_1}{I_{n_1}} n_Z \sin\omega t \\
\frac{N}{I_{n_1}} \dot{n}_Y + n_Z \ddot{n}_X - n_X \ddot{n}_Z + \Omega_X^2 n_X &= \pm \frac{k_2 s_2}{I_{n_1}} n_Z \cos\omega t \\
\frac{N}{I_{n_1}} \dot{n}_Z + n_X \ddot{n}_Y - n_Y \ddot{n}_X \quad &= \frac{k_1 s_1}{I_{n_1}} n_X \sin\omega t \mp \\
&\mp \frac{k_2 s_2}{I_{n_1}} n_Y \cos\omega t
\end{aligned}\right\} \qquad (14.22)$$

where $N = \omega_n I_n$ is the angular momentum component along the axis of symmetry of the body and $\Omega_X^2 = m g s / I_{n_1}$. If the body oscillates like a compound pendulum about the line of nodes Ω (without rotating), then Ω_X $(s > 0)$ is the circular frequency of these oscillations.

The above system can be considerably simplified if the angle between the axis of symmetry of the body (z-axis) and the Z-axis remains small. In such a case the unit sphere on which the endpoint of the vector $\vec{n}$ moves deviates slightly from the plane $Z = 1$, and therefore $n_Z \approx 1$.

If further, following M. SCHULER [1, 2], the components n_X and n_Y are replaced by the angles of rotation α and β of the symmetry axis of the body about the X- and Y-axis respectively, assuming that in the XZ-plane $\alpha = 0$ and in the YZ-plane $\beta = 0$, then

$$n_X \approx \sin\beta \approx \beta \quad \text{and} \quad n_Y \approx -\sin\alpha \approx -\alpha$$

Substituting these expressions into the system (14.22), we obtain a system of linear differential equations with constant coefficients

$$\begin{aligned} \ddot{\alpha} + \Omega_X^2\,\alpha + \frac{N}{I_{n_1}}\dot{\beta} &= -\frac{k_1 s_1}{I_{n_1}}\sin\omega t \\ \ddot{\beta} + \Omega_X^2\,\beta - \frac{N}{I_{n_1}}\dot{\alpha} &= \pm\frac{k_2 s_2}{I_{n_1}}\cos\omega t \end{aligned} \tag{14.23}$$

The solution of this system for the case of a linear force ($k_2 = 0$) is given by R. WIEBELITZ [1].

Let us now take the friction of the CARDAN suspension into account, assuming that the source of friction is in the bearings and that the friction is proportional to the angular velocity. Then, restricting ourselves to small angles, we must add supplementary terms $\varrho_1\dot{\alpha}$ and $\varrho_2\dot{\beta}$ respectively to the left-hand sides of system (14.23).

In addition, replacement of the moment of inertia I_{n_1} by two different principal moments of inertia I_X and I_Y (I_X, for example, is the moment of inertia of the body about the X-axis for the case where the symmetry axis of the body is restricted to the YZ-plane), results in the replacement of Ω_X by two different quantities Ω_X and Ω_Y in the above equations.

Thus we obtain the equations of motion in the form

$$\begin{aligned} \ddot{\alpha} + \Omega_X^2\alpha + \varrho_1\dot{\alpha} + a_1\dot{\beta} &= -b_1\sin\omega t \\ \ddot{\beta} + \Omega_Y^2\beta + \varrho_2\dot{\beta} - a_2\dot{\alpha} &= \pm b_2\cos\omega t \end{aligned} \tag{14.24}$$

where ϱ_1, ϱ_2; a_1, a_2; b_1, b_2 are certain constants.

WIEBELITZ also gives a physical discussion of the solutions of system (14.24) for the case where the friction is neglected ($\varrho_1 = \varrho_2 = 0$) and the case where friction is taken into account.

In the particular case of a circular periodic force ($k_1 = k_2 = k$, $s_1 = s_2 = s$), the nonlinear system (14.22) can be solved exactly, obtain and we the simple stationary solution

$$\begin{aligned} n_X &= \pm\frac{C_0}{\Omega_X^2 - C_0\omega^2 \pm (N/I_{n_1})\,\omega}\,\frac{k\,s}{I_{n_1}}\cos\omega t \\ n_Y &= \frac{C_0}{\Omega_X^2 - C_0\omega^2 \pm (N/I_{n_1})\,\omega}\,\frac{k\,s}{I_{n_1}}\sin\omega t, \qquad n_Z = C_0 \end{aligned} \tag{14.25}$$

C_0, being a constant of integration, cannot be chosen arbitrarily but must be determined from the condition

$$|\vec{n}|^2 = n_X^2 + n_Y^2 + n_Z^2 = 1$$

which expresses the fact that $\vec{n}$ is a unit vector. In this way we obtain for the determination of C_0 an equation of degree four, namely

$$\left(C_0 \frac{k s}{I_{n_1}}\right)^2 = (1 - C_0^2)\left(\Omega_X^2 - C_0 \omega^2 \pm \frac{N}{I_{n_1}} \omega\right)^2$$

When solved for ω, this equation assumes a simpler form

$$\omega = \pm \frac{N}{2 C_0 I_{n_1}} + \varepsilon_2 \sqrt{\frac{N^2}{4 C_0^2 I_{n_1}^2} + \frac{\Omega_X^2}{C_0} - \varepsilon_1 \frac{k s}{I_{n_1} \sqrt{1 - C_0^2}}} \tag{14.26}$$

where $\varepsilon_1, \varepsilon_2 = \pm 1$. WIEBELITZ has discussed this case in detail.

An approximate calculation can be carried out in the case where a periodic elliptical force is acting on a fast rotating body. In fact, for large values of ω_n the second term in formula (14.20) can be neglected so that

$$\vec{n} \approx \frac{\vec{I}}{\omega_n I_n} = \frac{\vec{I}}{N}$$

Substituting this expression for $\vec{n}$ into system (14.22), multiplying the equations by I_{n_1}, and introducing the notations

$$\frac{k_2 s_2}{m g s} = \bar{\alpha}, \qquad \frac{\Omega_X^2 I_{n_1}}{N} = \omega_2, \qquad \frac{k_1 s_1}{k_2 s_2} = \bar{\beta}$$

this system becomes

$$\left.\begin{aligned} \dot{I}_X - \omega_2 I_Y &= -\bar{\alpha}\bar{\beta}\, \omega_2 I_Z \sin\omega t \\ \dot{I}_Y + \omega_2 I_X &= \pm \bar{\alpha}\, \omega_2 I_Z \cos\omega t \\ \dot{I}_Z \qquad &= \bar{\alpha}\bar{\beta}\, I_X \omega_2 \sin\omega t \mp \bar{\alpha}\, I_Y \omega_2 \cos\omega t \end{aligned}\right\} \tag{14.27}$$

WIEBELITZ solves this system, following BRAUNBEK, by the method of successive approximations, putting

$$I_X = \sum p_n \bar{\alpha}^n, \qquad I_Y = \sum q_n \bar{\alpha}^n, \qquad I_Z = \sum r_n \bar{\alpha}^n \tag{14.28}$$

and obtaining for the determination of p_n, q_n, r_n systems of the form

$$\begin{aligned} \dot{p}_n - \omega_2 q_n &= -\bar{\beta}\, \omega_2 r_{n-1} \sin\omega t \\ \dot{q}_n + \omega_2 p_n &= \pm \omega_2 r_{n-1} \cos\omega t \\ \dot{r}_n \qquad &= \bar{\beta}\, \omega_2 p_{n-1} \sin\omega t \mp \omega_2 q_{n-1} \cos\omega t \qquad (n = 1, 2, \ldots) \end{aligned}$$

For $n = 0$ the right-hand sides vanish and we obtain that

$$p_0 = A_0 \cos(\omega_2 t + \delta), \qquad q_0 = -A_0 \sin(\omega_2 t + \delta), \qquad r_0 = C_0$$

The method of variation of constants, applied to the above homogeneous system, gives the recursion formulas

$$p_n = \omega_2 \int_0^t r_{n-1}[-\bar{\beta} \sin\omega x \cos\omega_2 (t-x) \pm \cos\omega x \sin\omega_2 (t-x)]\, dx + \\ + A_n \cos(\omega_2 t + \delta_n)$$

$$q_n = \omega_2 \int_0^t r_{n-1}[\bar{\beta} \sin\omega x \sin\omega_2 (t-x) \pm \cos\omega x \cos\omega_2 (t-x)]\, dx - \\ - A_n \sin(\omega_2 t + \delta_n)$$

$$r_n = \omega_2 \int_0^t [\bar{\beta}\, p_{n-1} \sin\omega x \mp q_{n-1} \cos\omega x]\, dx + C_n$$

The constants of integration A_n, δ_n and C_n $(n > 0)$ must be selected in such a way that the right-hand sides of the first two formulas do not contain the terms $\begin{smallmatrix}\cos\\ \sin\end{smallmatrix}(\omega_2 + \delta)$ and that the third formula does not contain any constants.

While the convergence of the series in the case of BRAUNBEK's approximations is doubtful, WIEBELITZ was able to show that series (14.28) converge for sufficiently small values of $\bar{\alpha}$ except at the point $\omega = 0$ and for values of ω for which resonance occurs.

14.3. Other types of torques. The motion of a heavy symmetric rigid body subject to a vertical displacement of its point of suspension has been discussed by F. WEIDENHAMMER [1], and N. V. BANKOVSKAYA [1], and the motion of an asymmetric heavy body with a vibrating suspension point by A. M. FEDORČENKO [1, 2].

Papers by G. O. FRIDLENDER [1], G. E. KUZMAK [1], A. YU. IŠLINSKII [8], D. R. MERKIN [1], YU. K. ŽBANOV [1] and YA. N. ROITENBERG [1, 2] have been concerned with the precession of a rigid body under the action of an external moment. The stability of a symmetric rigid body in the precessional motion has been studied by O. BOTTEMA [3]. For the random disturbances of the regular precession of a gyroscope see YU. V. LINNIK and V. S. NOVOSELOV [1].

A method for increasing the stability of a rapidly rotating symmetric rigid body has been given by A. M. FEDORČENKO [3].

V. N. KOŠLYAKOV [2] assumed that (i) the components of the external moment generated in a resisting medium are $-\lambda A p$, $-\lambda B q$, $-\lambda C r$, where λ is a coefficient of proportionality, (ii) the equatorial moments of inertia A and B differ only slightly from each other, say $B > A$, and (iii) $C > A$, where C is the axial moment of inertia, and sought a solution of the EULER equations in the form of power series in terms of the small parameter $\varepsilon = (B - A)/C$. The method of successive approximations was applied in such a way that the zero approximation corresponds to the symmetric case $A = B$.

In the case of a symmetric gyroscope and under the assumption that the components of the external moment are $-\lambda_1 p$, $-\lambda_2 q$, $-\lambda_3 r$, where λ_1, λ_2, λ_3 are certain constant coefficients of proportionality, the EULER equations are integrable in terms of BESSEL functions.

A. N. KRYLOV and YU. A. KRUTKOV [1], and B. V. BULGAKOV [1] have shown that if the resistance is proportional to the first power of the velocity, then the motion of the axis of the gyroscope depends on the ratio C/A. In the cases $C > A$ and $C < A$ the vector of the instantaneous angular velocity of the gyroscope tends to coincide with the axis of the largest principal moment of inertia.

Miss Maria Teresa VACCA [2] has studied the motion of a rigid body about a fixed point when each particle of the body is acted upon by a force which is a certain linear function of the coordinates of the particle referred to the principal axes of inertia of the body. The equations admit two first integrals corresponding to conservation of energy and angular momentum. If a certain relation holds between the parameters of the system, a third integral exists, which together with the other two permits complete integration by quadratures.

G. GRIOLI [5] gives a complete analytic description of the possible motions of a heavy rigid body about a fixed point under the assumption that the angular momentum with respect to the fixed point is the sum of two vectors, one of which has a fixed direction in space and the other has a fixed direction in the body.

PART II

Several coupled rigid bodies

Chapter IV

Gyrostats

§ 15. Permanent axes of rotation of a heavy gyrostat about a fixed point

15.1. Equations of motion of a gyrostat. Consider a mechanical system S which consists of an invariable part S_1 and other bodies S_2, variable or rigid, but not rigidly connected with S_1. If the external forces applied to the system are known, it is not possible to study the motion of the invariable part S_1 without taking into consideration the motion of S_2 relative to S_1. One of the simplest cases of such motions is that for which the motion of S_2 relative to S_1 does not change the mass geometry of system S. Such mechanical systems are called gyrostats.

Let the invariable part S_1 of a gyrostat be fixed at one of its points, O, which will be taken as the origin of two coordinate trihedrals: a fixed trihedral $OXYZ$ with vertically directed OZ-axis, and a moving trihedral $Oxyz$, the axes of which are directed along the principal axes of inertia of the gyrostat at O. On the basis of the theorem for the addition of velocities, the angular momentum of the gyrostat about the point O can be resolved into two components: the vector $\vec{I}$, due to the transport motion of the whole system S considered as a rigid body, and the vector $\vec{I}'$ due to the relative motion of S_2 with respect to S_1. Then the basic equation of dynamics assumes the form

$$\frac{d(\vec{I}+\vec{I}')}{dt}=\vec{M}$$

where differentiation refers to the fixed trihedral and $\vec{M}$ is the moment of the external forces with respect to the fixed point O. If $\vec{\omega}$ denotes the angular velocity of the invariable part S_1, then the above equation is equivalent to

$$\dot{\vec{I}}+\dot{\vec{I}}'+\vec{\omega}\times(\vec{I}+\vec{I}')=\vec{M}$$

where dots stand for the derivatives with respect to t, referred to the moving trihedral $O\,x\,y\,z$. The above equation determines the motion of system S provided that the gyrostatic moment $\vec{I}'$ is known, i.e. the motion of S_2 relative to S_1 is known. Therefore, following V. Volterra [1] and V. N. Drofa [1], let us consider the simple case where the gyrostatic moment $\vec{I}'$ is constant in magnitude and in direction relative to S_1 and the only force applied is gravity. Then the above equation reduces to

$$\dot{\vec{I}}+\vec{\omega}\times\vec{I}+\vec{\omega}\times\vec{I}'=m\,g(\vec{K}\times\vec{r}_0) \tag{15.1}$$

In addition, we have the Poisson equation

$$\dot{\vec{K}}+\vec{\omega}\times\vec{K}=0 \tag{15.2}$$

For applications to (i) the problem of the motion of the mass center of a system which is free of any external forces and the barycentric ellipsoid of inertia is an ellipsoid of revolution, (ii) the problem of the motion of the terrestrial poles caused by the angular momentum arising from the flow of water in the oceans, and (iii) the problem of variation of the latitude on the Earth's surface, the reader is referred to the papers by G. Peano [1, 2, 3], V. Volterra [1], V. A. Steklov [3], U. Cassina [1], and Miss Maria Teresa Vacca [1].

15.2. Permanent rotations of a gyrostat. Consider now under what conditions the gyrostat can rotate about an axis which is fixed in the gyrostat. If such an axis exists, then

$$\vec{\omega}=\omega\vec{\varkappa}$$

where $\vec{\varkappa}(a, b, c)$ is a unit vector of constant direction. In such a case by virtue of the fact that

$$\vec{I}(A\,a\,\omega,\, B\,b\,\omega,\, C\,c\,\omega) = \omega\vec{J}(A\,a,\, B\,b,\, C\,c)$$

$$\vec{I}\left(A\,a\frac{d\omega}{dt},\, B\,b\frac{d\omega}{dt},\, C\,c\frac{d\omega}{dt}\right) = \vec{J}\frac{d\omega}{dt}$$

equations (15.1) and (15.2) become

$$\vec{J}\frac{d\omega}{dt} + \omega^2\left(\vec{\varkappa}\times\vec{J}\right) + \omega\left(\vec{\varkappa}\times\vec{I'}\right) = m\,g\left(\vec{K}\times\vec{r_0}\right) \tag{15.3}$$

$$\dot{\vec{K}} + \omega\left(\vec{\varkappa}\times\vec{K}\right) = 0 \tag{15.4}$$

If we multiply equation (15.4) scalarly by $\vec{K}$, $\vec{\varkappa}$ respectively, we obtain two integrals

$$\vec{K}\cdot\vec{K} = K^2 = 1 \tag{15.5}$$

$$\vec{K}\cdot\vec{\varkappa} = \nu = \text{const.} \tag{15.6}$$

Similarly, if we multiply equation (15.3) scalarly by $\vec{K}$, $\vec{r_0}$, $\vec{I'}$ respectively, we obtain

$$\vec{J}\cdot\vec{K}\frac{d\omega}{dt} + \omega^2\left(\vec{\varkappa}\times\vec{J}\right)\cdot\vec{K} + \omega\left(\vec{\varkappa}\times\vec{I'}\right)\cdot\vec{K} = 0 \tag{15.7}$$

$$\vec{J}\cdot\vec{r_0}\frac{d\omega}{dt} + \omega^2\left(\vec{\varkappa}\times\vec{J}\right)\cdot\vec{r_0} + \omega\left(\vec{\varkappa}\times\vec{I'}\right)\cdot\vec{r_0} = 0 \tag{15.8}$$

$$\vec{J}\cdot\vec{I'}\frac{d\omega}{dt} + \omega^2\left(\vec{\varkappa}\times\vec{J}\right)\cdot\vec{I'} - m\,g\left(\vec{K}\times\vec{r_0}\right)\cdot\vec{I'} = 0 \tag{15.9}$$

Equations (15.7) and (15.8) for ω are similar; therefore their coefficients must be proportional, i.e.

$$\frac{\vec{J}\cdot\vec{K}}{L} = \frac{(\vec{\varkappa}\times\vec{J})\cdot\vec{K}}{M} = \frac{(\vec{\varkappa}\times\vec{I'})\cdot\vec{K}}{N}$$

where L, M, N denote the constant coefficients of equation (15.8).

Two independent equations of the above system are

$$M\left(\vec{J}\cdot\vec{K}\right) - L\left(\vec{\varkappa}\times\vec{J}\right)\cdot K = 0 \tag{15.10}$$

$$N\left(\vec{\varkappa}\times\vec{J}\right)\cdot\vec{K} - M\left(\vec{\varkappa}\times\vec{I'}\right)\cdot\vec{K} = 0 \tag{15.11}$$

Differentiating equation (15.10) with respect to the time t and making use of equation (15.4) and integral (15.6), we obtain the equation

$$-L\left(\vec{J}\cdot\vec{K}\right) - M\left(\vec{\varkappa}\times\vec{J}\right)\cdot\vec{K} = -L\,\nu\left(\vec{J}\cdot\vec{\varkappa}\right) \tag{15.12}$$

Solving the system consisting of equations (15.10), (15.11) and (15.12), we find that

$$\left.\begin{aligned}\vec{J}\cdot\vec{K} &= \frac{L^2\nu(\vec{J}\cdot\vec{\varkappa})}{L^2+M^2}\\ (\vec{\varkappa}\times\vec{J})\cdot\vec{K} &= \frac{LM\nu(\vec{J}\cdot\vec{\varkappa})}{L^2+M^2}\\ (\vec{\varkappa}\times\vec{I'})\cdot\vec{K} &= \frac{LN\nu(\vec{J}\cdot\vec{\varkappa})}{L^2+M^2}\end{aligned}\right\} \tag{15.13}$$

Thus for $\vec{K}(\alpha,\beta,\gamma)$ there are five algebraic equation (15.5), (15.6) and (15.13) with constant coefficients; four of these equations are linear in α, β, γ. In order that equations (15.6) and (15.13), linear relative to α, β, γ, ν, be consistent it is necessary that the determinant of this system, i.e.

$$D = (\vec{J}\cdot\vec{\varkappa})\left[(\vec{\varkappa}\times\vec{J})\cdot\vec{I'}\right]\left[1-\frac{L}{L^2+M^2}(\vec{J}\cdot\vec{\varkappa})(\vec{\varkappa}\cdot\vec{r_0})\right] \tag{15.14}$$

be zero. One condition for this is

$$(\vec{\varkappa}\times\vec{J})\cdot\vec{I'} = 0 \tag{15.15}$$

The other conditions yield nothing new compared to condition (15.15). If $D=0$, not all determinants of order three are zero. Consequently, we can find the values of α, β, γ from a system of nonhomogeneous linear equations with constant coefficients. Then equation (15.4) shows that $\vec{\varkappa}$ is parallel to $\vec{K}$, i.e.

$$\frac{a}{\alpha}=\frac{b}{\beta}=\frac{c}{\gamma}=n \tag{15.16}$$

In other words, the permanent axis must be vertical.

Because the vectors $\vec{\varkappa}$ and $\vec{K}$ are parallel, the cocfficients of ω and ω^2 in equation (15.7) vanish, and since, in general, $\vec{J}\cdot\vec{K}\neq 0$, then

$$\frac{d\omega}{dt}=0 \tag{15.17}$$

i.e. the gyrostat rotates with constant velocity about the vertical axis.

Taking conditions (15.15) and (15.17) into consideration, equation (15.9) yields

$$(\vec{K}\times\vec{r_0})\cdot\vec{I'}=0 \tag{15.18}$$

Finally equation (15.8) by virtue of (15.17) and (15.18) provides and additional condition

$$(\vec{\varkappa}\times\vec{J})\cdot\vec{r_0}=0 \tag{15.19}$$

While equation (15.19) gives the STAUDE-MLODZEEVSKII cone of permanent axes for a heavy rigid body, equation (15.15) is the equation of the cone formed by the permanent axes of the balanced gyrostat. Equation (15.18) is the equation for the plane determined by the vectors $\vec{I}'$ and $\vec{r}_0$. According to conditions (15.15), (15.18) and (15.19)

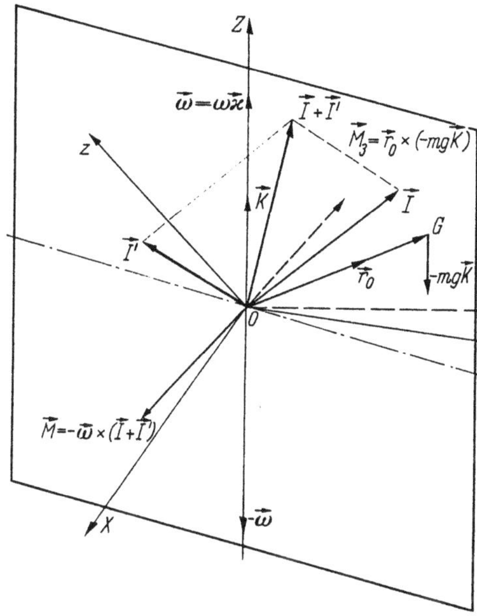

Fig. 37

the vectors $\vec{\omega} = \omega\vec{\varkappa}$, $\vec{I} = \omega\vec{J}$, $\vec{I}'$, $\vec{r}_0$, are all located in the vertical central plane (15.18) passing through the fixed point O and having an invariable orientation with respect to the gyrostat (Fig. 37).

The vertical straight line through O lies simultaneously on the above two cones and in the central vertical plane. Its equations in the moving coordinate trihedral $O\,x\,y\,z$ are

$$\frac{x}{\alpha} = \frac{y}{\beta} = \frac{z}{\gamma} \tag{15.20}$$

where

$$\alpha = \frac{1}{n}(C - B)\,R_y R_z$$

$$\beta = \frac{1}{n}(A - C)\,R_z R_x$$

$$\gamma = \frac{1}{n}(B - A)\,R_x R_y$$

$$n = \sqrt{(C-B)^2 R_y^2 R_z^2 + (A-C)^2 R_z^2 R_x^2 + (B-A)^2 R_x^2 R_y^2}$$

$$\vec{R} = \vec{I}' \times \vec{r}_0$$

and R_x, R_y, R_z are the projections of $\vec{R}$ on the axes of the moving coordinate trihedral $O\,x\,y\,z$.

The angular velocity ω of the permanent rotation of the gyrostat can be calculated from any one of the three identical scalar equations resulting from the vector equation (15.3), namely

$$\begin{aligned} n(A-C)(B-A)(C-B)\,R_x\,R_y\,R_z\,\omega^2 - \\ - n(A\,R_x\,I'_x + B\,R_y\,I'_y + C\,R_z\,I'_z)\,\omega + \\ + m\,g(A\,R_x\,x_0 + B\,R_y\,y_0 + C\,R_z\,z_0) = 0 \end{aligned} \qquad (15.21)$$

where I'_x, I'_y, I'_z are the components of $\vec{I}'$ in the $O\,x\,y\,z$-trihedral.

In general, of the two segments into which the vertical straight line (15.20) is subdivided by the fixed point O, only that segment can be the axis of permanent rotation for which the discriminant of equation (15.21) is always positive (replacement of one straight line segment by another implies replacement of both vectors $\vec{I}'$ and $\vec{r}_0$ by their opposites, which in turn implies the change of the sign of the free term in the above equation). The gyrostat can rotate about such a straight line segment in one or the other direction but the angular velocity in each case will be different. In order to explain these facts we shall assume that the vertical axis of rotation is given and we shall look for conditions under which a permanent rotation about the chosen axis is possible.

Since the weight moment $\vec{M}_3 = \vec{r}_0 \times (-m\,g\,\vec{K})$ and the gyrostatic moment $\vec{M} = -\vec{\omega} \times (\vec{I} + \vec{I}')$ are both directed along the straight line (but in opposite directions) through O perpendicular to the central vertical plane (15.18), then $\vec{M}_3$ can be counterbalanced by $\vec{M}$. On the other hand, the vertical axis of permanent rotation divides the central vertical plane into two half-planes. In order that the equality $\vec{M}_3 = -\vec{M}$ be satisfied, the center of mass G of the gyrostat must lie in the right central half-plane when looking from the endpoint of the vector $\vec{M}_1 = -(\vec{\omega} \times \vec{I})$. However, since the center of mass G of the gyrostat is given, then of the two straight line segments forming the axis of permanent rotation, that one must be directed upward relative to which the center of mass G lies in the right-hand half-plane when viewed from the endpoint of the vector $\vec{M}_1$.

The reversal of the direction of rotation about the given straight line segment changes the sign of the gyrostatic moment $\vec{M}_2 = -(\vec{\omega} \times \vec{I}')$. Rotation as before is possible but the angular velocity of rotation will now be different.

Let us mention two particular cases. If the ellipsoid of inertia of the gyrostat is an ellipsoid of revolution, say $A = B$, then the straight line (15.20) lies in the principal plane of inertia associated with the principal axis for which the moment of inertia is $C \neq A = B$. In such a case the equations of the straight line (15.20) reduce to

$$x R_x + y R_y = 0, \qquad z = 0$$

and the angular velocity of the gyrostat about this axis is $\omega = m g z_0 / n I_z'$.

In the case of a spherical ellipsoid of inertia equations (15.15) and (15.19) are satisfied identically, and the permanent axes of the gyrostat form the plane given by (15.18). Any straight line segment of this plane can serve as a permanent axis of rotation, but the rotation itself can take place in one direction only.

The angular velocity ω of the gyrostat, if calculated from the scalar equation obtained by projecting the vector equation (15.3) on the x-axis, is given by the formula

$$\omega = \frac{m g (\beta z_0 - \gamma y_0)}{b I_z' - c I_y'}$$

From here it follows that $\omega = 0$ for the axis passing through the center of mass G of the gyrostat, and $\omega = \infty$ for the axis coinciding with the vector $\vec{I'}$.

15.3. Other systems for which permanent rotations are possible. G. COLOMBO [2] has studied the motion under gravity of a system S consisting of two rigid bodies S_1 and S_2. Body S_1 by assumption has an absolutely fixed point while S_2 has a point fixed relative to S_1. Necessary and sufficient conditions are obtained in order that both bodies can rotate with uniform angular velocities about fixed axes.

G. CAPRIZ [1] has been concerned with a system which consists of (i) a rigid body S' with center of mass at G' and free to rotate about a fixed point $O \neq G'$, and (ii) a second rigid body S (a gyro) with center of mass at G and free to rotate about an axis of symmetry of S, fixed in S' and passing through O but not through G'. Motions of the system are determined which correspond to uniform rotations of S' about some fixed axis through O. V. V. RUMYANCEV [7] and A. ANČEV [1] have discussed the stability of permanent rotations of a gyrostat.

15.4. The existence of first integrals. I. A. KEIS [1] has discussed the existence of integrals which are linear and quadratic relative to the components of the angular velocity of a gyrostat. He also pointed out the gyrostatic analogues for certain known cases of integrability, such as those of LAGRANGE and POISSON, HESS and APPEL'ROT (loxodromic pendulum), STEKLOV and BOBYLEV, and KOVALEVSKAYA and DELONE, of the equations of motion of a heavy rigid body about a fixed point. In another paper KEIS [3] was concerned with the existence of algebraic integrals of the equations of motion of a gyrostat.

The introduction of LIE group theoretical methods has certain advantages over the LAGRANGE and ROUTH method for finding first integrals. Namely, by introduction of a suitable group of infinitesimal transformations, we are able to discover displacements of the given mechanical system which are more complicated than those described by the Lagrangian coordinates. On the other hand, the knowledge of cyclic displacements of a mechanical system, determined by the introduced group of infinitesimal transformations and the structure of the kinetic energy and the force function of the system, permits us to find certain first integrals of the equations of motion and thus to reduce the order of the system of equations. Regarding this topic the reader is referred to A. A. Bogoyavlenskii [4, 5], I. A. Keis [1] and L. N. Sretenskii [3].

15.5. Regular precession of a gyrostat. I. A. Keis [2] has discussed the existence of regular precession in the motion of a gyrostat about a fixed point under the action of gravity. An analogue of Grioli's case is obtained.

15.6. Equations of motion in terms of non-Eulerian angles. In many technical applications the gyroscope is an essential member of a system with several degrees of freedom, the stability of which with respect to a certain equilibrium position is of particular interest. In order to investigate the stability of such a system by the well known method of small oscillations, the motions of the different members of the system must be described in terms of convenient coordinates. As to the gyroscope itself, it is necessary for this purpose to replace the Euler angles by some other coordinates so that in terms of the new coordinates and their time derivatives the approximate equations of motion are linear.

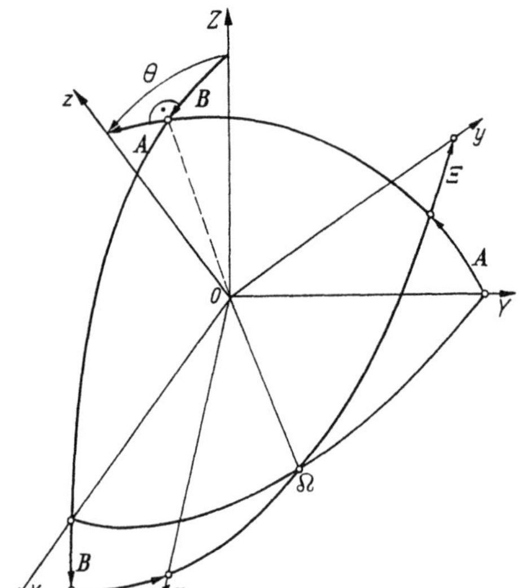

Fig. 38

E. Sponder [1], restricting himself to small displacements of the spin axis (z) from its equilibrium position, derived in a rigorous way the equations of motion in terms of three non-Eulerian angles A, B, Ξ. Angle A measures the angular distance of the spin axis (z) from the $X\,Z$-plane, angle B lies in the $X\,Z$-plane and is measured from the Z-axis. The third angle Ξ lies in the $x\,y$-plane and gives the position of the x-axis relative to the $X\,Z$-plane (Fig. 38). The non-Eulerian angles A and B together with the Eulerian angle Θ make up a spherical right triangle.

The components of the angular velocity $\vec{\omega}$ relative to the moving trihedral $O\,x\,y\,z$ are expressed in terms of the angles A, B, Ξ and their derivatives. After some sign changes and interchange of the symbols sin and cos these expressions become identical to those in terms of the Euler angles. Lagrange's equations are used for the derivation of the new equations which hold also for a slowly rotating gyroscope as well as for the limiting case of a spherical pendulum. See also F. Simonart [1].

§ 16. Asymmetric body subject to a self-excitement in the equatorial plane

16.1. Equations of motion. The technical problem underlying H. LEIPHOLZ's paper [1] is that of the motion of an aircraft with a rotor about its center of mass considered as a fixed point. Further, it is assumed that the aircraft moves above the stratosphere subject to forces applied at the center of mass, but no aerodynamical forces are taken into account. By virtue of the above assumptions the motion of the center of mass of the aircraft can be separated from that of the aircraft about its mass center.

By assumption the aircraft S consists of a rigid hollow body K and a frictionless rotor R, the axis of which passes through the mass centers G_R, G_K and G_S of R, K and S respectively and is at the same time one of the principal axes of inertia for the rotor at G_R as well as for the body K at G_K, and hence for the whole system S at G_S (Fig. 39).

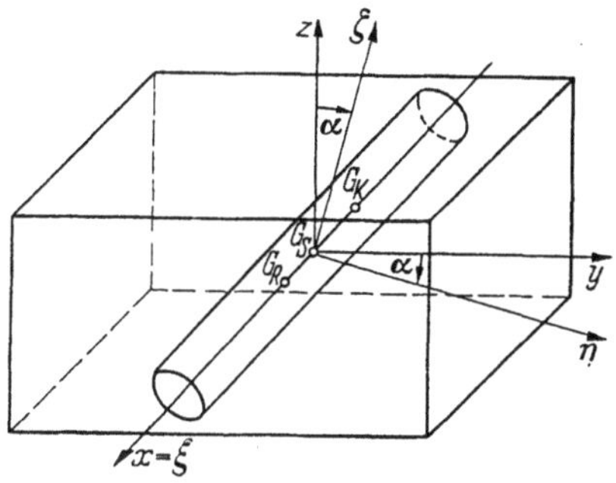

Fig. 39

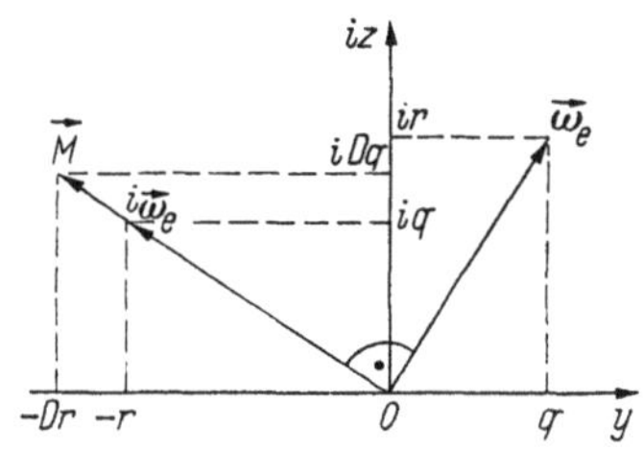

Fig. 40

Mathematically, the problem reduces to the motion of an asymmetric rigid body subject to forces, the moment vector $\vec{M}(0, -D\,r, D\,q)$ of which lies in the equatorial plane $(y\,z)$ of the body and is obtained from the projection $\vec{\omega}_e(0, q, r)$ of the angular velocity vector $\vec{\omega}(p, q, r)$ on the equator by rotation through a right angle and a magnification in the ratio $D:1$ (Fig. 40), i.e.

$$\vec{M} = D\,e^{i\,\pi/2}\,\vec{\omega}_e \tag{16.1}$$

where $\vec{\omega}_e = q + i\,r$, $\vec{M} = M_y + i\,M_z$ and D is a constant.

By virtue of the assumption that the resultant external force acting on the aircraft passes through its mass center G_S and no aerodynamic forces are present, the moment vector $\vec{M}(G_S)$ in the basic dynamic equation

$$\frac{d\vec{I}_S(G_S)}{dt} = \vec{M}(G_S) \tag{16.2}$$

is equal to zero, and hence equation (16.2) can be integrated, giving the motion of S about its center of mass G_S.

Two coordinate trihedrals with the common origin at G_S will be used: the $G_S\,x\,y\,z$-trihedral with unit vectors $\vec{i}$, $\vec{j}$, $\vec{k}$ along the axes rigidly connected with K and the $G_S\,\xi\,\eta\,\zeta$-trihedral with unit vectors $\vec{e}_1$, $\vec{e}_2$, $\vec{e}_3$ along the axes rigidly connected with the rotor. The x-axis and ξ-axis coincide by our assumption. The axes of both coordinate trihedrals are principal axes of inertia of the respective bodies. The principal moments of inertia of K in the $G_S\,x\,y\,z$-trihedral are denoted by A_K, B_K, C_K and those of the rotor R in the $G_S\,\xi\,\eta\,\zeta$-trihedral by A_R, B_R, C_R, whereby due to the symmetry of R with respect to the ξ-axis we have that $B_R = C_R$.

Let $\vec{\omega}(p, q, r)$ and $\vec{o}(o_x, o_y, o_z)$ be the angular velocity vectors of K and R respectively in terms of the components along the x, y, z-axes. From Fig. 39 it follows that the $\eta\,\zeta$-system rotates relative to the $y\,z$-system with the velocity

$$\dot{\alpha} = o_x - p$$

Hence

$$\vec{o} = \vec{\omega} + \dot{\alpha}\,\vec{i}$$

or, in terms of the components,

$$o_x = p + \dot{\alpha}$$
$$o_y = q$$
$$o_z = r$$

Since the angular momentum vector $\vec{I}_S(G_S)$ of the system is given by $\vec{I}_S(G_S) = \vec{I}_K(G_S) + \vec{I}_R(G_S)$ and since $\vec{M}(G_S) = 0$, the basic dynamic equation (16.2) gives

$$\frac{d\left[\vec{I}_K(G_S) + \vec{I}_R(G_S)\right]}{dt} = 0 \tag{16.3}$$

Keeping in mind that the reference point in what follows remains G_S, we shall drop G_S in all formulas. Then the last formula assumes the simple form

$$\frac{d\vec{I}_K}{dt} = -\frac{d\vec{I}_R}{dt} \tag{16.4}$$

With respect to the moving coordinate trihedral $G_S\,x\,y\,z$ the components of $\vec{I}_K$ and $\vec{I}_R$ are

$$\vec{I}_K = (I_{Kx}, I_{Ky}, I_{Kz}) = (A_K\,p,\ B_K\,q,\ C_K\,r)$$
$$\vec{I}_R = (I_{Rx}, I_{Ry}, I_{Rz}) = (A_R\,o_x,\ B_R\,q,\ B_R\,r)$$

Since

$$\frac{d\ldots}{dt} = \left(\frac{d\ldots}{dt}\right)_{\text{rel}} + \vec{\omega} \times \ldots$$

equation (16.4) is equivalent to the following three scalar equations

$$\left.\begin{aligned} A_K \dot{p} + (C_K - B_K)\, q\, r &= -A_R\, \dot{o}_x \\ B_K \dot{q} + (A_K - C_K)\, r\, p &= -B_R\, \dot{q} + B_R\, p\, r - A_R\, o_x\, r \\ C_K \dot{r} + (B_K - A_K)\, p\, q &= -B_R\, \dot{r} - B_R\, q\, p + A_R\, o_x\, q \end{aligned}\right\} \qquad (16.5)$$

We shall assume that

$$o_x = \text{const.}$$

In addition let us introduce the relations

$$B_S = B_K + B_R, \qquad C_S = C_K + C_R, \qquad D = A_R\, o_x = \text{const.}$$

Then the above system (16.5) assumes the final form

$$\left.\begin{aligned} A_K \dot{p} + (C_S - B_S)\, q\, r &= 0 \\ B_S \dot{q} + (A_K - C_S)\, r\, p &= -D\, r \\ C_S \dot{r} + (B_S - A_K)\, p\, q &= D\, q \end{aligned}\right\} \qquad (16.6)$$

Equations (16.6) can be interpreted as the EULER equations of an asymmetric "substitute" rigid body, the excitation of which depends upon the velocity of rotation $\vec{\omega}$. In fact, the moment vector $\vec{M}$, subject to which the substitute rigid body is moving, has components depending upon the vector $\vec{\omega}$, namely

$$\vec{M} = (0, -D\, r, D\, q)$$

where $\vec{M}$ is related to $\vec{\omega}_e$ by formula (16.1).

Since the motion of the system S about its mass center G_S can be separated from that of the mass center itself, equations (16.6) can be considered as the EULER equations of motion of an asymmetric rigid body, supported at its space-fixed mass center G_S and subject to the moment vector $\vec{M}$.

If we introduce a space-fixed coordinate trihedral $G_S\, X\, Y\, Z$, then the position of the body relative to such a coordinate trihedral is given by the EULER angles. They are related to the angular velocity vector $\vec{\omega}$ by formulas

$$\left.\begin{aligned} q &= \dot{\Theta}\cos\Phi + \dot{\psi}\sin\Theta\sin\Phi \\ r &= -\dot{\Theta}\sin\Phi + \dot{\psi}\sin\Theta\cos\Phi \\ p &= \dot{\psi}\cos\Theta + \dot{\Phi} \end{aligned}\right\} \qquad (16.7)$$

obtained from (1.8) by a circular permutation of the letters p, q, r.

16.2. Symmetric body K: $B_K = C_K$. In such a case $C_S - B_S = 0$ and the first equation of system (16.6) reduces to $A_K \dot{p} = 0$, and hence by integration

$$p = \text{const.} \tag{16.8}$$

Set

$$\nu = \frac{(A_K - B_S)\, p + D}{B_S} = \text{const.}$$

and let

$$\vec{\omega}_e = q + i\, r$$

be a complex representation of the projection $\vec{\omega}_e$ of the angular velocity vector $\vec{\omega}$ of K on the equatorial plane of the rigid body. Then the second and third equations of system (16.6) show that $\vec{\omega}_e$ satisfies the complex differential equation

$$\dot{\vec{\omega}}_e - i\, \nu\, \vec{\omega}_e = 0$$

the solution of which corresponding to the initial condition $\vec{\omega}_e(0) = q(0) + i\, r(0)$ is

$$\vec{\omega}_e = \vec{\omega}_e(0)\, e^{i \nu t}$$

Separating the real and imaginary parts, we obtain

$$\begin{aligned} q &= q(0) \cos \nu t - r(0) \sin \nu t \\ r &= q(0) \sin \nu t + r(0) \cos \nu t \end{aligned} \tag{16.9}$$

If we introduce the notations W_0, σ defined by

$$q(0) = W_0 \sin\sigma, \qquad r(0) = W_0 \cos\sigma$$

with

$$W_0 = \sqrt{q^2(0) + r^2(0)}, \qquad \tan\sigma = \frac{q(0)}{r(0)} = \text{const.}$$

then the above solution assumes the form

$$q = W_0 \sin(\sigma - \nu t), \qquad r = W_0 \cos(\sigma - \nu t) \tag{16.10}$$

Given the initial conditions, the vector $\vec{\omega}(p, q, r)$ is now completely determined. The expression for the square of its modulus

$$\omega^2 = p^2 + q^2 + r^2 = p^2 + W_0^2 = \text{const.}$$

shows that $\vec{\omega}$ is of constant length.

Consider now the angular momentum vector of the system $\vec{I}_S = \vec{I}_K + \vec{I}_R$ which can be rewritten in the form

$$\vec{I}_S = I_{Sx} \vec{i} + I_{Sy} \vec{j} + I_{Sz} \vec{k} = (A_K p + A_R\, \omega_x) \vec{i} + B_S q \vec{j} + C_S r \vec{k} \tag{16.11}$$

Further, introduce the vectors

$$\begin{aligned}\vec{I}_K^* &= I_{Kx}^* \vec{i} + I_{Ky}^* \vec{j} + I_{Kz}^* \vec{k} = A_K p \vec{i} + B_S q \vec{j} + C_S r \vec{k} \\ \vec{I}_R^* &= A_R o_x \vec{i} = D \vec{i} = D \vec{e}_1\end{aligned} \tag{16.12}$$

Then

$$\vec{I}_S = \vec{I}_K^* + \vec{I}_R^*$$

Since $d\vec{I}_S/dt = 0$ [see equation (16.3)], we conclude that $\vec{I}_S$ is a vector of constant length and fixed direction in space (since differentiation refers to an absolute coordinate trihedral).

Also the vectors $\vec{I}_K^*$ and $\vec{I}_R^*$, which are not fixed in space, are of constant length. Namely, observing the equality $C_S = B_S$, we obtain that

$$I_R^{*2} = A_R^2 o_x^2 = D^2 = \text{const.}$$

$$I_K^{*2} = A_K^2 p^2 + B_S^2 (q^2 + r^2) = A_K^2 p^2 + B_S^2 W_0^2 = \text{const.}$$

Further, the following relation holds:

$$\begin{aligned}I_S^2 &= \left(\vec{I}_K^* + \vec{I}_R^*\right)^2 = I_K^{*2} + 2\vec{I}_K^* \cdot \vec{I}_R^* + I_R^{*2} \\ &= I_K^{*2} + 2A_K p D + D^2 = \text{const.}\end{aligned} \tag{16.13}$$

Let us now refer the motion to the fixed inertial coordinate trihedral $G_S X Y Z$, the Z-axis of which coincides with the vector $\vec{I}_S$. Then the x-axis makes a constant angle Θ with the Z-axis. In fact,

$$\vec{I}_S \cdot \vec{i} = \left(\vec{I}_K^* + \vec{I}_R^*\right) \cdot \vec{i} = \left|\vec{I}_K^* + \vec{I}_R^*\right| \cos\Theta = A_K p + D = \text{const.}$$

Hence

$$\Theta = \Theta_0 = \text{const.} \tag{16.14}$$

Squaring and adding the first two equations of system (16.7) and using relations (16.10), we obtain that

$$W_0^2 = \dot{\Theta}^2 + \dot{\psi}^2 \sin^2\Theta$$

Since $\dot{\Theta} = 0$, it follows that

$$\dot{\psi}^2 = \frac{W_0^2}{\sin^2\Theta}$$

In order to calculate $\dot{\psi}$ we need $\sin\Theta$. Since

$$\cos\Theta = \frac{A_K p + D}{|\vec{I}_S|} = \frac{I_{Sz}}{|\vec{I}_S|}$$

then

$$\sin^2\Theta = 1 - \cos^2\Theta = \frac{B_S^2 W_0^2}{(A_K p + D)^2 + B_S^2 W_0^2}$$

and hence

$$\sin\Theta = \frac{B_S W_0}{|\vec{I}_S|}$$

In the above formulas for $\cos\Theta$ and $\sin\Theta$ the numerators are the x-component and the component orthogonal to the x-axis of the vector $\vec{I}_S = \vec{I}_K^* + \vec{I}_R^*$ respectively, while the denominator in both cases is the length of this vector.

Substituting the above expression of $\sin^2\Theta$ in the formula for $\dot{\psi}^2$ we obtain that

$$\dot{\psi}^2 = \frac{(A_K p + D)^2 + B_S^2 W_0^2}{B_S^2}$$

$$\dot{\psi} = \frac{|\vec{I}_S|}{B_S} = \text{const.} \tag{16.15}$$

and hence, by integration,

$$\psi(t) = \frac{|\vec{I}_S|}{B_S} t + \psi_0 \tag{16.16}$$

In order to calculate the third EULER angle Φ use the last formula of system (16.7), which gives

$$p = \frac{A_K p + D}{B_S} + \dot{\Phi}$$

Hence

$$\dot{\Phi} = \frac{(B_S - A_K) p - D}{B_S} = -\nu = \text{const.}$$

and, by integration, we obtain that

$$\Phi(t) = -\nu t + \Phi_0 \tag{16.17}$$

With reference to Fig. 39 this means that the body-fixed yz-plane rotates clockwise.

Thus the problem of motion of the rigid body is analytically solved: formulas (16.8) and (16.9) give the angular velocity vector $\vec{\omega}$, formula (16.11) the angular momentum vector $\vec{I}_S$, and formulas (16.14), (16.16) and (16.17) the EULER angles, and hence the position of the rigid body in space.

From the expressions of the EULER angles it follows that the general motion of a symmetric substitute rigid body is a regular precession with the velocity of precession given by formula (16.15).

A particular case of the motion takes place if

$$D = (B_S - A_K) p \tag{16.18}$$

Such a case can be realized by a proper choice of o_x, using the formula $D = A_K o_x$. Then $\nu = 0$ and

$$q = q(0) = \text{const.}, \qquad r = r(0) = \text{const.}$$

Hence the vectors $\vec{\omega}$ and $\vec{I}_K^*$ become body-fixed, and we have a permanent rotation about a body-fixed axis. Since for $\nu = 0$ we also have $\dot{\Phi} = 0$, we can assume that $\Phi_0 = 0$, i.e. the line of nodes coincides with the y-axis. Then also $\Phi = 0$ and $q = 0$, i.e. the vector $\vec{\omega}$ lies in the xZ-plane.

From formula (16.18) we obtain that $A_K p + D = B_S p$, and hence

$$\cos^2\Theta = \frac{B_S^2 p^2}{B_S^2(p^2 + W_0^2)} \quad \text{or} \quad \cos\Theta = \frac{p}{|\vec{\omega}|}$$

Consequently the vector $\vec{\omega}$ lies along the vertical Z-axis, which becomes the axis of permanent rotation. Since $q = 0$ it follows that the length of the self-exciting moment vector $\vec{M}$ is equal to Dr and $\vec{M}$ lies along the line of nodes which coincides with the y-axis.

Thus this case is analogous to that of permanent rotations of Staude and Mlodzeevskii of a heavy asymmetric rigid body.

16.3. Asymmetric body K: $B_K \neq C_K$. Putting

$$\left.\begin{aligned} \Omega_x &= \frac{B_S - C_S}{A_K} q r \\ \Omega_y &= \frac{C_S - A_K}{B_S} r p - \frac{D}{B_S} r \\ \Omega_z &= \frac{A_K - B_S}{C_S} p q + \frac{D}{C_S} q \end{aligned}\right\} \tag{16.19}$$

system (16.6) assumes the form

$$\frac{dp}{\Omega_x} = \frac{dq}{\Omega_y} = \frac{dr}{\Omega_z} = dt \tag{16.20}$$

which can be integrated as follows. From the above system it follows that

$$\frac{dq}{dp} = \frac{\Omega_y}{\Omega_x} = \frac{C_S - A_K}{B_S - C_S} \frac{A_K}{B_S} \frac{p}{q} - \frac{D A_K}{(B_S - C_S) B_S} \frac{1}{q}$$

and hence, by separation of variables, the first integral

$$q^2 = -\frac{A_K}{C_S - B_S}\left(\frac{C_S - A_K}{B_S} p^2 - \frac{2D}{B_S} p + \text{const.}\right) \tag{16.21}$$

is obtained. Similarly,

$$\frac{dr}{dp} = \frac{\Omega_z}{\Omega_x} = \frac{A_K - B_S}{B_S - C_S} \frac{A_K}{C_S} \frac{p}{r} + \frac{D A_K}{(B_S - C_S) C_S} \frac{1}{r}$$

and hence separating the variables we obtain the second integral

$$r^2 = \frac{A_K}{C_S - B_S}\left(\frac{B_S - A_K}{C_S}\, p^2 - \frac{2D}{C_S}\, p + \text{const.}\right) \tag{16.22}$$

If we eliminate q and r from Ω_x by means of integrals (16.21) and (16.22), then Ω_x is obtained as a function of p only, say $\bar{\Omega}_x(p)$. As a consequence

$$t = \int\limits_{p(0)}^{p} \frac{dp}{\bar{\Omega}_x(p)} \tag{16.23}$$

gives the third integral. This completes the integration of system (16.20) and therefore of system (16. 6).

By inverting the relation (16.23) we are able to express the components p, q, r of the angular velocity vector $\vec{\omega}$ as functions of the time t.

In order to obtain a mechanical interpretation of the above integrals we shall look for other integrals which are combinations of the ones already given. We recall that a relation of the form

$$f(p, q, r) = \text{const.}$$

is said to be an integral of system (16.20) provided that

$$\frac{df}{dp} = \frac{\partial f}{\partial p} + \frac{\partial f}{\partial q}\frac{dq}{dp} + \frac{\partial f}{\partial r}\frac{dr}{dp} = 0 \tag{16.24}$$

is identically satisfied if one substitutes for dq/dp and dr/dp their expressions obtained above.

Multiplying the equations of system (16.6) by p, q and r respectively, adding them, and integrating the equation so obtained, we obtain a first integral

$$f_1 = A_K p^2 + B_S q^2 + C_S r^2 = c_1 = \text{const.} \tag{16.25}$$

Similarly, multiplying the same equations by $A_K p$, $B_S q$ and $C_S r$ respectively, adding them, integrating the equation so obtained, and observing the relation

$$B_S - C_S = \frac{A_K \dot{p}}{q r}$$

[which follows from the first equation of system (16.6)], we obtain a second integral

$$f_2 = A_K^2 p^2 + B_S^2 q^2 + C_S^2 r^2 + 2A_K p D = c_2 = \text{const.} \tag{16.26}$$

Both integrals (16.25) and (16.26) identically satisfy equation (16.24) and have a mechanical meaning. In fact, comparing the expression for the kinetic energy T of the whole system,

$$T = \tfrac{1}{2}(A_K p^2 + B_S W_0^2 + D o_x)$$

with the expression (16.12) for I_K^*, we see that $f_1 = \text{const.}$ represents the energy integral

$$\vec{I}_K^* \cdot \vec{\omega} = 2T - D o_x = c_1 = \text{const.}$$

From equation (16.26) it follows that

$$(A_K p + D)^2 + B_S^2 q^2 + C_S^2 r^2 = c_2 + D^2 = \text{const.}$$

Comparing this with formulas (16.11) and (16.12) we conclude that

$$c_2 + D^2 = I_S^2$$

and hence $f_2 = \text{const.}$ is the angular momentum integral

$$f_2 = I_S^2 - D^2 = c_2$$

By means of the two integrals (16.25) and (16.26) we can express q and r in terms of p. Namely, multiplying equation (16.25) by B_S and ubtracting the result from equation (16.26), we obtain

$$C_S(C_S - B_S) r^2 + A_K(A_K - B_S) p^2 + 2 A_K p D = c_2 - B_S c_1$$

Putting

$$P_z = \frac{B_S - A_K}{C_S} p^2 - \frac{2D}{C_S} p + c_z$$
$$c_z = \frac{c_2 - B_S c_1}{C_S A_K} \tag{16.27}$$

we obtain

$$r = \sqrt{\frac{A_K}{C_S - B_S} P_z} \tag{16.28}$$

If we square the last formula, then equality (16.22) results. Similarly, multiplying equation (16.25) by C_S and subtracting the result from equation (16.26), we obtain

$$B_S(B_S - C_S) q^2 + A_K(A_K - C_S) p^2 + 2 A_K D p = c_2 - C_S c_1$$

Putting

$$P_y = \frac{C_S - A_K}{B_S} p^2 - \frac{2D}{B_S} p + c_y$$
$$c_y = \frac{c_2 - C_S c_1}{B_S A_K} \tag{16.29}$$

we obtain

$$q = i\sqrt{\frac{A_K}{C_S - B_S} P_y} \tag{16.30}$$

The square of the last equality is identical to equality (16.21).

From equalities (16.28), (16.30), and the first equality of system (16.19) it follows that

$$\bar{\Omega}_x(p) = i\sqrt{P_z P_y} = i\sqrt{W}$$

and

$$W = P_z P_y = -\bar{\Omega}_x^2$$

In what follows W will be called the gyroscopic function. In terms of these notations we have that

$$t = \int_{p(0)}^{p} \frac{dp}{i\sqrt{W}} \tag{16.31}$$

a relation which is equivalent to (16.23).

16.4. Gyroscopic function W. The behavior of the gyroscopic function W determines the nature of the components p, q, r of the angular velocity vector $\vec{\omega}$. Namely, formula (16.31) gives t as a function of p, and hence, by inversion, p as a function of the time t. Formulas (16.27) to (16.30) then determine q and r as functions of t.

The gyroscopic function W, being a product of two quadratic functions P_y and P_z, is a rational function of degree four in p. In order that formula (16.31) gives real values for t, it is necessary that $W \leqq 0$. In addition, we are interested in real values of q and r. In order to determine the nature of q and r, certain assumptions concerning the moments of inertia must be made. The following inequalities always hold

$$\begin{aligned} A_R + B_R &> C_R, & A_R + C_R &> B_R, & B_R + C_R &> A_R \\ A_K + B_K &> C_K, & A_K + C_K &> B_K, & B_K + C_K &> A_K \end{aligned} \tag{16.32}$$

In addition, without any loss of generality we shall assume in what follows that

$$C_S = C_K + C_R > B_S = B_K + B_R > A_K \tag{16.33}$$

By virtue of inequalities (16.33) we conclude from formulas (16.28) and (16.30) that q and r will assume real values for

$$P_z \geqq 0 \geqq P_y \tag{16.34}$$

These inequalities are also sufficient in order that $W \leqq 0$. Hence, in discussing the gyroscopic function W, we must always keep in mind inequalities (16.34), which guarantee the reality of the roots q and r as well as the reality of t and consequently the reality of p.

In drawing the graph of $W = W(p)$, it is also advisable to give the graphs of the generating parabolas P_y and P_z, which show the intervals in which inequalities (16.34) hold. By means of these three graphs it is an easy matter to determine intervals in which $W \leqq 0$. To such intervals then correspond real values for the three components of the angular velocity $\vec{\omega}$.

By analogy with the definition of the potential of a particle of mass one in a conservative force field, let us define the potential V of a function $f(t)$ by means of the integral

$$\int \ddot{f}\, df = V$$

Then

$$\int \ddot{f}\, df = \int \ddot{f} \frac{df}{dt}\, dt = \int \ddot{f} \dot{f}\, dt = \frac{1}{2} \dot{f}^2$$

and consequently

$$\dot{f}^2 = 2 \int \ddot{f}\, df$$

or

$$\dot{f}^2 = 2V \qquad (16.35)$$

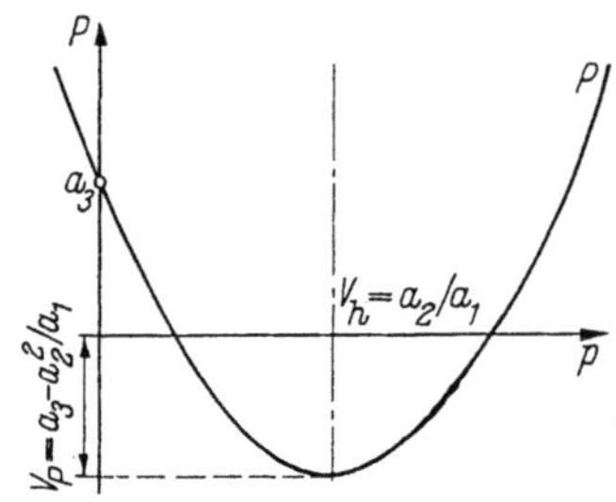

Fig. 41

If we put $f(t) = p(t)$, then comparison of formulas (16.35) and (16.31) shows that

$$-\tfrac{1}{2} W = V$$

i.e. the gyroscopic function $W(p)$ can be interpreted as a potential. If in addition to the graph $W = W(p)$ that of $\dot{p} = \sqrt{-W(p)}$ is also plotted on a separate plane, then an analysis of the phase-plane p, $\dot{p}$ allows us to make conclusions about the qualitative behavior of p.

Given the parabola

$$P = a_1 p^2 - 2a_2 p + a_3 \qquad (16.36)$$

the coordinates of the vertex and the P-intercept are seen in Fig. 41.

Comparing the above P-expression with the expressions of P_y and P_z respectively, we see that for the parabolas P_z and P_y we have respectively

$$a_{1z} = \frac{B_S - A_K}{C_S}, \quad a_{3z} = c_z, \quad V_{hz} = \frac{D}{B_S - A_K},$$

$$V_{P_z} = c_z - \frac{D^2}{C_S(B_S - A_K)}$$

and

$$a_{1y} = \frac{C_S - A_K}{B_S}, \quad a_{3y} = c_y, \quad V_{hy} = \frac{D}{C_S - A_K},$$

$$V_{P_y} = c_y - \frac{D^2}{B_S(C_S - A_K)}$$

From the above expressions for the coordinates of the vertices of the P_y- and P_z-parabolas we conclude that in the asymmetric case where

$C_S > B_S$ the vertices of P_y and P_z will never lie on a vertical line since $V_{hz} > V_{hy}$.

For the components of $\vec{\omega}$ the following types of solutions are possible:

T_1: elliptic functions
T_2: trigonometric functions
T_3: asymptotic solutions
T_4: stationary values

By asymptotic solutions is meant the case where the components of $\vec{\omega}$ approach finite limits as $t \to \infty$.

The possible types of solutions depend upon the nature of the roots of the gyroscopic function and the interval in which (16.34) holds. The gyroscopic function W can have:

W_1: four different real roots
W_2: two different real roots and a pair of complex conjugate roots
W_3: two pairs of complex conjugate roots
W_4: two different real roots and one double root
W_5: one simple real root and one triple real root
W_6: two double real roots
W_7: one quadruple real root

Cases W_3 and W_7 must be rejected. In case W_3 we have $W > 0$, $P_y > 0$ and $P_z > 0$ everywhere so that the double inequality (16.34) is nowhere satisfied. The case W_7 would require at least $V_{hy} = V_{hz}$ which in turn implies that $B_S = C_S$, a case which is to be excluded, or $D = 0$, the trivial case.

The remaining cases imply the following possibilities for the solutions:

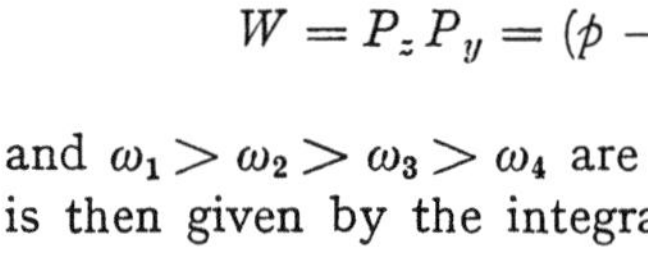

Fig. 42

$$\left.\begin{matrix} W_1 \\ W_2 \end{matrix}\right\} T_1, \quad W_4 \left\{\begin{matrix} T_2 \\ T_3 \\ T_4 \end{matrix}\right. \quad W_5 \left\{\begin{matrix} T_3 \\ T_4 \end{matrix}\right. \quad W_6 : T_4$$

16.5. Nonstationary solutions for $\vec{\omega}$.

In case W_1 the integral appearing on the right of formula (16.31) is an elliptic integral of the first kind, where

$$W = P_z P_y = (p - \omega_1)(p - \omega_2)(p - \omega_3)(p - \omega_4)$$

and $\omega_1 > \omega_2 > \omega_3 > \omega_4$ are the roots of $W = 0$ (Fig. 42). The time t is then given by the integral

$$t = \int_{p(0)}^{p} \frac{dp}{\sqrt{-(p - \omega_1)(p - \omega_2)(p - \omega_3)(p - \omega_4)}} \tag{16.37}$$

which can be transformed to the LEGENDRE normal form by means of the following transformations. For $\omega_4 \leqq p \leqq \omega_3$ put

$$p = \frac{\omega_4(\omega_3 - \omega_1) + \omega_1(\omega_4 - \omega_3)\sin^2\varphi}{\omega_3 - \omega_1 + (\omega_4 - \omega_3)\sin^2\varphi}, \qquad \sin^2\varphi = \frac{\omega_3 - \omega_1}{\omega_4 - \omega_3}\,\frac{p - \omega_4}{\omega_1 - p} \tag{16.38}$$

and for $\omega_2 \leqq p \leqq \omega_1$ let

$$p = \frac{\omega_2(\omega_3 - \omega_1) - \omega_3(\omega_2 - \omega_1)\sin^2\varphi}{\omega_3 - \omega_1 - (\omega_2 - \omega_1)\sin^2\varphi}, \qquad \sin^2\varphi = \frac{\omega_3 - \omega_1}{\omega_2 - \omega_1}\,\frac{p - \omega_2}{p - \omega_3} \tag{16.39}$$

Then

$$t = m\int_{\varphi_0}^{\varphi}\frac{d\varphi}{\sqrt{1 - k^2\sin^2\varphi}} \tag{16.40}$$

where

$$k^2 = \frac{\omega_4 - \omega_3}{\omega_4 - \omega_2}\,\frac{\omega_2 - \omega_1}{\omega_3 - \omega_1}$$

$$m = \frac{2}{\sqrt{(\omega_3 - \omega_1)(\omega_4 - \omega_2)}}$$

The choice between the above two transformations depends upon the intervals in which W is real.

If we introduce the usual notation

$$F(\varphi, k) = \int_0^{\varphi}\frac{d\varphi}{\sqrt{1 - k^2\sin^2\varphi}}$$

then

$$t = m[F(\varphi, k) - F(\varphi_0, k)]$$

where $F(\varphi_0, k)$ is a constant which can be replaced by t_0/m. Then

$$\frac{t + t_0}{m} = F(\varphi, k)$$

and hence by inversion

$$\sin\varphi = \operatorname{sn}\left(\frac{t + t_0}{m}, k\right)$$

The component p can be calculated from formulas (16.38) or (16.39). Using the formula

$$\operatorname{sn}^2 x = \frac{1 - \operatorname{cn} 2x}{1 + \operatorname{dn} 2x}$$

for $\omega_4 \leqq p \leqq \omega_3$ we obtain that

$$p = \frac{\omega_4(\omega_3 - \omega_1) + \omega_1(\omega_4 - \omega_3)\dfrac{1 - \operatorname{cn}\left(2\dfrac{t + t_0}{m}, k\right)}{1 + \operatorname{dn}\left(2\dfrac{t + t_0}{m}, k\right)}}{\omega_3 - \omega_1 + (\omega_4 - \omega_3)\dfrac{1 - \operatorname{cn}\left(2\dfrac{t + t_0}{m}, k\right)}{1 + \operatorname{dn}\left(2\dfrac{t + t_0}{m}, k\right)}}$$

and for $\omega_2 \leqq p \leqq \omega_1$ we have

$$p = \frac{\omega_2(\omega_3-\omega_1) - \omega_3(\omega_2-\omega_1)\dfrac{1-\operatorname{cn}\left(2\dfrac{t+t_0}{m}, k\right)}{1+\operatorname{dn}\left(2\dfrac{t+t_0}{m}, k\right)}}{\omega_3-\omega_1-(\omega_2-\omega_1)\dfrac{1-\operatorname{cn}\left(2\dfrac{t+t_0}{m}, k\right)}{1+\operatorname{dn}\left(2\dfrac{t+t_0}{m}, k\right)}}$$

Since q and r depend algebraically upon p, then these functions will also be given by elliptic functions of the time t.

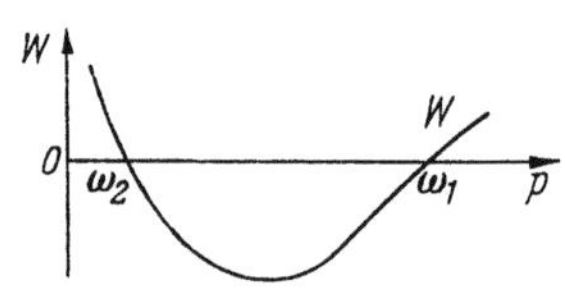

Fig. 43

In case W_2 the right-hand side of formula (16.31) is again an elliptic integral of the first kind which can be reduced to the form (16.37). Namely, if $\omega_1 > \omega_2$ are the real roots and $\omega_3 = b + i\,c$ and $\omega_4 = b - i\,c$ the complex conjugate roots (Fig. 43), then for $\omega_2 \leqq p \leqq \omega_1$, putting

$$p = \frac{\omega_1+\omega_2}{2} - \frac{\omega_1-\omega_2}{2}\,\frac{n-\cos\varphi}{1-n\cos\varphi}, \qquad \tan^2\frac{\varphi}{2} = \frac{\cos\sigma_1}{\cos\sigma_2}\,\frac{\omega_1-p}{p-\omega_2}$$

where

$$\tan\sigma_1 = \frac{\omega_1-b}{c}, \qquad \tan\sigma_2 = \frac{\omega_2-b}{c}, \qquad n = \tan\frac{\sigma_1-\sigma_2}{2}\tan\frac{\sigma_1+\sigma_2}{2}$$

$$\sigma_1, \sigma_2 < \pi/2$$

we obtain that

$$t = \bar{m}\int\limits_{\varphi_0}^{\varphi}\frac{d\varphi}{\sqrt{1-\bar{k}^2\sin^2\varphi}}$$

where

$$\bar{k}^2 = \sin^2\frac{\sigma_1-\sigma_2}{2}, \qquad \bar{m} = -\sqrt{\frac{\cos\sigma_1\cos\sigma_2}{c}}$$

By inversion it follows that

$$\sin\varphi = \operatorname{sn}\left(\frac{t+t_0}{\bar{m}}, \bar{k}\right)$$

Since

$$\cos\varphi = \operatorname{cn}\left(\frac{t+t_0}{\bar{m}}, \bar{k}\right)$$

then for the interval $\omega_2 \leqq p \leqq \omega_1$ the solution is

$$p = \frac{\omega_1+\omega_2}{2} - \frac{\omega_1-\omega_2}{2}\,\frac{n\operatorname{cn}\left(\dfrac{t+t_0}{\bar{m}}, \bar{k}\right)}{1-n\operatorname{cn}\left(\dfrac{t+t_0}{\bar{m}}, \bar{k}\right)}$$

In this case q and r will also be obtained as elliptic functions of the time t.

In case W_4 formula (16.31) assumes the form

$$t = \int_{p(0)}^{p} \frac{dp}{\sqrt{-(p-\omega_1)(p-\omega_2)(p-\omega_3)^2}}$$

and the graph of the gyroscopic function is one of the types seen in Fig. 44 and Fig. 45.

If we rewrite the above formula in the form

$$t = \int_{p(0)}^{p} \frac{dp}{(p-\omega_3)\sqrt{-p^2 + 2\frac{\omega_1+\omega_2}{2}p - \omega_1\omega_2}}$$

a cyclometric solution

$$t = -\frac{1}{\sqrt{-C}} \arcsin \frac{\left(\frac{\omega_1+\omega_2}{2} - \omega_3\right)p - \omega_1\omega_2 + \omega_3\frac{\omega_1+\omega_2}{2}}{(p-\omega_3)\sqrt{-\Delta}} + t_0$$

follows provided that

$$C = -\omega_3^2 + \omega_3(\omega_1+\omega_2) - \omega_1\omega_2 < 0 \tag{16.41}$$

$$\Delta = \omega_1\omega_2 - \frac{(\omega_1+\omega_2)^2}{4} < 0 \tag{16.42}$$

By inversion a trigonometric solution for p results, namely

$$p = \frac{\omega_1\omega_2 - \omega_3\frac{\omega_1+\omega_2}{2} + \sqrt{-\Delta}\,\omega_3 \sin[\sqrt{-C}(t-t_0)]}{\frac{\omega_1+\omega_2}{2} - \omega_3 + \sqrt{-\Delta}\sin[\sqrt{-C}(t-t_0)]}$$

Similarly, by virtue of the algebraic dependence of q and r upon p, trigonometric solutions will also be obtained for them.

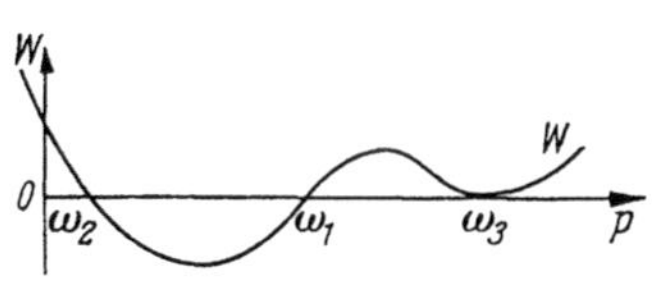

Fig. 44

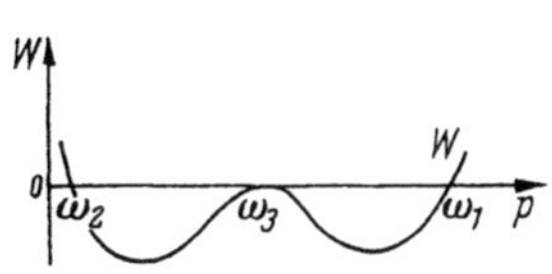

Fig. 45

Condition (16.42) is always satisfied by virtue of the fact that

$$\sqrt{\omega_1\omega_2} < \frac{\omega_1+\omega_2}{2}$$

for $\omega_1 \neq \omega_2$. Condition (16.41), however, is satisfied in the case of Fig. 44. In fact, let $\omega_3 = \omega_1 + \delta > \omega_1 > \omega_2 (\delta > 0)$. Then $C = \delta(\omega_2 - \omega_3) < 0$. If $\omega_3 = \omega_2 - \delta < \omega_2 < \omega_1$, then $C = \delta(\omega_3 - \omega_1) < 0$.

In the case of Fig. 45 we have $\omega_2 < \omega_3 = \omega_1 - \delta < \omega_1$, leading to $C = \delta(\omega_3 - \omega_2) > 0$, or $\omega_2 < \omega_3 = \omega_2 + \delta < \omega_1$, leading to $C = \delta(\omega_1 - \omega_3) > 0$. In such a case we have the asymptotic solution

$$t = -\frac{1}{\sqrt{C}} \operatorname{ar\,cosh} \left| \frac{\left(\frac{\omega_1 + \omega_2}{2} - \omega_3\right) p - \omega_1 \omega_2 + \omega_3 \frac{\omega_1 + \omega_2}{2}}{(p - \omega_3)\sqrt{-\Delta}} \right| + t_0$$

The above formula shows that

$$\lim_{t \to \infty} p = \omega_3$$

In case W_5 formula (16.31) assumes the form

$$t = \int\limits_{p(0)}^{p} \frac{dp}{i\sqrt{(p - \omega_1)(p - \omega_2)^3}} = \frac{2}{\omega_1 - \omega_2} \sqrt{\frac{p - \omega_1}{\omega_2 - p}} + t_0$$

the graph of $W = W(p)$ being given in Fig. 46.

By inversion of the last formula the expression

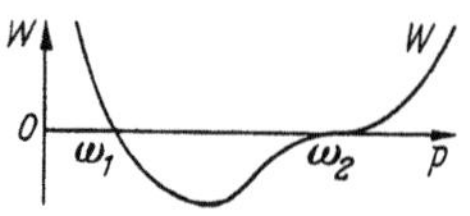

Fig. 46

$$p = \frac{4\omega_1 + \omega_2(\omega_1 - \omega_2)^2(t - t_0)^2}{4 + (\omega_1 - \omega_2)^2(t - t_0)^2}$$

is obtained which shows that

$$\lim_{t \to \infty} p = \omega_2$$

16.6. Determination of the position relative to a fixed coordinate trihedral. System (16.6) and the three equations for the direction cosines of the unit vector $\vec{K}$ along the fixed Z-axis, namely

$$\frac{d\alpha}{dt} = \beta r - \gamma q = \mathrm{A}$$

$$\frac{d\beta}{dt} = \gamma p - \alpha r = \mathrm{B}$$

$$\frac{d\gamma}{dt} = \alpha q - \beta p = \Gamma$$

constitute a system of six first order equations which can be written in the form

$$\frac{dp}{\Omega_x} = \frac{dq}{\Omega_y} = \frac{dr}{\Omega_z} = \frac{d\alpha}{\mathrm{A}} = \frac{d\beta}{\mathrm{B}} = \frac{d\gamma}{\Gamma} = \frac{dt}{1}$$

Eliminating the time t and introducing p as the independent variable, we obtain a system of five equations

$$\frac{dp}{\Omega_x} = \frac{dq}{\Omega_y} = \frac{dr}{\Omega_z} = \frac{d\alpha}{\mathrm{A}} = \frac{d\beta}{\mathrm{B}} = \frac{d\gamma}{\Gamma} \tag{16.43}$$

for the five quantities, $q, r, \alpha, \beta, \gamma$ as functions of p. Its integration requires five integrals $f_k = c_k$ $(k = 1, \ldots, 5)$, satisfying the equation

$$\frac{df_k}{dp} = \frac{\partial f_k}{\partial p} + \frac{\partial f_k}{\partial q}\frac{dq}{dp} + \frac{\partial f_k}{\partial r}\frac{dr}{dp} + \frac{\partial f_k}{\partial \alpha}\frac{d\alpha}{dp} + \frac{\partial f_k}{\partial \beta}\frac{d\beta}{dp} + \frac{\partial f_k}{\partial \gamma}\frac{d\gamma}{dp} = 0$$

where for the derivatives dq/dp, dr/dp, $d\alpha/dp$, $d\beta/dp$, $d\gamma/dp$ their expressions from system (16.43) are to be substituted.

So far three integrals are known, namely

$$f_1 = \vec{I}_K^* \cdot \vec{\omega} = c_1$$
$$f_2 = I_S^2 - D^2 = c_2$$
$$f_3 = \alpha^2 + \beta^2 + \gamma^2 = c_3 = 1$$

A simple calculation shows that

$$f_4 = (A_K p + D)\,\alpha + B_S\, q\, \beta + C_S\, r\, \gamma = c_4 = \text{const.}$$

Since

$$f_4 = I_{Sx}\,\alpha + I_{Sy}\,\beta + I_{Sz}\,\gamma = \vec{I}_S \cdot \vec{K} = I_{SZ} = c_4$$

the fourth integral expresses the fact that the Z-component of the angular momentum vector $\vec{I}_S$ is constant. This, however, is obvious since $\vec{I}_S$ is a vector of fixed length and direction in space.

For the determination of a fifth integral JACOBI's theory of the last multiplier can be used. Namely, since $M_J = 1$ is a multiplier, then a fifth integral can be put in the form

$$f_5 = \int \frac{\mathrm{A}\, dp - \Omega_x\, d\alpha}{\dfrac{\partial(f_1, f_2, f_3, f_4)}{\partial(q, r, \beta, \gamma)}} = \text{const.}$$

or

$$f_5 = \int \frac{\beta \dfrac{dp}{q} - \gamma \dfrac{dp}{r} + \dfrac{C_S - B_S}{A_K}\, d\alpha}{C_S\, r\, \beta - B_S\, q\, \gamma} = c_5 = \text{const.}$$

where q and r are given by formulas (16.30) and (16.28) respectively. In order to carry out the integration, we must express β and γ by means of the four integrals in terms of p and α. If the notations

$$\varphi_2 = \frac{c_4 - (A_K p + D)\,\alpha}{C_S^2 P_z - B_S^2 P_y}\,, \qquad \varphi_3 = \sqrt{\varphi_2^2 - \frac{A_K}{C_S - B_S}\,\frac{1 - \alpha^2}{C_S^2 P_z - B_S^2 P_y}}$$

are introduced, a calculation shows that $f_5 = c_5$ can be put in the form

$$f_5 = \int \frac{(C_S P_z - B_S P_y)\, dp}{i\sqrt{P_y P_z}\,(C_S^2 P_z + B_S^2 P_y)} - \int \frac{(C_S - B_S)\,\varphi_2\, dp}{i\,\varphi_3\,(C_S^2 P_z + B_S^2 P_y)}$$
$$+ \int \frac{d\alpha}{i\,\varphi_3\,(C_S^2 P_z + B_S^2 P_y)} = c_5 = \text{const.}$$

in which all the quadratures can be carried out since everything is expressed in terms of p and α. In addition, this is a real relation between p and α since $i\sqrt{P_y P_z}$ and $i\varphi_3$ are real.

By means of the integrals $f_k = c_k$ $(k = 1, \ldots, 5)$ we can now calculate the EULER angles. Namely, according to the preceding sections the components p, q, r can be expressed as functions of the time t. From the integral $f_5 = c_5$ we can express $\alpha = g_1[p(t)]$ as a function of t. Finally from the integrals $f_3 = c_3$, $f_4 = c_4$ and the already known functions $p(t)$ and $\alpha(t)$ we can also calculate $\beta = g_2(t)$ and $\gamma = g_3(t)$ in terms of the time t.

The EULER angles as functions of the time can be calculated from the relations

$$\alpha = \cos\Theta, \quad \beta = \sin\Theta\sin\Phi, \quad \gamma = \sin\Theta\cos\Phi$$

Then

$$\Theta = \operatorname{arc}\cos\alpha = \operatorname{arc}\cos g_1(t)$$

$$\Phi = \operatorname{arc}\tan(\beta/\gamma) = \operatorname{arc}\tan\frac{g_2(t)}{g_3(t)}$$

The third angle is given by the integral

$$\psi = \int \frac{p(t) - \dot{\Phi}}{\cos\Theta}\, dt$$

or, after substituting the above expressions for Θ and Φ,

$$\psi = \int \frac{p(t)\,[g_2^2(t) + g_3^2(t)] - g_3(t)\,\dot{g}_2(t) + g_2(t)\,\dot{g}_3(t)}{g_1(t)\,[g_2^2(t) + g_3^2(t)]}\, dt$$

Due to the complexity of the integrals involved in f_5 the relation $\alpha = g_1[p(t)]$ deduced from the integral $f_5 = c_5$ can be calculated only numerically, and hence the complete solution sketched above can be obtained only by numerical methods.

16.7. Permanent rotations. In the case of permanent rotations (STAUDE-MLODZEEVSKII) characterized by the fact that the angular momentum vector $\vec{I}_K^*$ of the substitute rigid body remains body-fixed, and hence $(d\vec{I}_K^*/dt)_{\text{rel}} = 0$, the EULER vector equation

$$\left(\frac{d\vec{I}_K^*}{dt}\right)_{\text{rel}} + \omega \times \vec{I}_K^* = \vec{M}$$

reduces to the equation

$$\vec{\omega} \times \vec{I}_K^* = \vec{M}$$

for $\vec{\omega}(p^0, q^0, r^0)$ or to the system of three scalar equations

$$\left.\begin{aligned}(C_S - B_S)\, q^0 r^0 &= 0\\ (A_K - C_S)\, r^0 p^0 &= -D\, r^0\\ (B_S - A_K)\, p^0 q^0 &= D\, q^0\end{aligned}\right\} \tag{16.44}$$

for which we seek a constant solution $p^0 = a$, $q^0 = b$, $r^0 = c$.

Since the fulfillment of system (16.44) is necessary for the existence of stationary solutions for $\vec{\omega}$, the following solutions given by Table 1 must be rejected

Table 1

p^0	q^0	r^0
a	b	c
0	b	c
0	b	0
0	0	c

(where a, b, c are non-zero constants)

Table 2

p^0	q^0	r^0	Equation
$a^* = D/(B_S - A_K)$	b	0	(16.45)
$a = D/(C_S - A_K)$	0	c	(16.46)
$\bar{a}$	0	0	(16.47)
0	0	0	(16.48)

On the other hand, expressions given in Table 2 satisfy system (16.44) and therefore are possible solutions.

In order to determine the position of the axes of permanent rotations of the substitute rigid body consider the scalar product

$$\vec{\omega}_e \cdot \left(\vec{\omega} \times \vec{I}_K^*\right) = \vec{M} \cdot \vec{\omega}_e = 0 \tag{16.49}$$

where the vectors $\vec{\omega}_e$, $\vec{\omega} \times \vec{I}_K^*$ and $\vec{M}$ have the components

$$\vec{\omega}_e(0, q, r), \quad \vec{\omega} \times \vec{I}_K^*[(C_S - B_S)\, q\, r,\ (A_K - C_S)\, r\, p,\ (B_S - A_K)\, p\, q],$$
$$\vec{M}(0, -D\, r, D\, q)$$

In terms of these components equation (16.49) reduces to the equation

$$p\, q\, r (B_S - C_S) = 0 \tag{16.50}$$

which is satisfied for $B_S = C_S$ (symmetric body K) or if at least one of the components p, q, r vanishes. If $B_S = C_S$, then p, q, r can all be different from zero and this case was discussed in Section 16.2. By a proper transformation (see H. LEIPHOLZ [1]) this case can always be reduced to the case (16.46). If $B_S \neq C_S$, then one of the components p, q or r must be zero. Hence the axes of permanent rotations are generators of planes. Stability considerations will show us that stable permanent axes of rotation are the generators of the p, r-plane cut along the r-axis. The STAUDE cone degenerates to a plane with a cut.

For the investigation of the stability of $\vec{\omega}$ relative to a perturbation of the initial values $p(0)$, $q(0)$, $r(0)$ two methods can be used. The first method is based upon the phase-plane analysis; the second is that of LYAPUNOV.

Let us investigate first the stability of solution (16.46). Let $\vec{\omega}_0 = (a, 0, c)$ be the unperturbed state of motion and $\vec{\omega}_0 + \vec{x} = (a + x_1, x_2, c + x_3)$ the perturbed one. If the latter expressions are substituted in system (16.6), then the perturbations $\vec{x}(x_1, x_2, x_3)$ satisfy the system

$$\begin{aligned} A_K \dot{x}_1 + (C_S - B_S)\, x_2 (c + x_3) &= 0 \\ B_S \dot{x}_2 + (A_K - C_S)\,(a + x_1)\,(c + x_3) &= -D(c + x_3) \\ C_S \dot{x}_3 + (B_S - A_K)\,(a + x_1)\, x_2 &= D\, x_2 \end{aligned}$$

which possesses two integrals

$$B_S x_2^2 + C_S (c + x_3)^2 + A_K (a + x_1)^2 = \text{const.} \tag{16.51}$$

$$B_S^2 x_2^2 + C_S^2 (c + x_3)^2 + A_K^2 \left[(a + x_1) + \frac{D}{A_K}\right]^2 = \text{const.} \tag{16.52}$$

Construct the function

$$\begin{aligned} V_1 = B_S(\lambda - B_S)\, x_2^2 + C_S(\lambda - C_S)\, x_3^2 + A_K(\lambda - A_K)\, x_1^2 + \\ + 2C_S\, c(\lambda - C_S)\, x_3 + 2A_K(\lambda\, a - A_K\, a - D)\, x_1 = \text{const.} \end{aligned}$$

If we choose for the constant factor λ the value

$$\lambda = C_S = \frac{A_K\, a + D}{a} \tag{16.53}$$

then V_1 becomes a positive definite function of x_1 and x_2. Since V_1 is a combination of the above two integrals and thus $d V_1/dt = 0$, it follows that under the assumption (16.53) which is equivalent to $a = D/(C_S - A_K)$ there is stability relative to x_1 and x_2. However, by virtue of the integral (16.51) there is stability also with respect to x_3. Hence condition (16.53), which as a consequence of the existence of solution (16.46) is satisfied, guarantees the stability of solution (16.46).

Next let us investigate the stability of solution (16.47). Let $\vec{\omega}_0 = (\bar{a}, 0, 0)$ be the unperturbed state of motion and $\vec{\omega}_0 + \vec{x} = (\bar{a} + x_1, x_2, x_3)$ the perturbed state. Then the perturbations $\vec{x}(x_1, x_2, x_3)$ satisfy the system

$$\begin{aligned} A_K \dot{x}_1 + (C_S - B_S)\, x_2\, x_3 &= 0 \\ B_S \dot{x}_2 + (A_K - C_S)\,(\bar{a} + x_1)\, x_3 &= -D\, x_3 \\ C_S \dot{x}_3 + (B_S - A_K)\,(\bar{a} + x_1)\, x_2 &= D\, x_2 \end{aligned}$$

which possesses the two integrals

$$B_S x_2^2 + C_S x_3^2 + A_K(\bar{a} + x_1)^2 = \text{const.}$$

$$B_S^2 x_2^2 + C_S x_3^2 + A_K^2\left[(\bar{a} + x_1) + \frac{D}{A_K}\right]^2 = \text{const.}$$

Construct the function

$$V_2 = B_S(\lambda - B_S)\, x_2^2 + C_S(\lambda - C_S)\, x_3^2 + A_K(\lambda - A_K)\, x_1^2 + \\ + 2A_K[\bar{a}(\lambda - A_K) - D]\, x_1 = \text{const.}$$

which is a combination of the above integrals. This function will be positive definite provided that

$$\lambda > B_S, \quad \lambda > C_S, \quad \lambda > A_K, \quad \lambda = (A_K \bar{a} + D)\sqrt{\bar{a}} \qquad (16.54)$$

Hence under assumptions (16.54) there is stability relative to x_1, x_2 and x_3. Together with $\vec{\omega}_0 = (\bar{a}, 0, 0)$ conditions (16.54), which are equivalent to

$$D > 0, \quad \bar{a} < \frac{D}{C_S - A_K} = V_{hy}, \quad \bar{a} < \frac{D}{B_S - A_K} = V_{hz} \qquad (16.55)$$

correspond to the case W_4. In fact, since $q = r = 0$, $p = \bar{a}$ is a root of the polynomial P_y as well as of P_z. Since by (16.55) $D > 0$, then V_{hy} and V_{hz} are both positive, and $p = \bar{a}$ lies to the left of the vertices of both parabolas. By virtue of (16.34) the P_z-parabola is flatter than the P_y-parabola (Fig. 47).

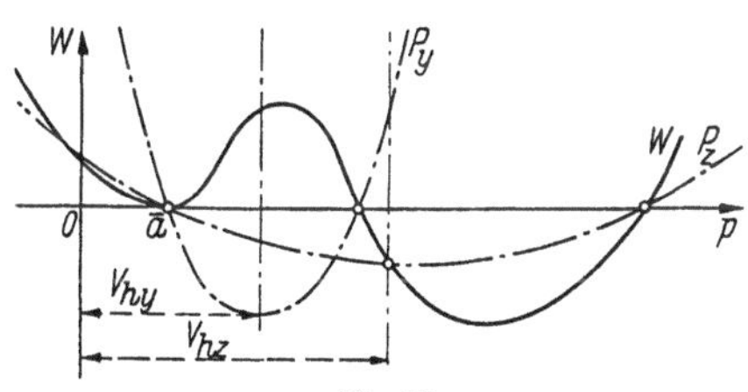

Fig. 47

So far it has been assumed that $D > 0$. Corresponding results are obtained for $D < 0$ if a reflection about the vertical straight line $p = \bar{a}$ is performed. This is seen from the LYAPUNOV V-function

$$V_2^* = B_S(B_S - \lambda)\, x_2^2 + C_S(C_S - \lambda)\, x_3^2 + A_K(A_K - \lambda)\, x_1^2 + \\ + 2A_K[a(A_K - \lambda) + D]\, x_1 = \text{const.}$$

constructed by means of the above two integrals. The stability conditions here are

$$D < 0, \quad \bar{a} > D/(C_S - A_K) = V_{hy}$$
$$\bar{a} > D/(B_S - A_K) = V_{hz}$$

In case (16.48) let the perturbed state of motion be represented by $\vec{\omega}_0 + \vec{x} = (x_1, x_2, x_3)$. Then system (16.6) leads to the following system of equations for the perturbations

$$A_K \dot{x}_1 + (C_S - B_S)\, x_2\, x_3 = 0$$
$$B_S \dot{x}_2 + (A_K - C_S)\, x_1\, x_3 = -D\, x_3$$
$$C_S \dot{x}_3 + (B_S - A_K)\, x_1\, x_2 = D\, x_2$$

This system possesses a positive definite integral

$$V_3 = A_K x_1^2 + B_S x_2^2 + C_S x_3^2 = \text{const.}$$

which can be used as a LYAPUNOV V-function since the condition $dV_3/dt = 0$ is automatically satisfied. Hence in this case the perturbed motion is stable (Fig. 48).

Solution (16.45) cannot be realized by virtue of conditions (16.33) and (16.34). In fact, for $p = a^* = D/(B_S - A_K)$ the polynomial P_z has a double root and simultaneously $P_y > 0$. The latter fact contradicts inequalities (16.34).

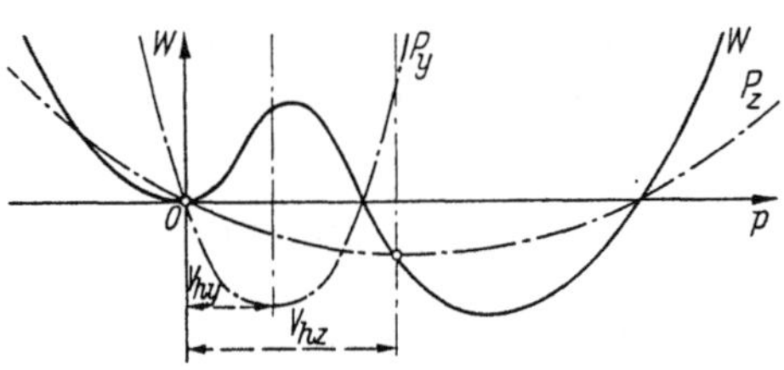

Fig. 48

Thus we see that under the assumptions (16.33) the stationary solutions (16.46) to (16.48) are stable. In these stable cases the minima on the graphs of the gyroscopic function W occur at the points corresponding to stationary solutions of p; the function W vanishes at these points. Since W can be interpreted as a certain potential function, the requirement that W be a minimum for stability is self-explanatory.

16.8. Motions corresponding to the stationary solutions of $\vec{\omega}$. In order to determine the motion in space corresponding to the stationary solution (16.46), $\vec{\omega}_0 = (a, 0, c)$, the system

$$\left.\begin{aligned} p &= a = \dot{\psi}\cos\Theta + \dot{\Phi} \\ q &= 0 = \dot{\psi}\sin\Theta\sin\Phi + \dot{\Theta}\cos\Phi \\ r &= c = \dot{\psi}\sin\Theta\cos\Phi - \dot{\Theta}\sin\Phi \end{aligned}\right\} \qquad (16.56)$$

must be solved. In the asymmetric case the kinetic energy is

$$T = \tfrac{1}{2}(A_K p^2 + B_S q^2 + C_S r^2 + D\, o_x)$$

Since in the present case $q = 0$, we can rewrite T in the form

$$T = \tfrac{1}{2}\,[A_K p^2 + C_S(q^2 + r^2) + D\, o_x]$$

or, by substitution of the above expressions for p, q, r,

$$T = \tfrac{1}{2}\,[A_K(\dot{\psi}\cos\Theta + \dot{\Phi})^2 + C_S(\dot{\Theta}^2 + \dot{\psi}^2\sin^2\Theta) + A_R(\dot{\psi}\cos\Theta + \dot{\Phi} + \dot{\alpha})^2] \qquad (16.57)$$

We observe that ψ and Φ are cyclic coordinates, and hence

$$\frac{\partial T}{\partial \dot{\Phi}} = A_K p + D = I_{Sx} = \text{const.} \tag{16.58}$$

$$\frac{\partial T}{\partial \dot{\psi}} = (A_K p + D) \cos\Theta + C_S \dot{\psi} \sin^2\Theta = I_{SZ} = \text{const.} \tag{16.59}$$

Solving the second integral for $\dot{\psi}$ and using the first integral we obtain

$$\dot{\psi} = \frac{I_{SZ} - (A_K p + D)\cos\Theta}{C_S \sin^2\Theta} = \frac{I_{SZ} - I_{Sx}\cos\Theta}{C_S \sin^2\Theta} \tag{16.60}$$

If we solve (16.57) for $\dot{\Theta}^2$ and make use of (16.60), the equation

$$\dot{\Theta}^2 + \frac{(I_{SZ} - I_{Sx}\cos\Theta)^2}{I_S^2 \sin^2\Theta} = \frac{2}{C_S}\left[T - \frac{1}{2}(A_K p^2 + D o_x)\right] = r^2$$

results. Putting $\cos\Theta = g$ and observing that as a consequence of $q = 0$ it follows that $I_{Se}^2 = C_S^2 r^2$, where I_{Se} is the equatorial component of $\vec{I}_S$, we obtain the equation

$$\dot{g}^2 + \frac{|\vec{I}_S|^2}{C_S^2} g^2 - 2\frac{I_{SZ} I_{Sx}}{C_S^2} g = \frac{I_{Se}^2 - I_{SZ}^2}{C_S^2} \tag{16.61}$$

which has a single-valued solution, reducible to $\Theta = \Theta_0 = \text{const.}$, for a particular choice of the fixed coordinate trihedral (see H. Leipholz [1]).

Substituting $\Theta = \text{const.}$ into equations (16.56), we can find a solution under the assumption $\dot{\Phi} = \Phi = 0$, namely,

$$\dot{\psi}^2 = p^2 + r^2 = \vec{\omega}_0^2, \quad \dot{\psi} = |\vec{\omega}_0| \tag{16.62}$$

Since

$$|\vec{\omega}_0| = \sqrt{p^2 + r^2} = \sqrt{a^2 + c^2} = \omega_0 = \text{const.}$$

then also

$$\cos\Theta = p/|\vec{\omega}_0| = a/\omega_0 = \text{const.}$$

and hence

$$\Theta = \arccos(p/\dot{\psi}) \tag{16.63}$$

Thus to the stable permanent rotation (16.46) of the substitute rigid body there corresponds a regular precession defined by formulas (16.62) and (16.63), the velocity of proper rotation $\dot{\Phi}$ about the symmetry axis x of the body being equal to zero (Fig. 49).

In the case of the stationary solution (16.47), $\vec{\omega}_0 = (\bar{a}, 0, 0)$, we obtain as before (since $q = r = 0$) the equation

$$\dot{\Theta}^2 + \frac{(I_{Sz} - I_{Sx}\cos\Theta)^2}{I_S^2 \sin^2\Theta} = 0$$

or

$$\dot{g}^2 + \frac{I_{Sx}^2}{C_S^2} g^2 - 2\frac{I_{Sz} I_{Sx}}{C_S^2} g = -\frac{I_{Sz}^2}{C_S^2}$$

which is of the same type as equation (16.61), and therefore possesses a solution $\Theta = \text{const}$. Since $r = 0$, then from the third equation of

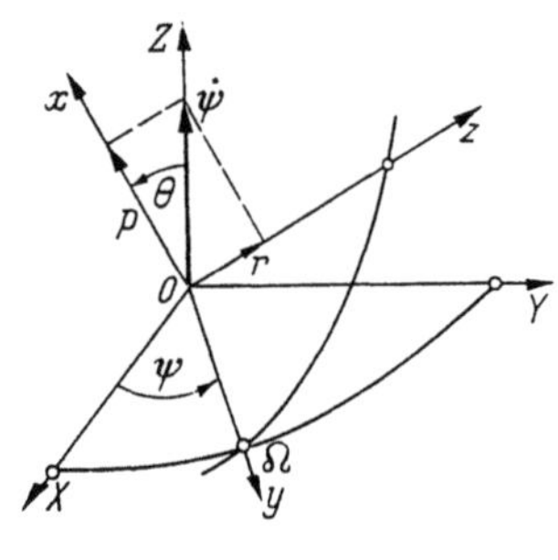

Fig. 49

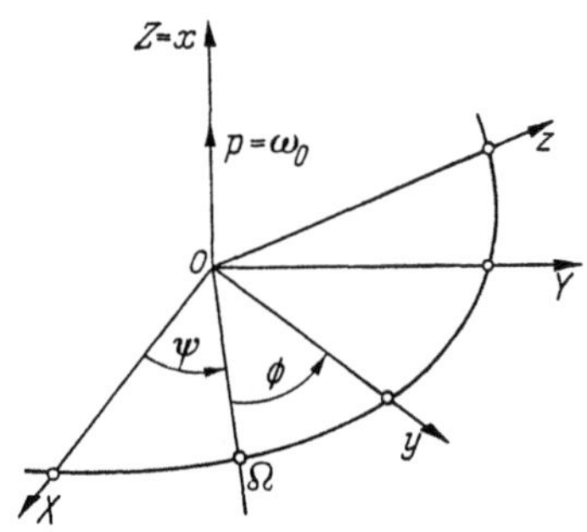

Fig. 50

system (16.56) we conclude that $\Theta \equiv 0$. From the first equation of system (16.56) it then follows that $p = \dot{\psi} + \dot{\Phi} = \dot{\chi} = \bar{a}$. We have a permanent rotation about the symmetry axis of the body (Fig. 50).

In the case of solution (16.47) there is another particular solution of system (16.56), namely

$$\Theta = \text{const.} \neq 0, \qquad \dot{\psi} = 0, \qquad \dot{\Phi} = p$$

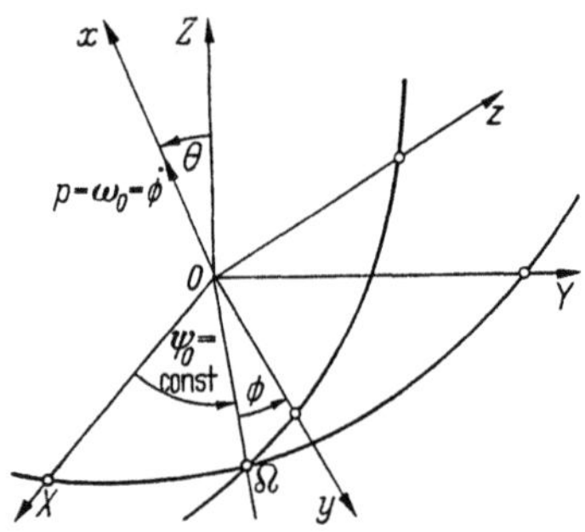

Fig. 51

This, however, does not differ essentially from the preceding solution, the only difference being that the space-fixed coordinate trihedral $OXYZ$ occupies a more general position relative to the body-fixed trihedral $Oxyz$ (Fig. 51).

To the stationary solution (16.48), $\vec{\omega}_0 = (0, 0, 0)$, corresponds the rest position of the rigid body with the trivial solution $\Theta = \text{const.}$, $\psi = \text{const.}$, $\Phi = \text{const.}$ of system (16.56).

Chapter V

Gyroscope in a Cardan suspension

§ 17. Aspects of the Cardan suspension of gyroscopes

17.1. Introduction. While the CARDAN suspension of a gyroscope has been used for about 150 years [F. BOHNENBERGER (1817), L. FOUCAULT (1852)], its effect on the motion and stability of the gyroscope has been investigated only during the last 25 years. See, for example, E. L. NIKOLAI [1, 2], H. PORITSKY [1], B. T. PLYMALE and R. GOODSTEIN [1], K. MAGNUS [1], R. GRAMMEL and H. ZIEGLER [1], N. G. ČETAEV [3], and others. E. L. NIKOLAI in 1939 investigated the inertial motion of a balanced symmetric gyroscope in a CARDAN suspension, taking into account the masses of the gimbals, and the stability of the motion in the case of a large angular velocity. He also investigated the effect of the magnitude and direction of the angular velocity of the outer gimbal on the stability of the axis of the gyroscope in the vertical position and discovered a curious phenomenon. K. MAGNUS [2] showed that a similar phenomenon also occurs in the case of a non-balanced symmetric gyroscope in a CARDAN suspension. In the same paper he obtained by means of a LYAPUNOV V-function sufficient conditions for the stability of rotation of the gyroscope about the vertical axis of the outer gimbal of a CARDAN suspension.

The stability of the regular precession of a gyroscope in a CARDAN suspension for the case where the nutation angle $\Theta_0 \neq 0$ has been investigated by V. N. SKIMEL' [2]. V. V. RUMYANCEV [5] derived a sufficient condition for the stability of the regular precession of a gyroscope in a CARDAN suspension by means of a LYAPUNOV V-function in the form of a linear bundle of first integrals of the motion. For $\Theta_0 = 0$ this condition gives a sufficient stability condition for the axis of the gyroscope in the vertical position. Also the necessity of this condition is shown, and the influence of dissipative forces on the stability of motion of a gyroscope is clarified. In another paper V. V. RUMYANCEV [6] investigated the stability of certain motions of a heavy symmetric gyroscope in a CARDAN suspension in the case where the fixed axis of the outer gimbal is horizontal.

Under the same assumptions as those by V. V. RUMYANCEV [6], A. A. BOGOYAVLENSKII [3] obtained a particular integral of the equations of motion. While K. MAGNUS [3] investigated a particular case of the motion of a heavy symmetric gyroscope in a CARDAN suspension when the axis of the outer gimbal coincides with the vertical, A. M. TABAROVSKII [1] considered the case where the axis of the outer gimbal makes an arbitrary angle with the vertical.

The influence on the stability of motion of a viscous friction at the bearings of the gimbal axes was investigated by V. V. RUMYANCEV [5]. On the other hand, L. I. KUZNECOV [1] and V. V. KREMENTULO [1, 2] have investigated the stability of the gyroscope, taking into account the dry friction on the axis of the inner gimbal.

B. V. BULGAKOV [1] studied the motion of the axis of the rotor in the case where the outer CARDAN ring is subject to a precession with constant angular velocity. The problem is solvable in terms of elliptic quadratures. Three types of motion are possible depending on whether the initial angular velocity of the inner CARDAN ring is smaller, equal or larger than a certain well-defined quantity. In the first case the axis of the rotor oscillates about that of the precession, in the second case the vertex of the gyroscope asymptotically approaches a definite limiting position while in the third case the inner Cardan ring together with the rotor both rotate in one and the same direction but with variable angular velocity. In the second part of the paper the air resistance is taken into account. See also A. V. Svetlov [1] and B. V. Bulgakov [2].

17.2. Statement of the problem. Consider a gyroscope suspended in two gimbals, as shown in Fig. 52, in such a way that the axis of

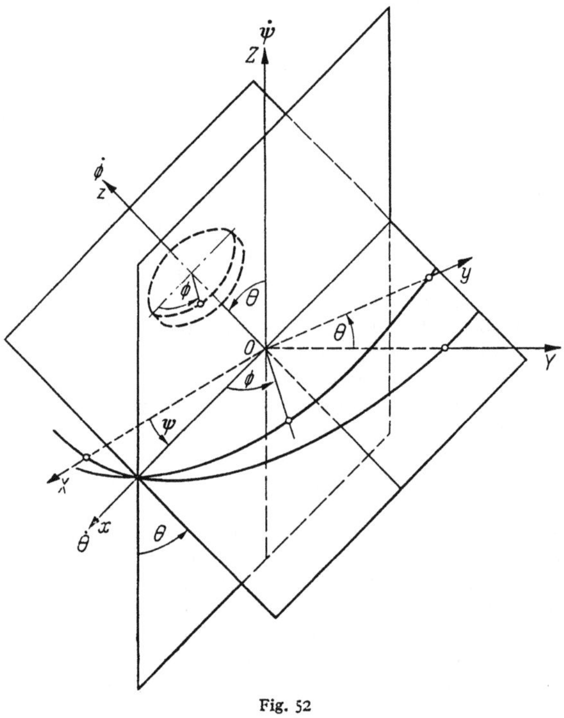

Fig. 52

the outer gimbal is vertical, i.e. it coincides with the direction of gravity. We also assume that friction at all bearings may be neglected. The system consisting of the rotor, the inner and outer gimbal rings has three degrees of freedom. Therefore its position can be determined by means of three coordinates such as the Euler angles ψ, Θ, Φ, where ψ is the angle of rotation of the outer Cardan ring about its space fixed axis (Z-axis), measured from a certain axis (X-axis) orthogonal to the Z-axis, Θ is the angle of rotation of the inner gimbal ring about its axis (x-axis), i.e. Θ is the angle showing the inclination of the inner gimbal plane relative to the outer gimbal plane, and Φ gives the position of the rotor relative to the inner gimbal ring.

Assume that in the initial position the planes of both CARDAN rings coincide, i.e. $\psi = \Theta = 0$. Further assume, in accordance with the case of LAGRANGE and POISSON, that the mass center of the system consisting of the rotor and the inner gimbal ring lies on the axis of the rotor at the distance s from the fixed point of suspension O. This distance is considered positive if in the initial position of the system the mass center is above the fixed point O.

Let the x, y, z-axes be the principal axes of inertia for the inner gimbal ring and, because of the assumed symmetry of the rotor, also for the latter. The principal axes of inertia for the outer gimbal ring need not be specified. Denote the principal moments of inertia about the above system of axes respectively for the rotor by $A_1 = B_1$, C_1 and for the inner gimbal ring by A_2, B_2, C_2. For the outer gimbal ring let C_3 be the moment of inertia about the Z-axis.

17.3. Equations of motion and their first integrals. We shall use the LAGRANGE method for finding the equations of motion of the gyroscope in a CARDAN suspension. For this purpose the kinetic and potential energies of the complete system are to be calculated.

The kinetic energy T of the system is equal to the sum of the kinetic energy of the outer gimbal, T_3, that of the inner gimbal, T_2, and the kinetic energy T_1 of the rotor. Since the outer gimbal rotates about the Z-axis with angular velocity $\dot{\psi}$, its kinetic energy is

$$T_3 = \tfrac{1}{2} C_3 \dot{\psi}^2$$

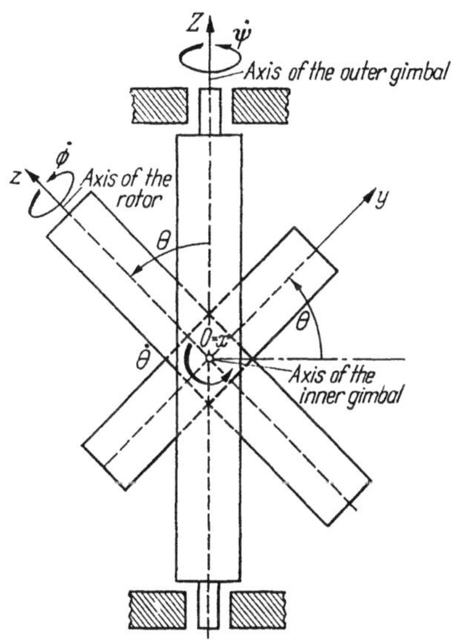

Fig. 53

The inner gimbal rotates about the x-axis with angular velocity $\dot{\Theta}$, and about the Z-axis together with the outer gimbal with angular velocity $\dot{\psi}$ (Figs. 53 and 54). Therefore the components of the angular velocity $\vec{\omega}_2$ (p_2, q_2, r_2) of rotation of the inner gimbal about the axes of the coordinate trihedral $O\,x\,y\,z$ are

$$p_2 = \dot{\Theta}, \qquad q_2 = \dot{\psi} \sin\Theta, \qquad r_2 = \dot{\psi} \cos\Theta$$

Hence

$$T_2 = \tfrac{1}{2} (A_2 \dot{\Theta}^2 + B_2 \dot{\psi}^2 \sin^2\Theta + C_2 \dot{\psi}^2 \cos^2\Theta)$$

The gyroscope rotates about the Z-axis with angular velocity $\dot{\psi}$, and together with the inner gimbal about the x-axis with angular velocity $\dot{\Theta}$, and about the z-axis with angular velocity $\dot{\Phi}$. Therefore

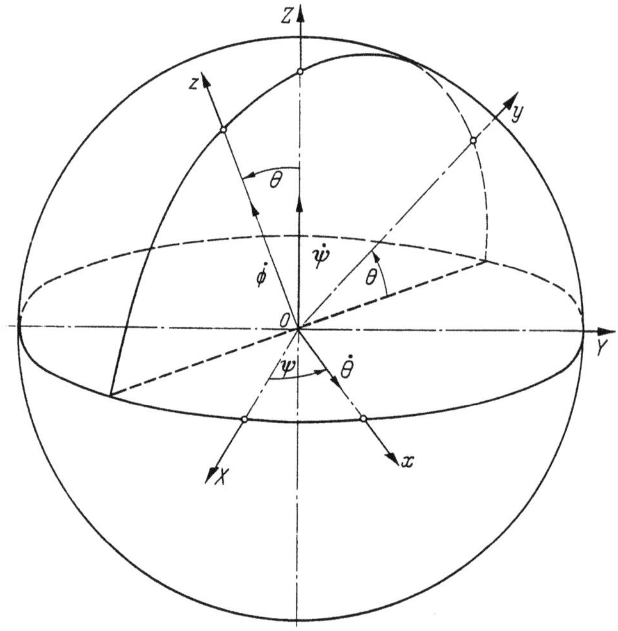

Fig. 54

the components of the angular velocity $\vec{\omega}_1(p_1, q_1, r_1)$ of rotation of the gyroscope on the axes of the trihedral $O\,x\,y\,z$ are

$$p_1 = \dot{\Theta}, \quad q_1 = \dot{\psi}\sin\Theta, \quad r_1 = \dot{\Phi} + \dot{\psi}\cos\Theta$$

Hence

$$T_1 = \tfrac{1}{2} A_1(\dot{\Theta}^2 + \dot{\psi}^2\sin^2\Theta) + \tfrac{1}{2} C_1(\dot{\Phi} + \dot{\psi}\cos\Theta)^2$$

Consequently

$$2T = A_1(\dot{\Theta}^2 + \dot{\psi}^2\sin^2\Theta) + C_1(\dot{\Phi} + \dot{\psi}\cos\Theta)^2 + \\ + A_2\dot{\Theta}^2 + B_2\dot{\psi}^2\sin^2\Theta + C_2\dot{\psi}^2\cos^2\Theta + C_3\dot{\psi}^2 \qquad (17.1)$$

or

$$2T = (A_1 + A_2)\dot{\Theta}^2 + [(A_1 + B_2 + C_3) - (A_1 + B_2 - C_2)\cos^2\Theta]\dot{\psi}^2 + \\ + C_1(\dot{\Phi} + \dot{\psi}\cos\Theta)^2 \qquad (17.2)$$

In order to calculate the potential energy U of the system the direction of the acceleration due to gravity is to be calculated. Without loss of generality we can assume that the vertical lies in the $X\,Z$-plane.

Then, as seen from Fig. 55, the angle Θ^* is to be calculated from the spherical triangle with side lengths Θ, Θ_0, Θ^*. Hence by the cosine law we can write that

$$\cos\Theta^* = \cos\Theta_0\cos\Theta + \sin\Theta_0\sin\Theta\sin\psi \qquad (17.3)$$

Let m be the mass of the rotor and the inner gimbal ring, and s the distance of the mass center G from the origin. Then the potential energy is

$$U = m g s\cos\Theta^* = m g s(\cos\Theta_0\cos\Theta + \sin\Theta_0\sin\Theta\sin\psi) \qquad (17.4)$$

The mass center of the outer gimbal ring lies by assumption on the Z-axis and therefore there is no change in the potential energy of the outer gimbal in the motion of the system.

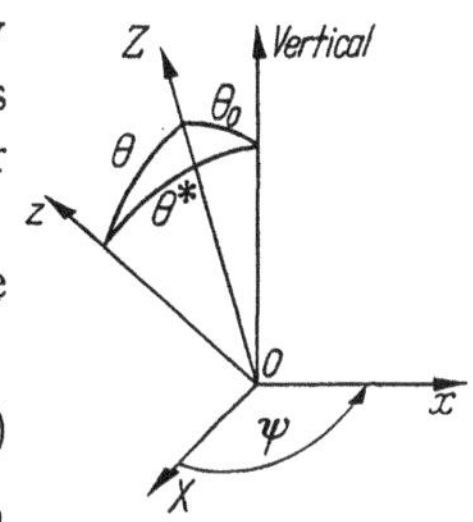

Fig. 55

Apply now the LAGRANGE equation of the second kind

$$\frac{d}{dt}\left(\frac{\partial T}{\partial\dot{q}}\right) - \frac{\partial T}{\partial q} = -\frac{\partial U}{\partial q} \qquad (17.5)$$

to our case by taking for the generalized coordinate q the EULER angles Φ, ψ and Θ in succession. Then we obtain the following equations of motion

$$\frac{d}{dt}[C_1(\dot\Phi + \dot\psi\cos\Theta)] = 0 \qquad (17.6)$$

$$\frac{d}{dt}\{[(A_1+B_2+C_3) - (A_1+B_2-C_2)\cos^2\Theta]\dot\psi + C_1(\dot\Phi + \dot\psi\cos\Theta)\cos\Theta\}$$
$$= -m g s\sin\Theta_0\sin\Theta\cos\psi \qquad (17.7)$$

$$(A_1+A_2)\ddot\Theta - (A_1+B_2-C_2)\dot\psi^2\sin\Theta\cos\Theta + C_1\dot\psi(\dot\Phi + \dot\psi\cos\Theta)\sin\Theta$$
$$= m g s(\cos\Theta_0\sin\Theta - \sin\Theta_0\cos\Theta\sin\psi) \qquad (17.8)$$

Since Φ is a cyclic coordinate, we obtain at once by integration of (17.6) a first integral

$$\dot\Phi + \dot\psi\cos\Theta = n = \text{const.} \qquad (17.9)$$

which expresses the fact that the angular velocity component of the rotor along the z-axis is constant.

On the other hand, since no dissipative forces are involved the energy integral exists in the form

$$2T + 2U = (A_1+A_2)\dot\Theta^2 + [(A_1+B_2+C_3) - (A_1+B_2-C_2)\cos^2\Theta]\dot\psi^2 +$$
$$+ C_1 n^2 + 2m g s(\cos\Theta_0\cos\Theta + \sin\Theta_0\sin\Theta\sin\psi) = E = \text{const.} \qquad (17.10)$$

In order to reduce the problem to quadratures a third integral, independent of the above two, is required. Such an integral can be derived from equation (17.7) when ψ is a cyclic coordinate, i.e. $\partial U/\partial \psi = 0$. This occurs in the following two cases:

(i) the case of a torque-free gyroscope ($s = 0$),

(ii) the case of a CARDAN suspension for which the axis of the outer gimbal ring coincides with that of the vertical ($\Theta_0 = 0$, $\Theta = \Theta^*$).

The first case has been discussed in the well-known books by F. KLEIN and A. SOMMERFELD [1], and R. GRAMMEL [4], and recently by K. MAGNUS [1]. These discussions will not be reproduced here. The second case was investigated by K. MAGNUS [3].

For $\Theta_0 = 0$ it follows from (17.7) by virtue of (17.9) that

$$[(A_1 + B_2 + C_3) - (A_1 + B_2 - C_2)\cos^2\Theta]\,\dot\psi + C_1 n \cos\Theta = D_z = \text{const.} \tag{17.11}$$

In addition, the energy integral reduces to

$$(A_1 + A_2)\,\dot\Theta^2 + [(A_1 + B_2 + C_3) - (A_1 + B_2 - C_2)\cos^2\Theta]\,\dot\psi^2 + C_1 n^2 + 2 m g s \cos\Theta = E \tag{17.12}$$

The integrals (17.9), (17.11) and (17.12), known for the case $\Theta_0 = 0$, correspond to the three known integrals in the case of LAGRANGE and POISSON of a heavy rigid body. They reduce to the latter if the masses of the CARDAN rings are neglected ($A_2 = B_2 = C_2 = C_3 = 0$).

While (17.12) is the energy integral and (17.9) expresses the fact that the component of the angular momentum of the rotor along its axis (z-axis) is constant, (17.11) expresses the fact that the component of the total angular momentum along the vertical (Z-axis) is constant.

17.4. Solution in the case where the axis of the outer gimbal ring is vertical. From equation (17.11) we obtain for the velocity of precession $\dot\psi$ the expression

$$\dot\psi = \frac{D_z - C_1 n \cos\Theta}{(A_1 + B_2 + C_3) - (A_1 + B_2 - C_2)\cos^2\Theta} \tag{17.13}$$

If the angle of nutation Θ is known as a function of the time t, then ψ is given by a quadrature.

Substitution of (17.13) into (17.12) yields a single differential equation of the first order for Θ which by means of the notations

$$\cos\Theta = u \tag{17.14}$$

$$\left.\begin{aligned} \frac{C_1 n}{A_1 + A_2} &= k_1, & \frac{E - C_1 n^2}{A_1 + A_2} &= k_4,\\ \frac{D_z}{A_1 + A_2} &= k_2, & \frac{A_1 + B_2 + C_3}{A_1 + A_2} &= k_5,\\ \frac{2 m g s}{A_1 + A_2} &= k_3, & \frac{A_1 + B_2 - C_2}{A_1 + A_2} &= k_6, \end{aligned}\right\} \tag{17.15}$$

assumes the form

$$\left(\frac{du}{dt}\right)^2 = (1-u^2)\left[(k_4 - k_3 u) - \frac{(k_2 - k_1 u)^2}{k_5 - k_6 u^2}\right] = U(u) \qquad (17.16)$$

The inverse function $t = t(u)$ of $u = u(t)$ is given by the integral

$$t = t_0 + \int \frac{du}{\sqrt{U(u)}} \qquad (17.17)$$

If by inversion of (17.17) $u = u(t)$ is determined and, in addition, $\Theta = \arccos u$ is known, then the angle ψ is obtained by integrating (17.13):

$$\psi = \psi_0 + \int \frac{k_2 - k_1 u}{k_5 - k_6 u^2}\, dt \qquad (17.18)$$

The angles Θ and ψ determine the position of the CARDAN suspension system and the axis of the rotor. The angle of rotation Φ of the rotor is of no interest but in case it is needed it can be determined from (17.9).

Contrary to the case of LAGRANGE and POISSON, for a heavy symmetric gyroscope in a CARDAN suspension the gyroscopic function $U(u)$ is a fractional rational function of u. Therefore, in general, the integrals (17.17) and (17.18) can be calculated only numerically.

17.5. Regular precessions. We shall investigate whether for a gyroscope in a CARDAN suspension regular precessions, corresponding to those in the case of LAGRANGE and POISSON, exist. In the latter case the axes of the possible precession cones are always vertical, i.e. they coincide with the direction of gravity. In the case of a heavy gyroscope in a CARDAN suspension there are two significant directions: that of gravity and that of the axis of the outer CARDAN ring. It is easy to see that each cone of a possible regular precession must have both directions as axes of symmetry simultaneously. Therefore one can expect to have the case of a regular precession only if the axis of the outer CARDAN ring is vertical, i.e. if it coincides with the direction of gravity.

If both directions do not coincide, then in the case of a precession symmetric relative to the vertical, non-uniform motions of the CARDAN system would result which in turn would lead to the corresponding non-uniform moments of reaction for the complete system. Hence a description of the precession cone with uniform velocity would be impossible.

Conversely, in the case of a precession symmetric with respect to the axis of the outer gimbal, the moment due to gravity would change during the course of motion not only the direction of the precessional axis but the magnitude of the precession as well. As a consequence the existence of a constant velocity of precession would be impossible.

Thus it is sufficient to investigate the case $\Theta_0 = 0$. The regular precessions are characterized by the conditions

$$\begin{aligned} \ddot{\psi} &= 0, \quad \dot{\psi} = \text{const.} \\ \dot{\Theta} &= 0, \quad \Theta = \text{const.} \end{aligned} \qquad (17.19)$$

Substituting (17.19) into the integrals (17.9) and (17.11) we obtain two equations for the determination of the constants n and D_z. Upon substitution of (17.19) into the third equation of motion (17.8) the equation

$$[(A_1 + B_2 - C_2)\, \dot{\psi}^2 \cos\Theta - C_1\, n\, \dot{\psi} + m\, g\, s] \sin\Theta = 0$$

or

$$(k_6\, u\, \dot{\psi}^2 - k_1\, \dot{\psi} + \tfrac{1}{2}\, k_3) \sin\Theta = 0 \qquad (17.20)$$

results. If we exclude the case $\sin\Theta = 0$, the quadratic equation (17.20) gives two values for the velocity of precession,

$$\left.\begin{matrix} \dot{\psi}_1 \\ \dot{\psi}_2 \end{matrix}\right\} = \frac{k_1}{2 k_6 u} \left(1 \pm \sqrt{1 - \frac{2 k_3 k_6 u}{k_1^2}}\right) \qquad (17.21)$$

Since usually $k_6 > 0$ (by construction of the inner CARDAN ring), the radicand would be negative in the case the product $k_3\, u$ is positive. This will happen if k_3 and u have the same sign; hence for $s > 0$, $0 \leqq \Theta \leqq \pi/2$ and for $s < 0$, $\pi/2 \leqq \Theta \leqq \pi$. Both cases correspond to an "elevated gyroscope". If, however, k_3 and u have opposite signs, we have the case of a "hanging gyroscope".

It follows from (17.21) that a hanging gyroscope always has two regular precessions. The sense of rotation in each case is different since both solutions in (17.21) have different signs. For an elevated gyroscope there are two different regular precessions only if the condition

$$k_1^2 > 2 k_3\, k_6\, u$$

or

$$C_1^2\, n^2 > 4 (A_1 + B_2 - C_2)\, m\, g\, s \cos\Theta \qquad (17.22)$$

is satisfied. Both solutions in (17.21) have in this case the same sign so that the regular precessions of the elevated gyroscope proceed in the same sense. In the limiting case they can coincide so that a double root for $\dot{\psi}$ exists. In such a case the inequality sign in (17.22) must be replaced by the equality sign while in (17.21) the radicand vanishes.

In the extreme case $\Theta = 0$ of the elevated gyroscope known as the standing gyroscope (the mass center is vertically above the point of intersection of the two CARDAN axes), (17.22) is a necessary condition for stability as well. We shall see, however, that, contrary to the case of LAGRANGE and POISSON, in the case of a gyroscope in a CARDAN suspension this condition is only necessary but not sufficient for the stability of the standing gyroscope.

17.6. Stability in the case where the position of the axis of the outer gimbal ring is vertical. This position is characterized by the conditions

$$\Theta = 0 (u = 1), \quad \dot{\Theta} = 0, \quad \dot{\Phi} = \dot{\Phi}_0, \quad \dot{\psi} = \dot{\psi}_0 \tag{17.23}$$

where $\dot{\Phi}_0$ and $\dot{\psi}_0$ are arbitrary constants. It is easy to see that (17.23) is a solution of equations (17.6), (17.7) and (17.8) for $\Theta_0 = 0$. In terms of the notations (17.14) and (17.15) integrals (17.9), (17.10) and the one obtained from equation (17.7) assume the form

$$\dot{\Phi} + \dot{\psi} u = n \tag{17.24}$$

$$(k_5 - k_6 u^2) \dot{\psi} + k_1 u = k_2 \tag{17.25}$$

$$\dot{\Theta}^2 + (k_5 - k_6 u^2) \dot{\psi}^2 + k_3 u = k_4 \tag{17.26}$$

Given (17.23), equations (17.24) to (17.26) determine the constants n, k_2 and k_4 which depend on the initial values. The constant k_1 also depends on the initial values but it is already determined by means of n.

Consider now the perturbed motion

$$\dot{\Theta} = 0 + x_1, \quad \dot{\Phi} = \dot{\Phi}_0 + x_2, \quad \dot{\psi} = \dot{\psi}_0 + x_3, \quad u = 1 - x_4 \tag{17.27}$$

for which the constants due to change of the initial values will assume the values

$$n = n_0 + R, \quad k_1 = k_{10} + K_1, \quad k_2 = k_{20} + K_2, \quad k_4 = k_{40} + K_4$$

Then the equations for the perturbations assume the integrals

$$x_2 + x_3 - \dot{\psi}_0 x_4 - x_3 x_4 = R \tag{17.28}$$

$$-k_6 x_3 x_4^2 + 2 k_6 x_3 x_4 - \dot{\psi}_0 k_6 x_4^2 + (k_5 - k_6) x_3 + \\ + (2 \dot{\psi}_0 k_6 - k_1) x_4 = K_2 - K_1 \tag{17.29}$$

$$x_1^2 - k_6 x_3^2 x_4^2 + 2 k_6 x_3^2 x_4 - 2 \dot{\psi}_0 k_6 x_3 x_4^2 + (k_5 - k_6) x_3^2 + 4 \dot{\psi}_0 k_6 x_3 x_4 - \\ - \dot{\psi}_0^2 k_6 x_4^2 + 2 \dot{\psi}_0 (k_5 - k_6) x_3 + (2 \dot{\psi}_0^2 k_6 - k_3) x_4 = K_4 \tag{17.30}$$

Let us seek a LYAPUNOV V-function in the form of a linear bundle of integrals (17.28) to (17.30) of the equations for the perturbations. For this purpose it is desirable to introduce a new variable as defined by the equation

$$x_5^2 = 2x_4 - x_4^2 = 1 - u^2 > 0 \tag{17.31}$$

Upon introduction of a new constant K_0, equation (17.31) can be rewritten in the form

$$x_4^2 + x_5^2 - 2x_4 = K_0 \neq 0 \tag{17.32}$$

Consider now a LYAPUNOV V-function of the form

$$V = K_4 + \alpha_1(K_2 - K_1) + \alpha_2 K_0 + \alpha_3 R^2 \tag{17.33}$$

where α_1, α_2, α_3 are constants to be determined later.

The time derivative of V is identically zero since the function V consists of constants only. If we can find conditions for which V is positive definite, then by a well known theorem of LYAPUNOV we have sufficient conditions for the stability of the unperturbed motion.

Substituting (17.28) to (17.30) and (17.32) into (17.33), we obtain for V the expression

$$\begin{aligned} V = {} & x_1^2 + \alpha_3 x_2^2 + (\alpha_3 + k_5 - k_6)\, x_3^2 + (\alpha_2 + \alpha_3 \dot\psi_0^2)\, x_4^2 + \\ & + (k_6 \dot\psi_0^2 + k_6 \alpha_1 \dot\psi_0 + \alpha_2)\, x_5^2 + 2\alpha_3 x_2 x_3 - 2\alpha_3 \dot\psi_0 x_2 x_4 - \\ & - 2\alpha_3 \dot\psi_0 x_3 x_4 + (2\dot\psi_0 + \alpha_1)(k_5 - k_6)\, x_3 - (k_3 + \alpha_1 k_1 + 2\alpha_2)\, x_4 + \\ & + 2\alpha_3 \dot\psi_0 x_3 x_4^2 - 2\alpha_3 x_3^2 x_4 + (2k_6 \dot\psi_0 + \alpha_1 k_6)\, x_3 x_5^2 - 2\alpha_3 x_2 x_3 x_4 + \\ & + \alpha_3 x_3^2 x_4^2 + k_6 x_3^2 x_5^2 \end{aligned} \tag{17.34}$$

The linear terms in (17.34) disappear for

$$\alpha_1 = -2\dot\psi_0, \qquad \alpha_2 = k_1 \dot\psi_0 - \tfrac{1}{2} k_3 \tag{17.35}$$

With these values for α_1 and α_2 the LYAPUNOV V-function assumes the form

$$\begin{aligned} V = {} & x_1^2 + \alpha_3 (x_2 + x_3)^2 + (k_5 - k_6)\, x_3^2 + (k_1 \dot\psi_0 - \tfrac{1}{2} k_3 + \alpha_3 \dot\psi_0^2)\, x_4^2 + \\ & + (k_1 \dot\psi_0 - \tfrac{1}{2} k_3 - k_6 \dot\psi_0^2)\, x_5^2 - 2\alpha_3 \dot\psi_0 (x_2 + x_3)\, x_4 + \\ & + 2\alpha_3 \dot\psi_0 x_3 x_4^2 - 2\alpha_3 x_3^2 x_4 - 2\alpha_3 x_2 x_3 x_4 + \alpha_3 x_3^2 x_4^2 + k_6 x_3^2 x_5^2 \end{aligned} \tag{17.36}$$

Thus V consists of a quadratic form in x_i $(i = 1, \ldots, 5)$ and terms of order three and four in the same variables.

The quadratic form is positive definite provided that the inequalities

$$\begin{gathered} \alpha_3 > 0, \quad k_5 - k_6 > 0, \quad k_1 \dot\psi_0 - \tfrac{1}{2} k_3 > 0, \\ k_1 \dot\psi_0 - \tfrac{1}{2} k_3 - k_6 \dot\psi_0^2 > 0 \end{gathered} \tag{17.37}$$

hold.

These conditions guarantee not only positive values for the coefficients of the quadratic terms but also the fulfillment of the SYLVESTER inequality for the quadratic form in terms of $x_2 + x_3$ and x_4, namely

$$\begin{vmatrix} \alpha_3 & \alpha_3 \dot{\psi}_0 \\ \alpha_3 \dot{\psi}_0 & k_1 \dot{\psi}_0 - \frac{1}{2} k_3 + \alpha_3 \dot{\psi}_0^2 \end{vmatrix} = \alpha_3 (k_1 \dot{\psi}_0 - \tfrac{1}{2} k_3) > 0 \qquad (17.38)$$

The first condition of (17.37) can be satisfied by a choice of α_3; the second condition is satisfied by virtue of (17.15) since $k_6 > 0$; the third condition is satisfied provided that the fourth holds. Thus (17.37) give sufficient conditions for the stability of the vertical position of the axis of the gyroscope relative to the variables Θ, $\dot{\Theta}$, $\dot{\Phi}$ and $\dot{\psi}$.

The bounds for the admissible range of $\dot{\psi}_0$ are

$$\left.\begin{matrix} \dot{\psi}_{01} \\ \dot{\psi}_{02} \end{matrix}\right\} = \frac{k_1}{2k_6} \left(1 \pm \sqrt{1 - \frac{2k_3 k_6}{k_1^2}}\right)$$

or, in terms of the original notations,

$$\left.\begin{matrix} \dot{\psi}_{01} \\ \dot{\psi}_{02} \end{matrix}\right\} = \frac{C_1 n}{2(A_1 + B_2 - C_2)} \left[1 \pm \sqrt{1 - \frac{(A_1 + B_2 - C_2)\, 4m g s}{C_1^2 n^2}}\right]$$

These solutions are real, and hence they correspond to real physical motions provided that the inequality

$$C_1^2 n^2 > (A_1 + B_2 - C_2)\, 4m g s$$

holds. For $B_2 = C_2 = 0$ this condition becomes the well known condition for stability in the case of LAGRANGE and POISSON (Sec. 5.5). While this condition is necessary and sufficient for the stability of the LAGRANGE gyroscope, it is only necessary for the stability of the gyroscope in a CARDAN suspension. In order that stability takes place, in addition, the condition

$$\dot{\psi}_{02} < \dot{\psi}_0 < \dot{\psi}_{01}$$

must be satisfied (under the assumption that $k_1 > 0$ and $k_6 > 0$).

Hence the velocity with which the system rotates about the vertical position of the axis of the gyroscope is a decisive factor for the stability of a gyroscope in a CARDAN suspension. Physically this result is easy to understand if one takes into account the fact that in the case of LAGRANGE and POISSON the quantities ψ and $\dot{\psi}$ are not determined for the vertical position of the axis of the gyroscope. The line of nodes and its azimuthal velocity of rotation are only computational quantities. In the case of a gyroscope in a CARDAN suspension, however, the angle ψ has a real meaning since it determines the motion of the whole system.

If we assume without loss of generality that $n > 0$, then in the case of a standing gyroscope ($s > 0$)

$$0 < \dot{\psi}_{02} < \dot{\psi}_{01}$$

This implies that a standing gyroscope cannot be stable unless a push on the gimbal ring is given in the direction of the rotor spin. This push, however, should not exceed the upper stability bound $\dot{\psi}_{01}$. For a hanging gyroscope ($s < 0$) the mass center is vertically below the point of intersection of the two CARDAN axes and we have

$$\dot{\psi}_{02} < 0 < \dot{\psi}_{01}$$

Hence a hanging gyroscope with $\dot{\psi}_0 = 0$ is always stable. The LAGRANGE hanging gyroscope ($s < 0$) is always stable; however, a hanging gyroscope in a CARDAN suspension could lose its stability if its CARDAN system is given a push either in the direction opposite to that of the rotor spin with $\dot{\psi}_0 < \dot{\psi}_{02} < 0$ or in the direction of the rotor spin with $\dot{\psi}_0 > \dot{\psi}_{10} > 0$. Hence stability holds only for a restricted range of $\dot{\psi}_0$-values. The bounds of this range coincide with those values of the angular velocity $\dot{\psi}$ which are given by (17.21) for the regular precessions with $\Theta \to 0$. The vertical position of the axis of the gyroscope is stable only for $\dot{\psi}_0$-values which lie between the velocities of the regular precession.

The stability range of a standing gyroscope is always smaller than that of a hanging one. For $k_1^2 = 2k_3 k_6$ this range reduces to zero since in such a case the necessary condition (17.22) for the stability of a standing gyroscope is no longer satisfied.

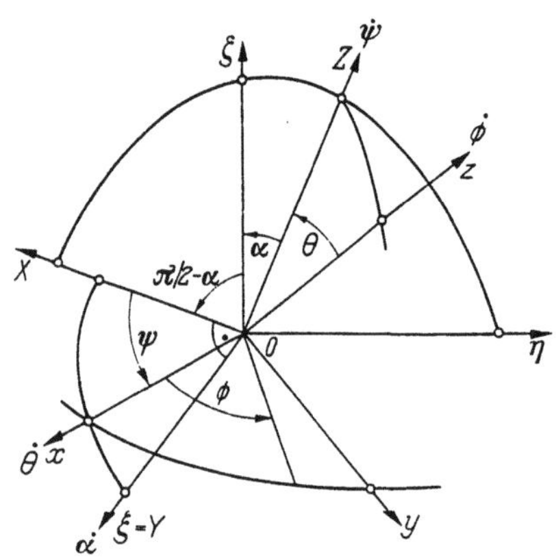

Fig. 56

17.7. Stability in the case where the position of the axis of the outer gimbal ring is inclined. Consider two fixed rectangular coordinate trihedrals $O\,\xi\,\eta\,\zeta$ and $O\,X\,Y\,Z$ with the common origin O at the fixed point of the gyroscope (Fig. 56). Let the $O\,\zeta$-axis coincide with the vertical, and let the $O\,\xi$-axis and $O\,\eta$-axis have an arbitrary position in the horizontal plane. Further, let the $O\,Z$-axis coincide with the axis of the outer gimbal, and let it lie in the $\zeta\,\eta$-plane making an angle α with the $O\,\zeta$-axis. Let the $O\,X$-axis be in the same plane, thus making an angle $\pi/2 - \alpha$ with the $O\zeta$-axis, and let the OY-axis coincide with the $O\xi$-axis. Consider also a body-fixed right-hand coordinate trihedral $Oxyz$, the Ox-axis of which coincides with the axis of the inner gimbal ring, and the Oz-axis with the axis of the rotor.

Case 1. Let us investigate the stability of the particular solution

$$\Theta = \alpha, \quad \psi = \frac{\pi}{2}, \quad \dot{\Theta} = 0, \quad \dot{\psi} = 0, \quad \dot{\Phi} = n \qquad (17.40)$$

of the equation of motion (17.6), (17.7) and (17.8). In such a case the axis of the inner gimbal is horizontal; the plane of the inner gimbal is vertical and the gyroscope rotates with constant angular velocity n about the vertical position of the axis of the rotor. For $\alpha = 0$ and $\alpha = \pi/2$ (the axis of the outer gimbal is vertical and horizontal respectively), the corresponding solutions have been investigated by V. V. RUMYANCEV. In what follows it will be assumed that $\alpha \neq 0$.

Let the perturbed motion be

$$\Theta = \alpha + \xi_1, \quad \dot{\Theta} = \dot{\xi}_1 = \eta_1, \quad \psi = \frac{\pi}{2} + \xi_2, \quad \dot{\psi} = \dot{\xi}_2 = \eta_2,$$
$$\dot{\Phi} = n + \eta_3$$

Then the equations for the perturbations assume the integrals

$$V_1 = (A_1 + A_2)\,\eta_1^2 + [C_3 + C_2 + (A_1 + B_2 - C_2)\sin^2\alpha]\,\eta_2^2 +$$
$$+ C_1\eta_3^2 + 2C_1 n\,\eta_3 - m g s(\xi_1^2 + \xi_2^2\sin^2\alpha) + \cdots = \text{const.}$$
$$V_2 = \eta_3 = \text{const.}$$

where in the first integral only terms up to the second order relative to the perturbations have been written out.

Consider the integral

$$V = V_1 - 2C_1 n V_2 = (A_1 + A_2)\,\eta_1^2 +$$
$$+ [C_3 + C_2 + (A_1 + B_2 - C_2)\sin^2\alpha]\,\eta_2^2 +$$
$$+ C_1\eta_3^2 - m g s(\xi_1^2 + \xi_2^2\sin^2\alpha) + \cdots = \text{const.}$$

For $s < 0$ and $\alpha \neq 0$ this is a positive definite function relative to the variables $\xi_1, \xi_2, \eta_1, \eta_2, \eta_3$. Therefore under the above assumptions by a well known theorem of LYAPUNOV the particular solution (17.40) is stable relative to $\Theta, \psi, \dot{\Theta}, \dot{\psi}, \dot{\Phi}$. For $s = 0$ the solution (17.40) is stable relative to the variables $\dot{\Theta}, \dot{\psi}, \dot{\Phi}$, the stability being secular.

For $s > 0$ the function V is no longer positive definite. Therefore in such a case solution (17.40) can have only gyroscopic stability. Let us find the conditions for this kind of stability.

The equations for the perturbations in the first approximation are

$$a\,\ddot{\xi}_1 = d\,\xi_1 - c\,\dot{\xi}_2, \qquad b\,\ddot{\xi}_2 = e\,\xi_2 + c\,\dot{\xi}_1 \tag{17.41}$$

where

$$A_1 + A_2 = a, \qquad C_3 + C_2 + (A_1 + B_2 - C_2)\sin^2\alpha = b$$

$$C_1\, n \sin\alpha = c, \qquad m\,g\,s = d, \qquad m\,g\,s\sin^2\alpha = e, \qquad (n \neq 0)$$

They assume a first integral indicated by N. G. ČETAEV [1],

$$\Gamma = 2(b\,d\,\xi_1\,\dot{\xi}_2 - a\,e\,\dot{\xi}_1\,\xi_2) - c(d\,\xi_1^2 + e\,\xi_2^2) + \\ + \frac{a e - b d}{2c}(b\,\dot{\xi}_2^2 - a\,\dot{\xi}_1^2 + d\,\xi_1^2 - e\,\xi_2^2) = \text{const.}$$

Consider the quadratic integral

$$\frac{1}{2} c V_1 - \Gamma - C_1 c n V_2 = \frac{c^2 - b d + a e}{2c} a\,\dot{\xi}_1^2 + 2 a e\,\dot{\xi}_1\,\xi_2 + \\ + \frac{c^2 - b d + a e}{2c} e\,\xi_2^2 + \frac{c^2 + b d - a e}{2c} b\,\dot{\xi}_2^2 - 2 b d\,\xi_1\,\dot{\xi}_2 + \\ + \frac{c^2 + b d - a e}{2c} d\,\xi_1^2 + \frac{1}{2} C_1 c\,\eta_3^2$$

For $s > 0$ this integral will be positive definite provided that

$$c^2 - \left(a e + b d + 2\sqrt{a b d e}\right) > 0 \tag{17.42}$$

Consequently, if the integral Γ holds for the equations of perturbations, then condition (17.42) is sufficient for the stability of the particular solution (17.40) relative to the variables Θ; ψ, $\dot{\Theta}$, $\dot{\psi}$, $\dot{\Phi}$. In the original notations inequality (17.42) assumes the form

$$C_1^2 n^2 \sin^2\alpha - m g s\left[C_3 + C_2 + (2A_1 + A_2 + B_2 - C_2)\sin^2\alpha + \right. \\ \left. + 2\sin\alpha\sqrt{(A_1 + A_2)\left[C_3 + C_2 + (A_1 + B_2 - C_2)\sin^2\alpha\right]}\right] > 0$$

which for $\alpha = \pi/2$ coincides with the condition for stability obtained by V. V. RUMYANCEV [6].

Let us investigate the necessity of condition (17.42). For this purpose consider the characteristic equation of the variational equations (17.41). This equation is

$$a\,b\,\lambda^4 + (c^2 - a e - b d)\,\lambda^2 + d e = 0 \tag{17.43}$$

If the particular solution (17.40) is stable, then the roots of (17.43) must have zero real parts. However, for $s > 0$ and $\alpha \neq 0$ equation (17.43) cannot have zero roots. Therefore its roots must be purely imaginary and the squares of the latter negative. This circumstance requires the simultaneous fulfillment of the inequalities

$$c^2 - a\,e - b\,d > 0, \qquad (c^2 - a\,e - b\,d)^2 - 4a\,b\,d\,e > 0$$

which in turn require the fulfillment of inequality (17.42). Thus for $\alpha \neq 0$ inequality (17.42) is necessary for the stability of the particular solution (17.40). For the case $\alpha = \pi/2$ the necessity of condition (17.42) for stability was obtained by ČZAN, SY-IN [1].

Case 2. Consider the particular solution

$$\Theta = 0, \quad \psi = 0, \quad \dot\Theta = 0, \quad \dot\psi = 0, \quad \dot\Phi = n \tag{17.44}$$

of equations (17.6), (17.7) and (17.8). In such a case the inner and outer gimbal planes coincide, and both are vertical while the axis of the rotor coincides with the axis of the outer gimbal.

Let the perturbed motion be

$$\Theta = \xi_1, \quad \dot\Theta = \dot\xi_1, \quad \psi = \xi_2, \quad \dot\psi = \dot\xi_2, \quad \dot\Phi = n + \eta_1$$

Then the equations for the perturbations, up to and including second order terms in the perturbations, assume the form

$$\begin{aligned} (A_1 + A_2)\,\ddot\xi_1 - m\,g\,s\,\xi_1 \cos\alpha + m\,g\,s\,\xi_2 \sin\alpha + C_1\,n\,\xi_1\,\dot\xi_2 + \cdots &= 0 \\ (C_3 + C_2)\,\ddot\xi_2 + m\,g\,s\,\xi_1 \sin\alpha + C_1\,\dot\eta_1 - C_1\,n\,\xi_1\,\dot\xi_1 + \cdots &= 0 \\ \dot\eta_1 &= 0 \end{aligned}$$

The characteristic equation corresponding to the first approximation is

$$\lambda^4 - \frac{m\,g\,s\cos\alpha}{A_1 + A_2}\,\lambda^2 - \frac{(m\,g\,s\sin\alpha)^2}{(A_1 + A_2)\,(C_3 + C_2)} = 0$$

Hence

$$\lambda^2 = \frac{m\,g\,s}{2(A_1 + A_2)}\left[\cos\alpha \pm \sqrt{\cos^2\alpha + \frac{4(A_1 + A_2)}{C_3 + C_2}\sin^2\alpha}\right]$$

For $s \neq 0$ the above formula shows that for $0 < \alpha < \pi/2$ the characteristic equation has a positive root. Consequently solution (17.44) is unstable on the basis of a theorem by LYAPUNOV on the instability of motion according to the first approximation. By the same argument there is instability of motion (17.44) for $\alpha = \pi/2$ as well as for $s > 0$, $\alpha = 0$.

Case 3. If the axis of the outer gimbal is horizontal, i.e. $\alpha = \pi/2$, then equations (17.6), (17.7) and (17.8) assume a particular solution

$$\psi = 0, \quad \dot\psi = 0, \quad \Theta = \frac{m\,g\,s}{C_1\,n}\,t + \Theta_0, \quad \dot\Theta = \frac{m\,g\,s}{C_1\,n},$$
$$\Phi = nt + \Phi_0, \quad \dot\Phi = n \tag{17.45}$$

The outer and inner gimbal planes are vertical while the inner gimbal plane rotates with a constant angular velocity about its vertical axis, and the gyroscope rotates with constant angular velocity about the axis of the rotor. The parameters of the system and the angular velocities satisfy the relation $C_1 \dot{\Phi} \dot{\Theta} = m g s$. The motion (17.45) has the nature of a regular precession for the angle of nutation equal to $\pi/2$.

17.8. Gyroscopes subject to various perturbing moments. The deviation of the gyrovertical as a result of changing the angular velocity of the rotor of the gyroscope has been studied by V. N. Košlyakov [1]. He showed that the angle of rotation of the inner CARDAN ring (gimbal) is determined by a linear second order differential equation and an additional quadrature.

If the angular velocity of the rotor decreases in such a way that the moment of the resisting forces varies proportionally to the first or the second power of the angular velocity, then V. N. Košlyakov [2] showed that the equation determining the angle of rotation of the inner gimbal can be integrated in terms of the BESSEL function of zero order or cylindrical functions respectively.

L. Ya. Roitenberg [1] has worked on the theory of gyroscope follow-up in the presence of random interference. A gyroscope follow-up is a gyro with three degrees of freedom, the axis of its outside CARDAN ring being vertical and that of its case horizontal. For the motion of a gyroscope in a CARDAN suspension and a nonaxially placed rotor see D. M. Klimov [1].

On the basis of an analysis of EULER's equations describing the motion of the CARDAN rings of a platform suspension, V. L. Rešetnikov [1] has obtained exact formulas for the moments of action on the platform by the CARDAN suspension.

R. Goodstein [1] applied a perturbation method to the nonlinear differential equations of motion of free and forced vibrations of a two-gimbal gyro.

V. V. Krementulo [1, 2] applied the direct method of Lyapunov, for the investigation of stability, to certain motions of a gyroscope in a CARDAN suspension when there is dry friction on the axis of the inner gimbal.

H. Ziegler [1] studied the motion of a symmetric gyroscope in a CARDAN suspension subject to a moment applied at the axis of the outer CARDAN ring.

G. A. Slomyanskii [1] was concerned with the integration of the equations of motion of a symmetric astatic gyroscope under various assumptions regarding the applied torque. See also A. M. Letov [1] and N. T. Kuzovkov [1].

A gyroscope rotating with a large angular velocity about its axis is subject to the moment of friction at the bearings and the moment of resistance of the surrounding medium. These moments as a rule exceed many times the moments of friction at the axes of the inner and outer rings of a CARDAN suspension. Therefore, in order to preserve a constant value for the angular momentum of the gyroscope, an electric drive is mounted in a CARDAN suspension. The purpose of this is to balance the damping moment of the resisting forces by the rotating moment of the electromagnetic forces.

E. L. Nikolai [2], D. M. Klimov [1], and others, investigating the motion of the above type of gyroscope, assumed that the angular velocity of the proper rotation remains constant. Such an assumption, however, is true only under certain restrictions imposed upon the moments originating from an electric drive. In addition, if the rings of a CARDAN suspension are not orthogonal, then the interaction between the nutational oscillations of the gyroscope and those relative to the axis of the proper rotation leads to the appearance of new phenomena considered by S. A. HARLAMOV [1, 2]. In another paper S. A. HARLAMOV [3] is concerned with the discussion of the influence of moments originating from an electric drive on the nutational oscillations and drift of the gyroscope, and with the investigation of conditions under which the angular velocity of the proper rotation of the gyroscope remains constant.

17.9. Gyroscopes on elastic foundations and moving bases. The influence of a moving support on the motion of a gyroscope, in particular, if the support is subject to periodic oscillations, the period of which is small in comparison with that of the precession of the axis, has been investigated by E. DE JONQUIÈRES [1] as early as 1886. Gyroscopes on elastically supported mass foundations have been the subject of investigations by A. TONDL [1] and YU. K. ŽBANOV [2]. ŽBANOV considered a two-gyro compass as a three-body system in which both rotors are elastically connected to its housing and studied resonance phenomena arising from periodic external excitements.

A gyroscope with an elastic rotor axis has been studied by V. V. KREMENTULO [3].

The motion of a gyrocompass and the stability of its motion have been investigated by A. YU. IŠLINSKII [2, 6], V. N. KOŠLYAKOV [3—6], and V. N. KOŠLYAKOV and V. F. LIAŠENKO [1]. Cases are considered where the base of the gyrocompass is fixed with respect to the Earth, or is moving relative to the latter.

A. A. SVEŠNIKOV [1] and YA. N. ROITENBERG [3, 4] have studied the motions of certain gyroscopic devices under the influence of random forces. Such a problem arises, for example, if the gyroscope is carried by a ship and if the rolling of the latter is taken into account.

V. S. NOVOSELOV [1] studied the motion of a stabilized gyroscopic system assuming an arbitrary dependence on time of the motion of the base and the mass of the gyroscopic system. Also the proper rotations of the gyroscopes are assumed to be nonstationary. The equations of motion obtained by a well known and frequently used method in the applied theory of gyroscopes do not account for the kinetic moments of the elements of suspension of the gyroscopic system nor for those of the gimbals of the gyroscopes nor for the equatorial components of the kinetic moments of the rotors and the kinetic moments of the motors. NOVOSELOV gave a justification for this method. See also L. I. KUZNECOV [2].

L. A. BUNATJAN [1] studied the motion of a free gyroscope on a non-stationary basis with allowance for the masses of the CARDAN suspension rings.

17.10. Application to inertial guidance systems. Among the many modern applications of the gyroscope let us mention inertial navigation.

In 1852 L. FOUCAULT [1—4] demonstrated by gyroscopic means the rotation of the Earth about its axis and showed that this rotation can be utilized to make a gyroscope seek out the local meridian of the Earth. Aside from demonstrating a principle his experiments raised anew the fundamental problem of defining inertial space in which NEWTON'S laws are valid.

All guidance systems perform three general functions: (i) determination of present vehicle position and motion, (ii) comparison of this information with desired position and motion, (iii) generation of correction commands for the vehicle control system. While there are various ways of reducing the last two functions to straightforward engineering procedures, the design of a guidance system which performs satisfactorily the first function is the most difficult part of the whole project. In all inertial guidance systems the reference space is provided by a reference member from which the angles essential for system operation are measured. This artificial reference plays the same role for automatic guidance that the celestial sphere plays in conventional navigation. The possibility of realizing such man-made orientational references is based on NEWTON's law of inertia, which expresses the property of matter that the momentum of each particle remains constant in magnitude and direction unless acted on by some force. When this principle is applied to a rigid wheel spinning at high speed about an axis of polar symmetry, gyroscopic effects appear. These effects cause a rotor to maintain the orientation of its spin axis with respect to inertial space if torque-free, or to respond by an angular velocity called precession of the spin axis towards the axis of the applied torque with a rate proportional to the magnitude of the applied torque. Utilization of these gyroscopic actions makes possible the design of gyroscopically controlled reference members that hold inertial space orientations with the accuracy required in practice. For information regarding the history and development of inertial guidance see J. M. KOOY and J. W. H. UYTENBOGAART [1], C. S. DRAPER, W. WRIGLEY and J. HOVORKA [1], and C. S. DRAPER [1].

A. YU. IŠLINSKII [5, 7, cf. also 1—4] reduced the problem of determining the position of a moving object by means of gyroscopes and accelerometers to the integration of a system of three nonlinear first order differential equations.

Also the discussion by C. FOX [1] regarding the mechanical determination of position and velocity on the Earth's surface is based upon the properties of gyroscopes.

PART III

Gyroscopes and artificial Earth satellites

Chapter VI

Rigid body in a central Newtonian field of forces

§ 18. Motion of a rigid body with a fixed point in a central Newtonian field of forces

18.1. Calculation of the force function. The force with which the mass μ situated at a point P attracts the mass m situated at M is given by NEWTON's law as

$$\vec{F} = -f\frac{m\mu}{r^2}\frac{\vec{r}}{r}$$

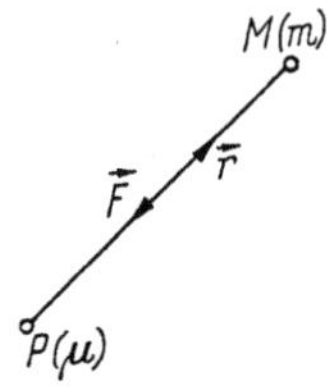

Fig. 57

where r is the distance between P and M, $\vec{r}$ is the radius vector of the point M (the origin of which is at the center of attraction P) and f is the constant of gravitation (Fig. 57). It follows from elementary considerations that

$$\Pi = -f\frac{m\mu}{r}$$

is the gravitational potential at the point M due to the mass μ situated at the point P. The function $U = -\Pi = f\frac{m\mu}{r}$ is called the mutual force function of the two masses.

Next consider a center of attraction $P(X, Y, Z)$ of mass μ and a body T of finite dimensions which is sufficiently far from the center P (Fig. 58). Let $M(X', Y', Z')$ be a point of T at which the mass dm is located. If $\delta(X', Y', Z')$ denotes the density of the body T at M, then $dm = \delta(X', Y', Z')\, d\tau$, where $d\tau$ is a volume element of T. The mutual force function of the body T and the particle P is then

$$U(X, Y, Z) = f\mu \int\limits_{(T)} \frac{dm}{\Delta} \qquad (18.1)$$

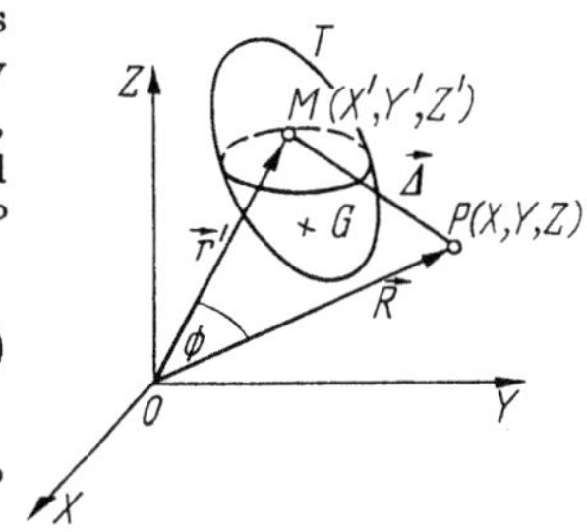

Fig. 58

where Δ is the distance between the points P and M, and hence

$$\Delta^2 = (X - X')^2 + (Y - Y')^2 + (Z - Z')^2$$

The integration in the right-hand side of formula (18.1) is to be extended over the whole volume T.

If we introduce spherical polar coordinates R, Θ, λ and r', Θ', λ' respectively, then

$$X = R \sin\Theta \cos\lambda, \qquad Y = R \sin\Theta \sin\lambda, \qquad Z = R\cos\Theta$$

$$X' = r' \sin\Theta' \cos\lambda', \qquad Y' = r' \sin\Theta' \sin\lambda', \qquad Z' = r' \cos\Theta'$$

and

$$\Delta^2 = R^2 + r'^2 - 2R\, r' \cos\varphi$$

where φ is the angle between the vectors $\vec{R} = \overrightarrow{OP}$ and $\vec{r}' = \overrightarrow{OM}$. Further we have that

$$\cos\varphi = \frac{1}{R\, r'}(X X' + Y Y' + Z Z') = \cos\Theta \cos\Theta' + \sin\Theta \sin\Theta' \cos(\lambda - \lambda')$$

If $r' < R$, then

$$\frac{1}{\Delta} = \frac{1}{R}\, \frac{1}{\sqrt{1 - 2\left(\frac{r'}{R}\right)\cos\varphi + \left(\frac{r'}{R}\right)^2}} = \frac{1}{R}\sum_{n=0}^{\infty}\left(\frac{r'}{R}\right)^n P_n(\cos\varphi)$$

$$= \sum_{n=0}^{\infty} \frac{r'^n}{R^{n+1}} P_n(\cos\varphi) \qquad (18.2)$$

where $P_n(\cos\varphi)$ is the LEGENDRE polynomial of degree n. For $n = 0, 1, 2, 3$ we have the following expressions for the LEGENDRE polynomials:

$$P_0(\cos\varphi) = 1, \qquad P_1(\cos\varphi) = \cos\varphi, \qquad P_2(\cos\varphi) = \tfrac{1}{2}(3\cos^2\varphi - 1)$$

$$P_3(\cos\varphi) = \tfrac{1}{2}(5\cos^3\varphi - 3\cos\varphi)$$

It is well known that series (18.2) converges absolutely and uniformly in a certain domain for arbitrary values of the angle φ.

Upon the substitution of the expression (18.2) into formula (18.1) and term-by-term integration we obtain that

$$U(R, \Theta, \lambda) = f \mu \sum_{n=0}^{\infty} \frac{1}{R^{n+1}} \int_{(T)} r'^n P_n(\cos\varphi)\, dm$$

Since by the addition formula of spherical functions

$$P_n(\cos\varphi) = P_n(\cos\Theta) P_n(\cos\Theta') + 2 \sum_{m=1}^{n} \frac{(n-m)!}{(n+m)!} P_n^m(\cos\Theta') P_n^m(\cos\Theta) \cos m(\lambda - \lambda')$$

is a spherical function of order n with respect to Θ and λ, then, integrating the expression $r'^n P_n(\cos\varphi)$ with respect to r', Θ', λ', we obtain again a spherical function of order n, say

$$Y_n(\Theta, \lambda) = \int_{(T)} r'^n P_n(\cos\varphi)\, dm \tag{18.3}$$

Hence

$$U(R, \Theta, \lambda) = f \mu \sum_{n=0}^{\infty} \frac{Y_n(\Theta, \lambda)}{R^{n+1}} \tag{18.4}$$

Any homogeneous polynomial of degree n, $U_n(X, Y, Z)$, when expressed in terms of spherical polar coordinates R, Θ, λ, can be represented in the form of a product of two functions, one of which depends only on R, while the other depends on Θ and λ, namely

$$U_n(X, Y, Z) = R^n Y_n(\Theta, \lambda) \tag{18.5}$$

where

$$Y_n(\Theta, \lambda) = \sum_{m_1+m_2+m_3=n} C_n^{(m_1, m_2, m_3)} \sin^{m_1+m_2}\Theta \cos^{m_3}\Theta \cos^{m_1}\lambda \sin^{m_2}\lambda$$

Substituting $Y_n(\Theta, \lambda)$ calculated from (18.5) into (18.4), we obtain the following expansion of the force function U,

$$U(X, Y, Z) = f \mu \sum_{n=0}^{\infty} \frac{U_n(X, Y, Z)}{R^{2n+1}} \tag{18.6}$$

The series on the right-hand side of (18.6) converges absolutely and uniformly in a certain domain for which $R > \overline{R}$ provided that all points of T satisfy the inequality $r' \leq \overline{R}$.

For the calculation of the polynomials U_n we can proceed as follows. By NEWTON's binomial formula we have that

$$(\nu^2 - 1)^n = \sum_{s=0}^{n} (-1)^s C_n^s \nu^{2n-2s}$$

and hence

$$\frac{d^n (\nu^2 - 1)^n}{d\nu^n} = \sum_{s=0}^{[n/2]} (-1)^s C_n^s (2n-2s)(2n-2s-1) \cdots [2n-2s-(n-1)] \nu^{n-2s}$$

where $[n/2]$ denotes the greatest integer less than or equal to $n/2$. On the other hand by RODRIGUES' formula

$$P_n(\nu) = \frac{1}{2^n n!} \frac{d^n(\nu^2-1)^n}{d\nu^n}$$

Hence

$$P_n(\nu) = \sum_{s=0}^{[n/2]} P_{ns} \nu^{n-2s} \tag{18.7}$$

where

$$P_{ns} = (-1)^s \frac{(2n-2s)!}{2^n s!\,(n-s)!\,(n-2s)!}$$

Substituting $\nu = \cos\varphi$ into formula (18.7) we obtain

$$P_n(\cos\varphi) = \sum_{s=0}^{[n/2]} P_{ns} \cos^{n-2s}\varphi$$

Multiplying both sides by $(R\,r')^n$ we obtain the equality

$$(R\,r')^n P_n(\cos\varphi) = \sum_{s=0}^{[n/2]} P_{ns}(X\,X' + Y\,Y' + Z\,Z')^{n-2s} R^{2s} r'^{2s} \qquad (18.8)$$

Each term on the right is a homogeneous polynomial of degree n with respect to the coordinates of the point P as well as of the point M. Consequently the whole sum is also a polynomial of the same type, and we can write that

$$R^n r'^n P_n(\cos\varphi) = \bar{P}_n(P, M) = \sum_{m_1+m_2+m_3=n} P_n^{(m_1, m_2, m_3)}(X', Y', Z')\, X^{m_1} Y^{m_2} Z^{m_3}$$

each coefficient being a homogeneous polynomial of degree n in terms of the coordinates X', Y', Z' of the point M. By formula (18.3) it follows now that

$$Y_n(\Theta, \lambda) = \int_{(T)} r'^n P_n(\cos\varphi)\, dm = \frac{1}{R^n} \int_{(T)} \bar{P}_n(P, M)\, dm = \frac{1}{R^n} U_n(X, Y, Z)$$

Hence

$$U_n(X, Y, Z) = \int_{(T)} \bar{P}_n(P, M)\, dm \qquad (18.9)$$

Next let us calculate the first four functions U_n for $n = 0, 1, 2, 3$. For this purpose let us first calculate the corresponding four functions $\bar{P}_n(P, M)$ defined by (18.8).

We have that

$$\bar{P}_0(P, M) = P_0(\cos\varphi) = 1$$

Further, for $n = 1$ we obtain that

$$\bar{P}_1(P, M) = R\,r'\,P_1(\cos\varphi) = R\,r'\cos\varphi = X\,X' + Y\,Y' + Z\,Z'$$

For $n = 2$ we find that

$$\begin{aligned}\bar{P}_2(P, M) &= R^2 r'^2 P_2(\cos\varphi) = R^2 r'^2(\tfrac{3}{2}\cos^2\varphi - \tfrac{1}{2}) \\ &= \tfrac{3}{2}(X\,X' + Y\,Y' + Z\,Z')^2 - \tfrac{1}{2} r'^2(X^2 + Y^2 + Z^2)\end{aligned}$$

or

$$\begin{aligned}\bar{P}_2(P, M) = \frac{X^2}{2}(3X'^2 - r'^2) + \frac{Y^2}{2}(3\,Y'^2 - r'^2) + \\ + \frac{Z^2}{2}(3Z'^2 - r'^2) + 3X\,Y\,X'\,Y' + 3X\,Z\,X'\,Z' + 3\,Y\,Z\,Y'\,Z'\end{aligned}$$

For $n = 3$ we obtain that

$$\begin{aligned}\bar{P}_3(P, M) &= R^3 r'^3 P_3(\cos\varphi) = R^3 r'^3(\tfrac{5}{2}\cos^3\varphi - \tfrac{3}{2}\cos\varphi) \\ &= \tfrac{5}{2}(R\,r'\cos\varphi)^3 - \tfrac{3}{2}(R\,r'\cos\varphi)\,R^2 r'^2\end{aligned}$$

or

$$\bar{P}_3(P, M) = \tfrac{5}{2}(X\,X' + Y\,Y' + Z\,Z')^3 - \tfrac{3}{2}r'^2(X\,X' + Y\,Y' + Z\,Z')\,(X^2 + Y^2 + Z^2)$$

or

$$\bar{P}_3(P, M) = X^3\left(\frac{5}{2}X'^3 - \frac{3}{2}X' r'^2\right) + X^2 Y\left(\frac{15}{2}Y' X'^2 - \frac{3}{2}Y' r'^2\right) +$$

$$+ X^2 Z\left(\frac{15}{2}Z' X'^2 - \frac{3}{2}Z' r'^2\right) + X Y^2\left(\frac{15}{2}X' Y'^2 - \frac{3}{2}X' r'^2\right) +$$

$$+ 15 X Y Z X' Y' Z' + X Z^2\left(\frac{15}{2}X' Z'^2 - \frac{3}{2}X' r'^2\right) +$$

$$+ Y^3\left(\frac{5}{2}Y'^3 - \frac{3}{2}Y' r'^2\right) + Y^2 Z\left(\frac{15}{2}Z' Y'^2 - \frac{3}{2}Z' r'^2\right) +$$

$$+ Y Z^2\left(\frac{15}{2}Y' Z'^2 - \frac{3}{2}Y' r'^2\right) + Z^3\left(\frac{5}{2}Z'^3 - \frac{3}{2}Z' r'^2\right)$$

Next, let us calculate the four functions $U_n (n = 0, 1, 2, 3)$. We have that

$$U_0(X, Y, Z) = \int_{(T)} dm = m$$

m being the mass of the body. Further,

$$U_1(X, Y, Z) = X \int_{(T)} X' \, dm + Y \int_{(T)} Y' \, dm + Z \int_{(T)} Z' \, dm = m(X \bar{X} + Y \bar{Y} + Z \bar{Z})$$

where $\bar{X}$, $\bar{Y}$, $\bar{Z}$ are the coordinates of the mass center G of the body. If the origin O of the coordinate trihedral $O X Y Z$ is taken at the mass center G of the body T, then $\bar{X} = \bar{Y} = \bar{Z} = 0$, and hence $U_1(X, Y, Z) \equiv 0$.

For $n = 2$ we have that

$$U_2(X, Y, Z) = \frac{X^2}{2} \int_{(T)} (3X'^2 - r'^2) \, dm + \frac{Y^2}{2} \int_{(T)} (3Y'^2 - r'^2) \, dm +$$

$$+ \frac{Z^2}{2} \int_{(T)} (3Z'^2 - r'^2) \, dm + 3X Y \int_{(T)} X' Y' \, dm +$$

$$+ 3X Z \int_{(T)} X' Z' \, dm + 3 Y Z \int_{(T)} Y' Z' \, dm$$

If we introduce the moments of inertia of the body T with respect to the coordinate axes, A, B, C, and the products of inertia, D, E, F, then

$$U_2(X, Y, Z) = \frac{X^2}{2}(B + C - 2A) + \frac{Y^2}{2}(C + A - 2B) + \frac{Z^2}{2}(A + B - 2C) +$$

$$+ 3D X Y + 3E X Z + 3F Y Z \tag{18.10}$$

If, in particular, the coordinate axes coincide with the principal axes of inertia of the body T at O, then $D = E = F = 0$, and the last formula assumes the form

$$U_2(X, Y, Z) = \frac{X^2}{2}(B + C - 2A) + \frac{Y^2}{2}(C + A - 2B) + \frac{Z^2}{2}(A + B - 2C)$$

The polynomial U_2 given by formula (18.10) can also be put in the form

$$U_2(X, Y, Z) = \frac{R^2}{2}(A + B + C - 3J)$$

where

$$J(X, Y, Z) = A\left(\frac{X}{R}\right)^2 + B\left(\frac{Y}{R}\right)^2 + C\left(\frac{Z}{R}\right)^2 - 2D\frac{X}{R}\frac{Y}{R} - 2E\frac{X}{R}\frac{Z}{R} - 2F\frac{Y}{R}\frac{Z}{R}$$

is the moment of inertia of the body with respect to the straight line OP.

Finally, for $n = 3$ we have that

$$\begin{aligned} U_3(X, Y, Z) = {} & U_3^{(3,0,0)} X^3 + U_3^{(2,1,0)} X^2 Y + U_3^{(2,0,1)} X^2 Z + \\ & + U_3^{(1,2,0)} X Y^2 + U_3^{(1,1,1)} X Y Z + U_3^{(1,0,2)} X Z^2 + \\ & + U_3^{(0,3,0)} Y^3 + U_3^{(0,2,1)} Y^2 Z + U_3^{(0,1,2)} Y Z^2 + U_3^{(0,0,3)} Z^3 \end{aligned}$$

where the coefficients are given by the following integrals

$$\begin{aligned} U_3^{(3,0,0)} &= \tfrac{1}{2}\int_{(T)} (5X'^2 - 3r'^2)\, X'\, dm \\ U_3^{(2,1,0)} &= \tfrac{1}{2}\int_{(T)} (15X'^2 - 3r'^2)\, Y'\, dm \\ U_3^{(2,0,1)} &= \tfrac{1}{2}\int_{(T)} (15X'^2 - 3r'^2)\, Z'\, dm \\ U_3^{(1,2,0)} &= \tfrac{1}{2}\int_{(T)} (15Y'^2 - 3r'^2)\, X'\, dm \\ U_3^{(1,1,1)} &= 15\int_{(T)} X' Y' Z'\, dm \\ U_3^{(1,0,2)} &= \tfrac{1}{2}\int_{(T)} (15Z'^2 - 3r'^2)\, X'\, dm \\ U_3^{(0,3,0)} &= \tfrac{1}{2}\int_{(T)} (5Y'^2 - 3r'^2)\, Y'\, dm \\ U_3^{(0,2,1)} &= \tfrac{1}{2}\int_{(T)} (15Y'^2 - 3r'^2)\, Z'\, dm \\ U_3^{(0,1,2)} &= \tfrac{1}{2}\int_{(T)} (15Z'^2 - 3r'^2)\, Y'\, dm \\ U_3^{(0,0,3)} &= \tfrac{1}{2}\int_{(T)} (5Z'^2 - 3r'^2)\, Z'\, dm \end{aligned}$$

Introducing the notations

$$J_{xyz} = \int_{(T)} X' Y' Z'\, dm, \qquad J_{xxy} = \int_{(T)} X'^2 Y'\, dm, \ \ldots$$

we obtain for $U_3(X, Y, Z)$ the expression

$$\begin{aligned} U_3(X, Y, Z) = {} & -\tfrac{3}{2}[(J_{xxx} + J_{xyy} + J_{xzz}) X + \\ & + (J_{yyy} + J_{yzz} + J_{yxx}) Y + (J_{zzz} + J_{zxx} + J_{zyy}) Z] + \\ & + \tfrac{5}{2}[J_{xxx} X^3 + J_{yyy} Y^3 + J_{zzz} Z^3 + 6 J_{xyz} X Y Z + \\ & + 3(J_{yxx} Y X^2 + J_{zxx} Z X^2 + J_{xyy} X Y^2 + J_{xzz} X Z^2 + J_{zyy} Z Y^2 + J_{yzz} Y Z^2)] \end{aligned}$$

Let

$$g = \frac{f\mu}{R^2}$$

denote the acceleration due to gravity at the distance R from the center of attraction P, and let us introduce instead of the coordinates X, Y, Z of P the direction cosines α, β, γ of $\vec{R} = \overrightarrow{OP}$ by putting

$$\alpha = \frac{X}{R}, \quad \beta = \frac{Y}{R}, \quad \gamma = \frac{Z}{R}$$

Then U, considered as a function of α, β, γ, assumes the form

$$\begin{aligned} U &= m R g + \frac{g(A+B+C)}{2R} + m g(\bar{X}\alpha + \bar{Y}\beta + \bar{Z}\gamma) - \\ &- \frac{3}{2}\frac{g}{R}(A\alpha^2 + B\beta^2 + C\gamma^2) - \frac{3}{2}\frac{g}{R^2}[(J_{xxx} + J_{xyy} + J_{xzz})\alpha + \\ &+ (J_{yyy} + J_{yzz} + J_{yxx})\beta + (J_{zzz} + J_{zxx} + J_{zyy})\gamma] + \\ &+ \frac{5}{2}\frac{g}{R^2}[J_{xxx}\alpha^3 + J_{yyy}\beta^3 + J_{zzz}\gamma^3 + 6J_{xyz}\alpha\beta\gamma + \\ &+ 3(J_{yxx}\alpha^2\beta + J_{zxx}\alpha^2\gamma + J_{xyy}\alpha\beta^2 + J_{xzz}\alpha\gamma^2 + \\ &+ J_{zyy}\beta^2\gamma + J_{yzz}\beta\gamma^2)] + \cdots \end{aligned} \tag{18.11}$$

Assume now that the origin O is a fixed point of the body T, and that the axes of the trihedral $OXYZ$ coincide with the principal axes of inertia of the body T at O. Henceforth the trihedral $OXYZ$ will be called $Oxyz$. Further, let $\overrightarrow{OP}$ be the direction of the negative OZ-axis. These assumptions amount to replacing α, β, γ by $-\alpha$, $-\beta$, $-\gamma$ in formula (18.11). Hence, except for a constant term, U assumes the expression

$$\begin{aligned} U &= -m g(x_0\alpha + y_0\beta + z_0\gamma) - \frac{3}{2}\frac{g}{R}(A\alpha^2 + B\beta^2 + C\gamma^2) + \\ &+ \frac{3}{2}\frac{g}{R^2}[(J_{xxx} + J_{xyy} + J_{xzz})\alpha + (J_{yyy} + J_{yzz} + J_{yxx})\beta + \\ &+ (J_{zzz} + J_{zxx} + J_{zyy})\gamma] - \\ &- \frac{5}{2}\frac{g}{R^2}[J_{xxx}\alpha^3 + J_{yyy}\beta^3 + J_{zzz}\gamma^3 + 6J_{xyz}\alpha\beta\gamma + \\ &+ 3(J_{yxx}\alpha^2\beta + J_{zxx}\alpha^2\gamma + J_{xyy}\alpha\beta^2 + \\ &+ J_{xzz}\alpha\gamma^2 + J_{zyy}\beta^2\gamma + J_{yzz}\beta\gamma^2)] + \cdots \end{aligned} \tag{18.12}$$

18.2. Calculation of the resultant force and moment. The orientation of the body T with respect to the fixed coordinate trihedral $OXYZ$ is determined by the orientation of a trihedral $Gxyz$ with respect to $OXYZ$, where G is an arbitrary point rigidly connected with T. In particular, G can be the mass center of the body T (Fig. 59). The orientation of the trihedral $Gxyz$ with respect to $OXYZ$ is determined by the EULER angles ψ, Φ, Θ, since the nine direction cosines of the axes of the trihedral $G\,x\,y\,z$ with respect to $O\,X\,Y\,Z$ can be expressed in terms of them. Thus the position and orientation of the body T with respect to the trihedral $O\,X\,Y\,Z$ is uniquely determined by six parameters

$$\bar{X},\ \bar{Y},\ \bar{Z},\ \psi,\ \Phi,\ \Theta$$

which vary as the body T moves.

If α_{ik} denote the direction cosines of the moving axes $G\,x\,y\,z$ relative to the fixed trihedral $O\,X\,Y\,Z$ according to the scheme

	x	y	z
X	α_{11}	α_{12}	α_{13}
Y	α_{21}	α_{22}	α_{23}
Z	α_{31}	α_{32}	α_{33}

then

$$\left.\begin{aligned} X' &= \bar{X} + \alpha_{11}\,x' + \alpha_{12}\,y' + \alpha_{13}\,z' \\ Y' &= \bar{Y} + \alpha_{21}\,x' + \alpha_{22}\,y' + \alpha_{23}\,z' \\ Z' &= \bar{Z} + \alpha_{31}\,x' + \alpha_{32}\,y' + \alpha_{33}\,z' \end{aligned}\right\} \tag{18.13}$$

where

$$\left.\begin{aligned} \alpha_{11} &= \cos\psi\cos\Phi - \sin\psi\sin\Phi\cos\Theta \\ \alpha_{12} &= -\cos\psi\sin\Phi - \sin\psi\cos\Phi\cos\Theta \\ \alpha_{13} &= \sin\psi\sin\Theta \\ \alpha_{21} &= \sin\psi\cos\Phi + \cos\psi\sin\Phi\cos\Theta \\ \alpha_{22} &= -\sin\psi\sin\Phi + \cos\psi\cos\Phi\cos\Theta \\ \alpha_{23} &= -\cos\psi\sin\Theta \\ \alpha_{31} &= \sin\Phi\sin\Theta \\ \alpha_{32} &= \cos\Phi\sin\Theta \\ \alpha_{33} &= \cos\Theta \end{aligned}\right\} \tag{18.14}$$

Consider now an element of mass dm of the body T located at the point $M(X',\ Y',\ Z')$ (Fig. 60). The center $P(X,\ Y,\ Z)$ attracts this element with the

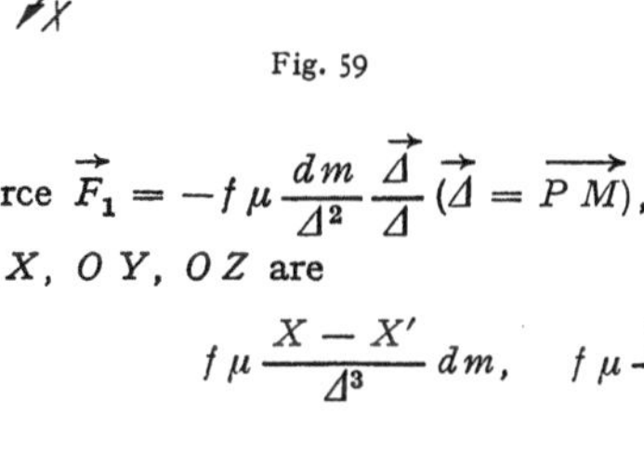

Fig. 59

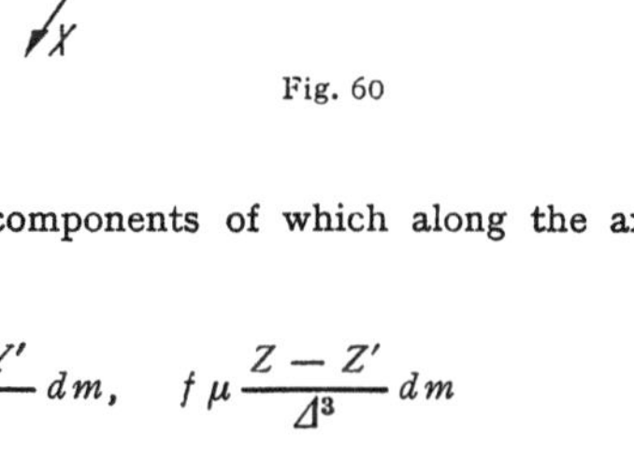

Fig. 60

force $\vec{F}_1 = -f\,\mu\dfrac{dm}{\Delta^2}\dfrac{\vec{\Delta}}{\Delta}$ $(\vec{\Delta} = \overrightarrow{P\,M})$, the components of which along the axes $O\,X,\ O\,Y,\ O\,Z$ are

$$f\,\mu\frac{X - X'}{\Delta^3}\,dm, \quad f\,\mu\frac{Y - Y'}{\Delta^3}\,dm, \quad f\,\mu\frac{Z - Z'}{\Delta^3}\,dm$$

and the moment of which with respect to the mass center G, $\vec{L}^G = \overrightarrow{GM} \times \vec{F}_1$, has the components

$$f\mu \frac{(Y' - \overline{Y})(Z - Z') - (Z' - \overline{Z})(Y - Y')}{\Delta^3} dm$$

$$f\mu \frac{(Z' - \overline{Z})(X - X') - (X' - \overline{X})(Z - Z')}{\Delta^3} dm$$

$$f\mu \frac{(X' - \overline{X})(Y - Y') - (Y' - \overline{Y})(X - X')}{\Delta^3} dm$$

Hence in a central field of forces with the center of attraction at P, a body of finite dimensions is subject to forces of attraction, the resultant force $\vec{F}$ and the moment $\vec{L}^G$ of which with respect to the mass center G of the body have respectively the components

$$\Xi = f\mu \int\limits_{(T)} \frac{X - X'}{\Delta^3} dm, \quad \mathrm{H} = f\mu \int\limits_{(T)} \frac{Y - Y'}{\Delta^3} dm, \quad \mathrm{Z} = f\mu \int\limits_{(T)} \frac{Z - Z'}{\Delta^3} dm \tag{18.15}$$

and

$$\left.\begin{aligned} L^G_{X'} &= f\mu \int\limits_{(T)} \frac{(Y' - \overline{Y})(Z - Z') - (Z' - \overline{Z})(Y - Y')}{\Delta^3} dm \\ L^G_{Y'} &= f\mu \int\limits_{(T)} \frac{(Z' - \overline{Z})(X - X') - (X' - \overline{X})(Z - Z')}{\Delta^3} dm \\ L^G_{Z'} &= f\mu \int\limits_{(T)} \frac{(X' - \overline{X})(Y - Y') - (Y' - \overline{Y})(X - X')}{\Delta^3} dm \end{aligned}\right\} \tag{18.16}$$

Considering in the above formulas X, Y, Z as constants and X', Y', Z' as variables, we have six functions in terms of the coordinates X', Y', Z' of the point M.

Introduce the mutual force function of the body T and the particle P,

$$U = f\mu \int\limits_{(T)} \frac{dm}{\Delta} \tag{18.17}$$

Then

$$\vec{F}(\Xi, \mathrm{H}, \mathrm{Z}) = \operatorname{grad} U \qquad \text{(with respect to } X', Y', Z')$$

If in formulas (18.15), (18.16) and (18.17) instead of X', Y', Z' new variables of integration x', y', z' are introduced by means of transformation formulas (18.13), then the force function U, the components of the resultant force $\vec{F}$, and those of the moment $\vec{L}^G$ become functions of the parameters $\overline{X}$, $\overline{Y}$, $\overline{Z}$; ψ, Φ, Θ (and of the coordinates X, Y, Z of the center of attraction P).

In addition we have that

$$\vec{F}(\Xi, \mathrm{H}, \mathrm{Z}) = \operatorname{grad} U \qquad \text{(with respect to } \overline{X}, \overline{Y}, \overline{Z})$$

and

$$\vec{L^G}(L^G_\psi, L^G_\Phi, L^G_\Theta) = \operatorname{grad} U \qquad \text{(with respect to } \psi, \Phi, \Theta)$$

Let $U^{(n)}(\alpha, \beta, \gamma)$ be the expression which results from $U(\alpha, \beta, \gamma)$ given by formula (18.12) when all terms of the order $n+1$ and higher in $1/R$ are omitted. Then in a moving coordinate trihedral $O\,x\,y\,z$, the origin O of which is at a fixed point of the body T and the axes of which are along the principal axes of inertia at O, the body T is subject to forces of attraction, whose resultant force $\vec{F}$ and resultant moment $\vec{L^0}$ with respect to O are

$$\vec{F} = \operatorname{grad} U^{(n)} \tag{18.18}$$

$$\begin{aligned}\vec{L^0} &= -\vec{R}\times\vec{F} = -\vec{R}\times\operatorname{grad} U^{(n)}\\ &= \left(\gamma\frac{\partial U^{(n)}}{\partial\beta} - \beta\frac{\partial U^{(n)}}{\partial\gamma}\right)\vec{i} + \left(\alpha\frac{\partial U^{(n)}}{\partial\gamma} - \gamma\frac{\partial U^{(n)}}{\partial\alpha}\right)\vec{j} + \left(\beta\frac{\partial U^{(n)}}{\partial\alpha} - \alpha\frac{\partial U^{(n)}}{\partial\beta}\right)\vec{k}\end{aligned} \tag{18.19}$$

Hence the components of $\vec{L^0}$ have the expressions

$$\left.\begin{aligned} L^0_x &= \gamma\frac{\partial U^{(n)}}{\partial\beta} - \beta\frac{\partial U^{(n)}}{\partial\gamma}\\ L^0_y &= \alpha\frac{\partial U^{(n)}}{\partial\gamma} - \gamma\frac{\partial U^{(n)}}{\partial\alpha}\\ L^0_z &= \beta\frac{\partial U^{(n)}}{\partial\alpha} - \alpha\frac{\partial U^{(n)}}{\partial\beta}\end{aligned}\right\} \tag{18.20}$$

18.3. Equations of motion and their first integrals. From the above approximation we obtain the approximate equations of motion of a rigid body about a fixed point in a NEWTONian central force field in the form

$$\left.\begin{aligned} A\frac{dp}{dt} + (C-B)\,q\,r &= \gamma\frac{\partial U^{(n)}}{\partial\beta} - \beta\frac{\partial U^{(n)}}{\partial\gamma}\\ B\frac{dq}{dt} + (A-C)\,r\,p &= \alpha\frac{\partial U^{(n)}}{\partial\gamma} - \gamma\frac{\partial U^{(n)}}{\partial\alpha}\\ C\frac{dr}{dt} + (B-A)\,p\,q &= \beta\frac{\partial U^{(n)}}{\partial\alpha} - \alpha\frac{\partial U^{(n)}}{\partial\beta}\end{aligned}\right\} \tag{18.21}$$

to which the POISSON equations for the direction cosines α, β, γ of the $O\,Z$-axis must be appended, namely

$$\frac{d\alpha}{dt} = r\beta - q\gamma, \quad \frac{d\beta}{dt} = p\gamma - r\alpha, \quad \frac{d\gamma}{dt} = q\alpha - p\beta \tag{18.22}$$

For $n = 0$, equations (18.21) reduce to the equations of motion of a heavy rigid body about a fixed point in a uniform gravitational field.

The right-hand sides of equations (18.21) are polynomials and the first integrals of these equations, the energy integral

$$A p^2 + B q^2 + C r^2 - 2 U^{(n)}(\alpha, \beta, \gamma) = \text{const.}$$

the angular momentum integral

$$A p \alpha + B q \beta + C r \gamma = \text{const.}$$

and the trivial integral

$$\alpha^2 + \beta^2 + \gamma^2 = 1$$

are algebraic. Since the system of equations (18.21) and (18.22) does not contain the time t explicitly and has the last JACOBI multiplier equal to one, the problem of integration of this system reduces to quadratures provided that four time independent first integrals are known.

There is a well known theorem to the effect that the equations of motion of a heavy rigid body about a fixed point ($n = 0$) assume a fourth algebraic integral in those and only those cases (EULER and POINSOT, LAGRANGE and POISSON, and KOVALEVSKAYA) for which the general solutions for $p, q, r; \alpha, \beta, \gamma$ are single-valued functions of the time t in the whole complex t-plane.

The question is whether the above theorem applies to equations (18.21) and (18.22) for $n > 0$.

Let us consider first the case $n = 1$.

In the case of kinetic symmetry of the body we have $A = B = C$, and systems (18.21) and (18.22), for $n = 1$, assume the fourth integral

$$x_0 p + y_0 q + z_0 r = \text{const.}$$

If the body has kinetic symmetry about the z-axis and the center of mass G lies on that axis, then

$$A = B, \qquad x_0 = y_0 = 0$$

(case analogous to that of LAGRANGE and POISSON), and systems (18.21) and (18.22), for $n = 1$, assume the fourth integral

$$r = \text{const.}$$

In the last case the reduction of the integration problem to quadratures was carried out by V. V. BELECKII [1].

If the fixed point coincides with the center of mass G of the body, i.e. if $x_0 = y_0 = z_0 = 0$, then systems (18.21) and (18.22), for $n = 1$, assume as the fourth integral the angular momentum integral

$$A^2 p^2 + B^2 q^2 + C^2 r^2 - \frac{3g}{R}(B C \alpha^2 + A C \beta^2 + A B \gamma^2) = \text{const.}$$

(case analogous to that of EULER and POINSOT). This integral was first discovered by F. DE BRUN [1], and the integration problem reduced to quadratures by G. KOBB [2], and later by Mrs. E. I. HARLAMOVA [3] using a different method. Thus complete integration of systems (18.21) and (18.22) is possible in the above two cases (besides the case of kinetic symmetry of the body).

Are there any other cases for which the general integrals of systems (18.21) and (18.22), for $n = 1$, are single-valued functions of the time t? YU. A. ARHANGEL'SKII [3] showed that such an investigation does not lead to any new cases other than the two aforementioned. The method used by him is that of a small parameter due to POINCARÉ.

Consider the substitution

$$p = \mu p_1, \quad q = \mu q_1, \quad r = \mu r_1, \quad \alpha = \mu^2 \alpha_1, \quad \beta = \mu^2 \beta_1,$$

$$\gamma = \mu^2 \gamma_1, \quad t = t_0 + \frac{t_1}{\mu}$$

by means of which system (18.21), for $n = 1$, assumes the form

$$A \frac{dp_1}{dt_1} + (C - B) q_1 r_1 = m g (z_0 \beta_1 - y_0 \gamma_1) + \mu^2 \frac{3g}{R} (C - B) \beta_1 \gamma_1$$

$$B \frac{dq_1}{dt_1} + (A - C) r_1 p_1 = m g (x_0 \gamma_1 - z_0 \alpha_1) + \mu^2 \frac{3g}{R} (A - C) \gamma_1 \alpha_1$$

$$C \frac{dr_1}{dt_1} + (B - A) p_1 q_1 = m g (y_0 \alpha_1 - x_0 \beta_1) + \mu^2 \frac{3g}{R} (B - A) \alpha_1 \beta_1$$

These equations contain a small parameter μ^2 and reduce for $\mu = 0$ to the system of equations which describes the motion of a heavy rigid body about a fixed point ($n = 0$). Concerning this system it is known that it has single-valued integrals only in the following three cases:

$x_0 = y_0 = z_0 = 0$ (EULER and POINSOT) (i)

$A = B, \quad x_0 = y_0 = 0$ (LAGRANGE and POISSON) (ii)

$A = B = 2C, \quad y_0 = z_0 = 0$ (KOVALEVSKAYA) (iii)

Therefore, in order to show the truth of the above statement it is sufficient to show that in case (iii) systems (18.21) and (18.22), for $n = 1$, have multiple-valued integrals.

Under the assumptions (iii) systems (18.21) and (18.22) reduce to the form

$$\left.\begin{aligned} &\frac{dp}{dt} - \frac{1}{2} q r = - b^2 \beta \gamma & &\frac{d\alpha}{dt} = \beta r - \gamma q \\ &\frac{dq}{dt} + \frac{1}{2} p r = \frac{c}{2} \gamma + b^2 \alpha \gamma & &\frac{d\beta}{dt} = \gamma p - \alpha r \\ &\frac{dr}{dt} \qquad\quad = - c \beta & &\frac{d\gamma}{dt} = \alpha q - \beta p \end{aligned}\right\} \qquad (18.23)$$

where

$$b^2 = \frac{3g}{2R}, \qquad c = \frac{m g x_0}{C}$$

If we introduce a small parameter μ by putting

$$p = \frac{p_1}{\mu}, \quad q = \frac{q_1}{\mu}, \quad r = \frac{r_1}{\mu}, \quad \alpha = \frac{\alpha_1}{\mu}, \quad \beta = \frac{\beta_1}{\mu},$$
$$\gamma = \frac{\gamma_1}{\mu}, \quad t = t_0 + \mu t_1$$

Then equations (18.23) assume the form

$$\left.\begin{aligned} &\frac{dp_1}{dt_1} - \frac{1}{2} q_1 r_1 = -b^2 \beta_1 \gamma_1 & &\frac{d\alpha_1}{dt_1} = r_1 \beta_1 - q_1 \gamma_1 \\ &\frac{dq_1}{dt_1} + \frac{1}{2} p_1 r_1 = b^2 \alpha_1 \gamma_1 + \mu \frac{c}{2} \gamma_1 & &\frac{d\beta_1}{dt_1} = p_1 \gamma_1 - r_1 \alpha_1 \\ &\frac{dr_1}{dt_1} \qquad = -\mu c \beta_1 & &\frac{d\gamma_1}{dt_1} = q_1 \alpha_1 - p_1 \beta_1 \end{aligned}\right\} \qquad (18.24)$$

For $\mu = 0$ the above system reduces to the simpler system

$$\left.\begin{aligned} &\frac{dp_1}{dt_1} - \frac{1}{2} q_1 r_1 = -b^2 \beta_1 \gamma_1 & &\frac{d\alpha_1}{dt_1} = r_1 \beta_1 - q_1 \gamma_1 \\ &\frac{dq_1}{dt_1} + \frac{1}{2} p_1 r_1 = b^2 \alpha_1 \gamma_1 & &\frac{d\beta_1}{dt_1} = p_1 \gamma_1 - r_1 \alpha_1 \\ &\frac{dr_1}{dt_1} \qquad = 0 & &\frac{d\gamma_1}{dt_1} = q_1 \alpha_1 - p_1 \beta_1 \end{aligned}\right\} \qquad (18.25)$$

which has the particular integral

$$p_1 = \frac{i}{t_1}, \quad q_1 = r_1 = \alpha_1 = 0, \quad \beta_1 = \frac{1}{b t_1}, \quad \gamma_1 = \frac{i}{b t_1} \qquad (i = \sqrt{-1})$$

Assuming that the integrals of system (18.24) can be expanded in powers of the parameter μ, we can write that

$$\begin{aligned} &p_1 = \frac{i}{t_1} + \mu p_2 + \cdots, \quad q_1 = \mu q_2 + \cdots, \quad r_1 = \mu r_2 + \cdots \\ &\alpha_1 = \mu \alpha_2 + \cdots, \ \beta_1 = \frac{1}{b t_1} + \mu \beta_2 + \cdots, \ \gamma_1 = \frac{i}{b t_1} + \mu \gamma_2 + \cdots \end{aligned} \qquad (18.26)$$

Substituting expansions (18.26) into system (18.24) and equating the coefficients of μ, we obtain equations for the determination of p_2, q_2, r_2, α_2, β_2 and γ_2. For example, we have that

$$\frac{dr_2}{dt_1} = \frac{c}{b t_1} \qquad (18.27)$$

Inasmuch as the quotient $c/b \neq 0$ it follows from equation (18.27) that r_2 has a movable critical logarithmic point. Hence system (18.23) has multiple, valued integrals. This fact proves the above statement.

Next YU. A. ARHANGEL'SKII [4] showed that a necessary condition for the existence of an additional algebraic integral (besides the three classical integrals) of systems (18.21) and (18.22), for $n = 1$, is that the ellipsoid of inertia about the fixed point be an ellipsoid of revolution. In other words, the validity of POINCARÉ's theorem for a heavy rigid body can be extended to the case of motion of a rigid body of a NEWTONian central force field for $n = 1$. The proof consists of a slight modification of HUSSON's proof of POINCARÉ's theorem.

Next consider the system of equations (18.21) for $n = 2$. YU. A. ARHANGEL'SKII [5] showed that in this case the theorem quoted on p. 266 no longer applies. This can be seen as follows. Take, for example, a rigid body the mass distribution of which satisfies the conditions

$$A = B, \quad x_0 = y_0 = 0$$

$$J_{zxx} = J_{zyy}, \quad J_{zzz} - 3J_{zxx} = l \neq 0$$

$$J_{xxx} = J_{yyy} = J_{xyz} = J_{yxx} = J_{xyy} = J_{yzz} = J_{xzz} = 0$$

In such a case the function $U^{(2)}$ reduces, except for a constant term, to

$$U^{(2)}(\gamma) = -m g z_0 \gamma - \frac{3}{2}\frac{g}{R}(C - A)\gamma^2 + \frac{3}{2}\frac{g}{R^2} l \gamma - \frac{5}{2}\frac{g}{R^2} l \gamma^3 \tag{18.28}$$

and the equations of system (18.21) for $n = 2$ become

$$\left.\begin{aligned} \frac{dp}{dt} - m q r &= -\varphi(\gamma)\beta \\ \frac{dq}{dt} + m p r &= \varphi(\gamma)\alpha \\ \frac{dr}{dt} \qquad &= 0 \end{aligned}\right\} \tag{18.29}$$

where

$$m = \frac{A - C}{A}, \quad \varphi(\gamma) = \frac{1}{A}\frac{\partial U^{(2)}}{\partial \gamma}$$

System (18.29) has the fourth integral $r = r_0$ which, together with the three first integrals, rewritten in the form

$$p^2 + q^2 - \frac{2U^{(2)}}{A} = h, \quad p\alpha + q\beta - (m - 1) r_0 \gamma = k,$$

$$\alpha^2 + \beta^2 + \gamma^2 = 1 \tag{18.30}$$

reduce the determination of γ to the equation

$$\left(\frac{d\gamma}{dt}\right)^2 = (1 - \gamma^2)\left[h + \frac{2U^{(2)}(\gamma)}{A}\right] - [k + (m - 1) r_0 \gamma]^2 \equiv P(\gamma) \tag{18.31}$$

Separating the variables in equation (18.31) and integrating we obtain

$$t - t_0 = \int \frac{d\gamma}{\sqrt{P(\gamma)}} \tag{18.32}$$

Inversion of integral (18.32) gives γ as a function of the time t which would be the general solution for γ. Since $P(\gamma)$ is a fifth degree polynomial in γ, integral (18.32) is a hyperelliptic integral, the inversion of which is not a single-valued function of t.

Thus, when $n = 2$, the existence of a fourth algebraic integral does not imply its single-valuedness. By similar reasoning it is easy to see that this will be the case for all $n > 2$ provided that $U^{(n)} = U^{(n)}(\gamma)$.

Consequently the theorem mentioned on p. 266 no longer holds when $n \geqq 2$. See also YU. A. ARHANGEL'SKII [5, 8].

18.4. Stability of rotation of a body fixed at its mass center. For $n = 1$ the equations of motions (18.21) and (18.22) assume the integrals

$$A p^2 + B q^2 + C r^2 + \frac{3g}{R}(A\alpha^2 + B\beta^2 + C\gamma^2) = h \tag{18.33}$$

$$A p \alpha + B q \beta + C r \gamma = k \tag{18.34}$$

$$A^2 p^2 + B^2 q^2 + C^2 r^2 - \frac{3g}{R}(B C \alpha^2 + A C \beta^2 + A B \gamma^2) = l \tag{18.35}$$

$$\alpha^2 + \beta^2 + \gamma^2 = 1 \tag{18.36}$$

A particular solution of these equations is

$$p = q = 0, \quad r = r_0; \quad \alpha = \beta = 0, \quad \gamma = 1$$

which represents a rotation of the body about the principal axis of inertia Oz with constant angular velocity r_0.

Let us investigate the stability of such a rotation (see V. V. BELECKII [2]). For this purpose consider the perturbed motion

$$p,\ q,\ r = r_0 + \zeta; \quad \alpha,\ \beta,\ \gamma = 1 + \delta \tag{18.37}$$

having in the first approximation the characteristic equation

$$\lambda^2(\lambda^4 + m\lambda^2 + n) = 0 \tag{18.38}$$

where

$$\begin{aligned} m &= -3\omega^2(a + b) + r_0^2(1 + a b) \\ n &= (3\omega^2 - r_0^2)^2 a b, \quad a = \frac{C - B}{A}, \quad b = \frac{C - A}{B}, \quad \omega^2 = \frac{g}{R} \end{aligned} \tag{18.39}$$

If at least one of the inequalities $n < 0$, $m < 0$, $m^2 - 4n < 0$ is satisfied, then among the roots of equation (18.38) there is at least one with a positive real part. Hence the corresponding unperturbed motion is unstable.

Let the rotation take place about the middle axis of inertia, i.e. let $A > C > B$ or $B > C > A$. Then $n < 0$ and for an arbitrary value of r_0 the rotation about this axis is unstable.

Further, let the body rotate about the largest axis of the ellipsoid of inertia (which corresponds to the axis of the smallest principal moment of inertia). Then

$$C < A, \quad C < B \tag{18.40}$$

From integrals (18.33), (18.35) and (18.36) it is easy to derive the following two integrals:

$$A p^2 + B q^2 + C r^2 + 3\omega^2[(A - C)\alpha^2 + (B - C)\beta^2] = V_1^0 \tag{18.41}$$

$$A(C - A) p^2 + B(C - B) q^2 + 3\omega^2(A - C)(C - B)(\alpha^2 + \beta^2) = V_2^0 \tag{18.42}$$

V_1^0 and V_2^0 being constants.

By virtue of inequalities (18.40) the quadratic form V_2^0 is of fixed sign with respect to the variables p, q, α, β. Consequently the unperturbed motion is stable relative to these variables. Then stability relative to r, γ follows from integrals (18.41) and (18.36) respectively. Thus rotation about the largest axis of the ellipsoid of inertia is stable for arbitrary values of r_0.

Finally, let the body rotate about the smallest axis of the ellipsoid of inertia (which corresponds to the axis of the largest principal moment of inertia), i.e. let $C > A$, $C > B$. Then $a > 0$ and $b > 0$.

Let us seek a sufficient condition for stability of rotation. By means of integrals (18.33) to (18.36) we can derive the following integrals for the perturbed motion (18.37):

$$A p^2 + B q^2 + C \zeta^2 + 2C r_0 \zeta + 3\omega^2[(A - C)\alpha^2 + (B - C)\beta^2] = V_1^0$$

$$2A p\alpha + 2B q\beta + 2C\zeta - C r_0(\alpha^2 + \beta^2 + \delta^2) - \\ - 2C\zeta(\alpha^2 + \beta^2 + \delta^2) = V_2^0$$

$$A^2 p^2 + B^2 q^2 + C^2 \zeta^2 + 2C^2 r_0 \zeta - \\ - 3\omega^2[B(C - A)\alpha^2 + A(C - B)\beta^2] = V_3^0$$

Let us seek a LYAPUNOV V-function in the form

$$V = V_1^0 - r_0(1 + \varkappa C) V_2^0 + \varkappa V_3^0$$

where $\varkappa$ is an arbitrary factor which will be determined later. Then V can be put in the form $V = L + W$, where L is a quadratic form and W contains terms of the third order only. If L is of fixed sign, then V will also be of fixed sign. The quadratic form

$$\begin{aligned} L = &(1+\varkappa A) A p^2 - 2r_0(1+\varkappa C) A p \alpha + \\ &+ \alpha^2[3\omega^2(A-C)(1+\varkappa B) + C r_0^2(1+\varkappa C)] + (1+\varkappa B) B q^2 - \\ &- 2r_0(1+\varkappa C) B q \beta + \beta^2[3\omega^2(B-C)(1+\varkappa A) + C r_0^2(1+\varkappa C)] + \\ &+ (1+\varkappa C) C \zeta^2 + C r_0^2(1+\varkappa C)\delta^2 \end{aligned}$$

will be positive definite if

$$\varkappa > -\frac{1}{C}, \quad r_0^2 > 3\omega^2 \frac{(1+\varkappa A)(1+\varkappa B)}{(1+\varkappa C)}$$

The function

$$\psi(\varkappa) = \frac{(1+\varkappa A)(1+\varkappa B)}{1+\varkappa C}$$

assumes at the point $\varkappa_1 = -\frac{1}{C}(1-\sqrt{a b}) > -\frac{1}{C}$ the minimum

$$\psi(\varkappa_1) = \left(\frac{\sqrt{a}+\sqrt{b}}{1+\sqrt{a b}}\right)^2$$

Taking $\varkappa_1$ instead of $\varkappa$ in the expression of V, we arrive at the conclusion that rotation about the smallest axis of the ellipsoid of inertia is stable for

$$r_0^2 > 3\omega^2\left(\frac{\sqrt{a}+\sqrt{b}}{1+\sqrt{a b}}\right)^2 \tag{18.43}$$

Let us show that for

$$r_0^2 < 3\omega^2\left(\frac{\sqrt{a}+\sqrt{b}}{1+\sqrt{a b}}\right)^2 \tag{18.44}$$

the rotation is unstable. Since $a < 1$ and $b < 1$, then

$$\frac{a+b}{1+a b} < \left(\frac{\sqrt{a}+\sqrt{b}}{1+\sqrt{a b}}\right)^2 \tag{18.45}$$

If

$$r_0^2 < 3\omega^2 \frac{a+b}{1+a b}$$

then $m < 0$, and the rotation is unstable.

Let

$$3\omega^2\frac{a+b}{1+a b} < r_0^2 < 3\omega^2\left(\frac{\sqrt{a}+\sqrt{b}}{1+\sqrt{a b}}\right)^2$$

Then

$$r_0^2(1+a b) > 3\omega^2(a+b) \tag{18.46}$$

By virtue of inequality (18.44) we have that $r_0^2 < 3\omega^2$, and by (18.46) we can write that

$$r_0^2(1 - \sqrt{a\,b})^2 > 3\,\omega^2(\sqrt{a} - \sqrt{b})^2 \tag{18.47}$$

The expression $m^2 - 4n$, rewritten in the form

$$m^2 - 4n = \left[r_0^2(1 - \sqrt{a\,b})^2 - 3\,\omega^2(\sqrt{a} - \sqrt{b})^2\right] \times \\ \times \left[r_0^2(1 + \sqrt{a\,b})^2 - 3\,\omega^2(\sqrt{a} + \sqrt{b})^2\right]$$

by virtue of inequalities (18.44) and (18.47) turns out to be negative. Hence the unperturbed motion is unstable.

The angular velocity

$$r_* = \sqrt{3}\,\omega\,\frac{\sqrt{a} + \sqrt{b}}{1 + \sqrt{a\,b}}$$

will be called the critical velocity.

If $r_0^2 \geqq 3\,\omega^2$, then the stability condition (18.43) is satisfied *a fortiori*.

Thus a body, fixed at its mass center and lying in a NEWTONian central force field, may be in equilibrium position only if one of its principal inertia axes is directed toward the center of attraction. The equilibrium will be unstable if this is the smallest or middle principal inertia axis, and stable if the direction to the center of attraction coincides with the direction of the largest principal axis of inertia at the fixed point. The conditions for a stable equilibrium are given by (18.40).

Finally, making a comparison of the case under consideration with that of EULER and POINSOT in a uniform gravitational field we arrive at the following conclusions.

1. In the classical problem a stationary rotation about any principal inertia axis is possible for any position of this axis in space.

In the NEWTONian central force field a stationary rotation about an axis of inertia is possible only if the direction of this axis passes through the center of attraction.

2. In the classical problem a rotation about the largest or smallest principal axis of inertia is stable and about the middle axis of inertia unstable for any value of the velocity of rotation.

In the case under consideration a rotation about the largest principal inertia axis is stable for arbitrary values of the velocity of rotation while a rotation about the smallest axis of inertia is stable only under the condition that the velocity of rotation does exceed a certain critical value. In the contrary case the rotation is unstable. Rotation about the middle principal inertia axis is always unstable.

18.5. Stability of rotation in the case $A = B$ and $U = U(\gamma)$. For $n = 1$ and $A = B$, $U = U(\gamma)$ equations (18.21) and (18.22) possess a particular solution

$$p = q = 0, \quad r = r_0; \quad \alpha = \beta = 0, \quad \gamma = 1 \tag{18.48}$$

which represents a rotation about the axis of kinetic symmetry provided that its direction coincides with that of the center of attraction. The differential equations for the perturbed motion

$$p, q, r = r_0 + \zeta; \quad \alpha, \beta, \gamma = 1 + \delta$$

possess the integrals

$$V_1 = A(p^2 + q^2) + C(\zeta^2 + 2\zeta) - 2\left(\frac{\partial U}{\partial \gamma}\right)_{\gamma=1} \delta - 2\left(\frac{\partial^2 U}{\partial \gamma^2}\right)_{\gamma=1} \delta^2 + W(\delta^3, \ldots)$$

$$V_2 = A(p\,\alpha + q\,\beta) + C(\zeta\,\delta + \zeta + r_0\,\delta)$$

$$V_3 = \alpha^2 + \beta^2 + \delta^2 + 2\delta = 0$$

$$V_4 = \zeta$$

Following N. G. ČETAEV [1] let us seek a LYAPUNOV V-function in the form of a bundle of integrals

$$V = V_1 + 2\varkappa V_2 - \left[-\left(\frac{\partial U}{\partial \gamma}\right)_{\gamma=1} + C r_0 \varkappa\right] V_3 - 2C(r_0 + \varkappa) V_4 + \\ + \frac{C(C-A)}{A} V_4^2 + \left(\frac{\partial^2 U}{\partial \gamma^2}\right)_{\gamma=1} V_3\,\delta$$

the derivative of which

$$\frac{dV}{dt} = \left(\frac{\partial^2 U}{\partial \gamma^2}\right)_{\gamma=1} V_3 \frac{d\delta}{dt}$$

is equal to zero since $V_3 = 0$. Therefore the unperturbed motion is stable if V is of fixed sign.

The function V can be put in the form $V = L + W^*$, where L is a quadratic form and W^* contains small terms of order higher than two. The function V is of fixed sign if L is of fixed sign, where

$$L = A\,p^2 + 2\varkappa A\,p\,\alpha - \left[-\left(\frac{\partial U}{\partial \gamma}\right)_{\gamma=1} + C r_0 \varkappa\right] \alpha^2 \\ + A\,q^2 + 2\varkappa A\,q\,\beta - \left[-\left(\frac{\partial U}{\partial \gamma}\right)_{\gamma=1} + C r_0 \varkappa\right] \beta^2 \\ + \frac{C}{A}\zeta^2 + 2\varkappa C\,\zeta\,\delta - \left[-\left(\frac{\partial U}{\partial \gamma}\right)_{\gamma=1} + C r_0 \varkappa\right] \delta^2$$

If the condition

$$C^2 r_0^2 + 4A\left(\frac{\partial U}{\partial \gamma}\right)_{\gamma=1} > 0 \tag{18.49}$$

is satisfied, then there exists a number $\varkappa$ such that L is a positive definite quadratic form.

Thus condition (18.49) is sufficient for stability of the unperturbed motion (18.48).

If

$$C^2 r_0^2 + 4A\left(\frac{\partial U}{\partial \gamma}\right)_{\gamma=1} < 0$$

the unperturbed motion is unstable. Consequently condition (18.49) is a necessary and sufficient condition for stability.

For $U = -m g z_0 \gamma$ condition (18.49) reduces to the condition of stability of rotation for a heavy rigid body in the case of LAGRANGE and POISSON:

$$C^2 r_0^2 > 4A\, m g z_0 \tag{18.50}$$

In the NEWTONian central force field, for $n = 1$, $U(\gamma)$ has the expression

$$U = -m g z_0 \gamma - \frac{3}{2} \frac{g}{R} (C - A) \gamma^2 \tag{18.51}$$

and therefore condition (18.49) assumes the form

$$C^2 r_0^2 > 4A \left[m g z_0 + \frac{3g}{R} (C - A) \right] \tag{18.52}$$

For $C > A$, this condition is stronger than (18.50), and for $C < A$, weaker than (18.50) in the sense that in the first case the perturbation forces tend to destabilize the motion while in the second case they contribute to its stabilization.

For $z_0 = 0$, $r_0 \neq 0$ we obtain a condition for the stability of rotation about the mass center:

$$r_0^2 > 4 \frac{A}{C} \frac{C - A}{C} 3 \frac{g}{R}$$

This condition coincides with that of (18.43) for $A = B$.

18.6. Stability of permanent rotations. Permanent rotations of an asymmetric rigid body and their stability in a NEWTONian central force field have been investigated by G. K. POŽARICKII [1]. For $n = 1$ the HAMILTONian function assumes the form

$$H = T - U = \frac{1}{2} (A p^2 + B q^2 + C r^2) + m g (x_0 \alpha + y_0 \beta + z_0 \gamma) + \\ + \frac{3g}{2R} (A \alpha^2 + B \beta^2 + C \gamma^2)$$

If for $p, q, r; \alpha, \beta, \gamma$ their expressions in terms of the EULER angles are substituted, then

$$\begin{aligned} H = \frac{1}{2} [&A (\dot\psi \sin\Theta \sin\Phi + \dot\Theta \cos\Phi)^2 + B (\dot\psi \sin\Theta \cos\Phi - \dot\Theta \sin\Phi)^2 + \\ &+ C (\dot\psi \cos\Theta + \dot\Phi)^2] + \\ &+ m g (x_0 \sin\Theta \sin\Phi + y_0 \sin\Theta \cos\Phi + z_0 \cos\Theta) + \\ &+ \frac{3g}{2R} (A \sin^2\Theta \sin^2\Phi + B \sin^2\Theta \cos^2\Phi + C \cos^2\Theta) \end{aligned}$$

Since ψ is a cyclic coordinate, the corresponding integral is $(L = T + U)$

$$\begin{aligned} P = \frac{\partial L}{\partial \dot\psi} = \frac{\partial T}{\partial \dot\psi} = \frac{\partial H}{\partial \dot\psi} &= A (\dot\psi \sin\Theta \sin\Phi + \dot\Theta \cos\Phi) \sin\Theta \sin\Phi + \\ &+ B (\dot\psi \sin\Theta \cos\Phi - \dot\Theta \sin\Phi) \sin\Theta \cos\Phi + C (\dot\psi \cos\Theta + \dot\Phi) \cos\Theta \\ &= A p \alpha + B q \beta + C r \gamma = \text{const.} = P_0 \end{aligned} \tag{18.53}$$

If

$$\dot{\Phi} = \dot{\Theta} = 0, \quad \Phi = \Phi_0, \quad \Theta = \Theta_0, \quad \dot{\psi} = \dot{\psi}_0$$

is a solution of the equations

$$\frac{\partial H}{\partial \dot{\psi}} = P_0, \quad \frac{\partial H}{\partial \dot{\Phi}} = \frac{\partial H}{\partial \dot{\Theta}} = \frac{\partial H}{\partial \Phi} = \frac{\partial H}{\partial \Theta} = 0 \tag{18.54}$$

then the equations of motion of a rigid body assume a particular solution

$$\dot{\Phi} = \dot{\Theta} = 0, \quad \Theta = \Theta_0, \quad \Phi = \Phi_0, \quad \dot{\psi} = \dot{\psi}_0, \quad \psi = \dot{\psi}_0 t + \psi_0 \tag{18.55}$$

known as a permanent rotation.

The last two equations of system (18.54) which determine the angles Φ_0 and Θ_0 under the assumption that $P = P_0$, $\dot{\Phi} = \dot{\Theta} = 0$ are

$$\begin{aligned} &\frac{1}{2} l \sin 2\Theta \left(\frac{3g}{R} + \frac{P_0^2}{K^2}\right) + \\ &\qquad + m g (x_0 \cos\Theta \sin\Phi + y_0 \cos\Theta \cos\Phi - z_0 \sin\Theta) = 0 \\ &\frac{1}{2} (A - B) \sin 2\Phi \sin\Theta \left(\frac{3g}{R} + \frac{P_0^2}{K^2}\right) + \\ &\qquad + m g (x_0 \cos\Phi - y_0 \sin\Phi) = 0 \end{aligned} \tag{18.56}$$

where

$$l = A \sin^2\Phi + B \cos^2\Phi - C$$

Eliminating the expression $\frac{3g}{R} + \frac{P_0^2}{K^2}$ in parenthesis in equations (18.56), we obtain the equation for the STAUDE cone in terms of the variables Φ, Θ.

If by means of the integral (18.53) the cyclic velocity component $\dot{\psi}$ is eliminated from T, then the kinetic energy becomes

$$T^* = R_2 + R_0$$

where R_2 is a positive definite quadratic form relative to the non-cyclic velocity components $\dot{\Theta}$, $\dot{\Phi}$, and R_0 is a constant. The energy integral assumes the form

$$R_2 + V = \text{const.} \tag{18.57}$$

where

$$V = \frac{P_0^2}{2K} + m g (x_0 \sin\Theta \sin\Phi + y_0 \sin\Theta \cos\Phi + z_0 \cos\Theta) + \frac{3g}{2R} K$$

is the modified potential energy of the system, and

$$K = A \sin^2\Theta \sin^2\Phi + B \sin^2\Theta \cos^2\Phi + C \cos^2\Theta$$

Hence the particular solution (18.55) will be stable relative to the variables $\dot{\Phi}$, $\dot{\Theta}$, $\dot{\psi}$, Φ, Θ provided that the quadratic form

$$\frac{\partial^2 V}{\partial \Phi^2}\,\delta\Phi^2 + 2\frac{\partial^2 V}{\partial\Phi\,\partial\Theta}\,\delta\Phi\,\delta\Theta + \frac{\partial^2 V}{\partial\Theta^2}\,\delta\Theta^2$$

is positive definite relative to the variations $\delta\Theta$, $\delta\Phi$. This will be the case if and only if

$$\frac{\partial^2 V}{\partial\Theta^2} > 0, \qquad \frac{\partial^2 V}{\partial\Phi^2}\,\frac{\partial^2 V}{\partial\Theta^2} - \left(\frac{\partial^2 V}{\partial\Phi\,\partial\Theta}\right)^2 > 0 \tag{18.58}$$

Written out in full these conditions are

$$l\left(\frac{3g}{R} - \frac{P_0^2}{K^2}\right)\cos 2\Theta + \frac{P_0^2 l^2}{K^3}\sin^2 2\Theta -$$
$$- m\,g\,(x_0 \sin\Theta\sin\Phi + y_0\sin\Theta\cos\Phi + z_0\cos\Theta) > 0$$

$$\left[l\left(\frac{3g}{R} - \frac{P_0^2}{K^2}\right)\cos 2\Theta + \frac{P_0^2 l^2}{K^3}\sin^2 2\Theta -\right.$$
$$\left.- m\,g\,(x_0\sin\Theta\sin\Phi + y_0\sin\Theta\cos\Phi + z_0\cos\Theta)\right]\times$$
$$\times\left[(A-B)\left(\frac{3g}{R} - \frac{P_0^2}{K^2}\right)\cos 2\Phi\sin^2\Theta + \frac{(A-B)^2 P_0^2\sin^4\Theta\sin^2 2\Phi}{K^3} -\right.$$
$$\left.- m\,g\sin\Theta\,(x_0\sin\Phi + y_0\cos\Phi)\right] -$$
$$-\left\{(A-B)\sin 2\Theta\sin 2\Phi\left[\frac{1}{2}\left(\frac{3g}{R} - \frac{P_0^2}{K^2}\right) + \frac{l P_0^2}{K^3}\sin^2\Theta\right] +\right.$$
$$\left.+ m\,g\cos\Theta\,(x_0\cos\Phi - y_0\sin\Phi)\right\}^2 > 0$$

These two inequalities together with equations (18.56) single out from the totality of permanent axes of rotation those which are stable.

18.7. Motion and stability of a gyroscope in a Cardan suspension. Consider two rectangular coordinate trihedrals $O\,X\,Y\,Z$ and $O\,x\,y\,z$ with the common origin O at the point of intersection of the axes of suspension (Fig. 61). The trihedral $O\,X\,Y\,Z$ is fixed in space with the $O\,Z$-axis along the axis of the outer gimbal. The trihedral $O\,x\,y\,z$ is rigidly connected with the inner gimbal, the $O\,x$-axis being directed along the axis of the latter and the $O\,z$-axis along the axis of the rotor. The $O\,x$-axis lies in the $O\,X\,Y$-plane.

The position of the gyroscope relative to the fixed coordinate trihedral $O\,X\,Y\,Z$ is determined by the EULERian angles ψ, Θ, Φ (Fig. 61). With our previous notations (see Sec. 17.3) the expression for the double kinetic energy is

$$2T = (A_1 + A_2)\,\dot{\Theta}^2 + [C_3 + C_2 + (A_1 + B_2 - C_2)\sin^2\Theta]\,\dot{\psi}^2 + \\ + C_1(\dot{\Phi} + \dot{\psi}\cos\Theta)^2$$

Further, assume that the center of attraction O_1 lies on the negative Z-axis and that the distance $R = \overline{O_1 O}$ is large in comparison with the dimensions of the rotor and the CARDAN rings. Let the mass center of the system consisting of the inner gimbal and the gyroscope be on the z-axis at the distance $l \geqq 1$ from the origin O, and denote the mass of this system by M. Then the force function U has the expression

$$U = -M\,g\,l\cos\Theta + \frac{3g}{2R}(A_1 + B_2 - C_1 - C_2)\cos^2\Theta \qquad (18.59)$$

If the center of attraction is at infinity ($R \to \infty$), then the second term on the right-hand side disappears, and U reduces to the expression of U for the homogeneous force field of gravitation. This will also be the case if the moments of inertia of the gyroscope and the inner gimbal satisfy the relation

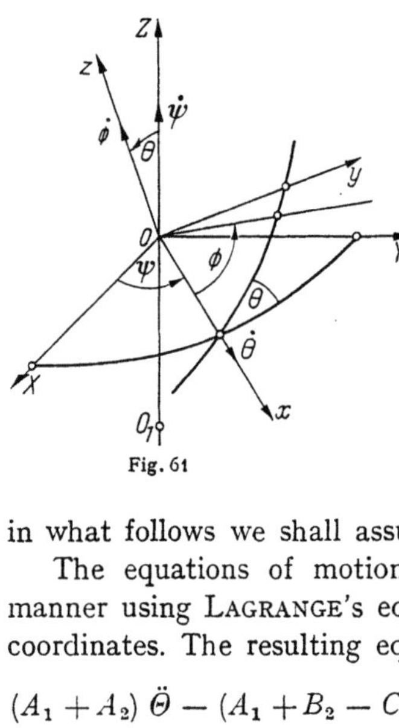

Fig. 61

$$A_1 + B_2 - C_1 - C_2 = 0 \qquad (18.60)$$

Consequently, if relation (18.60) holds, then the motion of the gyroscope under consideration will be identical to that of a gyroscope with the axis of the outer CARDAN ring vertical and lying in a homogeneous force field of gravitation, discussed earlier. Therefore, in what follows we shall assume that relation (18.60) is not satisfied.

The equations of motion can be obtained in a straightforward manner using LAGRANGE's equations with Θ, ψ and Φ as generalized coordinates. The resulting equations are

$$\left.\begin{aligned} &(A_1 + A_2)\,\ddot{\Theta} - (A_1 + B_2 - C_2)\,\dot{\psi}^2 \sin\Theta\cos\Theta + \\ &\qquad + C_1(\dot{\Phi} + \dot{\psi}\cos\Theta)\,\dot{\psi}\sin\Theta - M\,g\,l\sin\Theta + \\ &\qquad + \frac{3g}{R}(A_1 + B_2 - C_1 - C_2)\sin\Theta\cos\Theta = 0 \\ &\frac{d}{dt}\{[C_3 + C_2 + (A_1 + B_2 - C_2)\sin^2\Theta]\,\dot{\psi} + \\ &\qquad + C_1(\dot{\Phi} + \dot{\psi}\cos\Theta)\cos\Theta\} = 0 \\ &\qquad \frac{d}{dt}[C_1(\dot{\Phi} + \dot{\psi}\cos\Theta)] = 0 \end{aligned}\right\} \qquad (18.61)$$

If the masses of the CARDAN rings are neglected, then equations (18.61) reduce to those obtained by V. V. BELECKII [1] for the case analogous to that of LAGRANGE and POISSON.

The equations of motion (18.61) have the following first integrals

$$\left.\begin{aligned} &(A_1 + A_2)\,\dot{\Theta}^2 + [C_3 + C_2 + (A_1 + B_2 - C_2)\sin^2\Theta]\,\dot{\psi}^2 + \\ &\qquad + C_1(\dot{\Phi} + \dot{\psi}\cos\Theta)^2 + 2Mgl\cos\Theta - \\ &\qquad - \frac{3g}{R}(A_1 + B_2 - C_1 - C_2)\cos^2\Theta = h \\ &[C_3 + C_2 + (A_1 + B_2 - C_2)\sin^2\Theta]\,\dot{\psi} + \\ &\qquad + C_1(\dot{\Phi} + \dot{\psi}\cos\Theta)\cos\Theta = k \\ &\qquad \dot{\Phi} + \dot{\psi}\cos\Theta = n \end{aligned}\right\} \quad (18.62)$$

the first one being the energy integral, the last two corresponding to cyclic coordinates ψ and Φ; h, k and n are constants of integration.

The actual integration of equations (18.61) by reduction to quadratures has been carried out by A. M. TABAROVSKII [1]. He also gave stability conditions for the regular precessions.

18.8. Gyrostat in a central Newtonian force field. N. N. KOLESNIKOV [1] has studied the motion of a gyrostat consisting of a rigid body T_1 and rotors T_2, the axes of which are fixed relative to T_1. Friction on the axes of rotors and other dissipative effects are neglected. Sufficient conditions of stability for a particular solution of the equations of motion are derived by the direct method of LYAPUNOV.

Chapter VII

Motion of an artificial Earth satellite about its mass center

§ 19. The problem of separating the general motion of mutually attracting rigid bodies into translations of their mass centers and rotations about the latter

The study of the rotation of an artificial satellite about its mass center is of the greatest importance. From the mathematical point of view, however, this problem is very difficult.

In classical celestial mechanics the motions of celestial bodies are decomposed into translations of their mass centers and rotations about the latter, considering the coordinates of the mass centers as known functions of the time. In this way the rotational effect of the body on the motion of its mass center is neglected. However, in a non-homogeneous gravitational field the translations of the mass centers are no longer independent of the rotations about the latter. Therefore there arises the general problem of motion of celestial bodies as well as the problem of decomposing it into the above two component motions.

Certain particular cases of the second problem have been discussed by V. T. KONDURAR' in a series of papers [1—7]. The general case of n bodies has been the subject of investigations by G. N. DUBOŠIN [1—4].

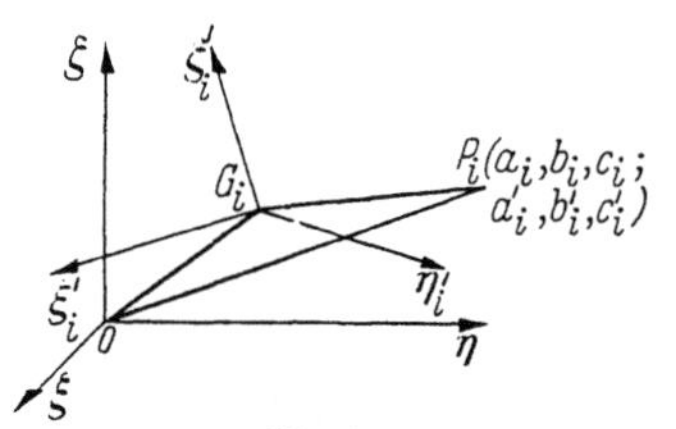

Fig. 62

19.1. Resultant force and moment. Consider a system of $n+1$ rigid bodies M_i $(i = 0, 1, \ldots, n)$ attracting each other according to NEWTON's law. Let ξ_i, η_i, ζ_i be the coordinates of the mass center G_i of M_i relative to a fixed, rectangular coordinate trihedral $O\,\xi\,\eta\,\zeta$.

Consider now an element of mass dm_i of the body M_i located at the point $P_i(a_i, b_i, c_i)$, and an element of mass dm_j of the body M_j located at the point $P_j(a_j, b_j, c_j)$. Then P_j attracts P_i with the force

$$\vec{F}_1^{(i,j)} = -f\,\frac{dm_i\,dm_j}{\Delta_{ji}^2}\,\frac{\vec{\Delta}_{ji}}{\Delta_{ji}} = f\,\frac{dm_i\,dm_j}{\Delta_{ij}^2}\,\frac{\vec{\Delta}_{ij}}{\Delta_{ij}} \qquad \left(\vec{\Delta}_{ji} = \overrightarrow{P_j P_i};\ i \neq j\right)$$

the moment of which with respect to the mass center G_i is

$$\vec{L}_1^{(i,j)} = \overrightarrow{G_i P_i} \times \vec{F}_1^{(i,j)}$$

Hence the resultant force $\vec{F}^{(i,j)}$ and the resultant moment $\vec{L}^{(i,j)}$ with respect to G_i have respectively the components

$$F_\xi^{(i,j)} = f \int\limits_{(M_i)} dm_i \int\limits_{(M_j)} \frac{a_j - a_i}{\Delta_{ij}^3}\,dm_j, \ldots, \ldots \tag{19.1}$$

and

$$L_\xi^{(i,j)} = f \int\limits_{(M_i)} dm_i \int\limits_{(M_j)} \frac{(b_i - \eta_i)(c_j - c_i) - (c_i - \zeta_i)(b_j - b_i)}{\Delta_{ij}^3}\,dm_j, \ldots, \ldots \tag{19.2}$$

Integrations in formulas (19.1) and (19.2) must be extended over the masses of the whole bodies M_i and M_j.

Let us associate with each body M_i a rigidly connected rectangular coordinate trihedral $G_i\,\xi_i'\,\eta_i'\,\zeta_i'$, the axes of which are directed along the principal inertia axes at G_i. Then with reference to Fig. 62 the coordinates a_i, b_i, c_i of P_i relative to $O\,\xi\,\eta\,\zeta$ in terms of the coordinates a_i', b_i', c_i' of P_i relative to $G_i\,\xi_i'\,\eta_i'\,\zeta_i'$, and the direction cosines of the second trihedral relative to the first one are given by the formulas

	ξ_i'	η_i'	ζ_i'
ξ	$a_{11}^{(i)}$	$a_{12}^{(i)}$	$a_{13}^{(i)}$
η	$a_{21}^{(i)}$	$a_{22}^{(i)}$	$a_{23}^{(i)}$
ζ	$a_{31}^{(i)}$	$a_{32}^{(i)}$	$a_{33}^{(i)}$

$$\left.\begin{aligned} a_i &= \xi_i + a_{11}^{(i)} a_i' + a_{12}^{(i)} b_i' + a_{13}^{(i)} c_i' \\ b_i &= \eta_i + a_{21}^{(i)} a_i' + a_{22}^{(i)} b_i' + a_{23}^{(i)} c_i' \\ c_i &= \zeta_i + a_{31}^{(i)} a_i' + a_{32}^{(i)} b_i' + a_{33}^{(i)} c_i' \end{aligned}\right\} \tag{19.3}$$

where the direction cosines $a_{ks}^{(i)}$ are given in terms of the EULER angles by means of formulas [see transformation (1.5)]

$$\left.\begin{aligned} a_{11}^{(i)} &= \cos\psi_i \cos\Phi_i - \sin\psi_i \sin\Phi_i \cos\Theta_i \\ a_{12}^{(i)} &= -\cos\psi_i \sin\Phi_i - \sin\psi_i \cos\Phi_i \cos\Theta_i \\ &\dots \end{aligned}\right\} \tag{19.4}$$

By means of formulas (19.3) and (19.4) and similar formulas for the body M_j, the force components (19.1) and moment components (19.2) become functions of 12 variables

$$\xi_i, \eta_i, \zeta_i; \quad \psi_i, \Phi_i, \Theta_i; \quad \xi_j, \eta_j, \zeta_j; \quad \psi_j, \Phi_j, \Theta_j$$

19.2. Equations of motion. Let m_k, A_k, B_k, C_k be the mass and the principal moments of inertia of the body M_k, and p_k, q_k, r_k the components of the angular velocity vector $\vec{\omega}_k$ of the body M_k relative to the trihedral $G_k \xi_k' \eta_k' \zeta_k'$. Then

$$\left.\begin{aligned} p_k &= \dot\psi_k \sin\Phi_k \sin\Theta_k + \dot\Theta_k \cos\Phi_k \\ q_k &= \dot\psi_k \cos\Phi_k \sin\Theta_k - \dot\Theta_k \sin\Phi_k \\ r_k &= \dot\psi_k \cos\Theta_k + \dot\Phi_k \end{aligned}\right\} \tag{19.5}$$

Let

$$T_k = \tfrac{1}{2} m_k (\dot\xi_k^2 + \dot\eta_k^2 + \dot\zeta_k^2) + \tfrac{1}{2}(A_k p_k^2 + B_k q_k^2 + C_k r_k^2)$$

be the kinetic energy of the body M_k, and

$$T = \sum_{k=0}^{n} T_k$$

the total kinetic energy of the system.

Further, introduce the mutual force function of the bodies M_i and M_k,

$$U_{ik} = f \int\limits_{(M_i)} dm_i \int\limits_{(M_k)} \frac{dm_k}{\Delta_{ik}} \qquad (i \neq k)$$

and the force function of the whole system

$$U = \sum_{i,k} U_{ik} \tag{19.6}$$

Then the equations for the combined motions of the bodies, consisting of translations of their mass centers and rotations about the latter, are

$$\left.\begin{aligned}
&\frac{d}{dt}\left(\frac{\partial T}{\partial \dot{\xi}_i}\right) = \frac{\partial U}{\partial \xi_i} &\quad& \frac{d}{dt}\left(\frac{\partial T}{\partial \dot{\psi}_i}\right) = \frac{\partial U}{\partial \psi_i}\\
&\frac{d}{dt}\left(\frac{\partial T}{\partial \dot{\eta}_i}\right) = \frac{\partial U}{\partial \eta_i} && \frac{d}{dt}\left(\frac{\partial T}{\partial \dot{\Phi}_i}\right) - \frac{\partial T}{\partial \Phi_i} = \frac{\partial U}{\partial \Phi_i}\\
&\frac{d}{dt}\left(\frac{\partial T}{\partial \dot{\zeta}_i}\right) = \frac{\partial U}{\partial \zeta_i} && \frac{d}{dt}\left(\frac{\partial T}{\partial \dot{\Theta}_i}\right) - \frac{\partial T}{\partial \Theta_i} = \frac{\partial U}{\partial \Theta_i}\\
&&& (i = 0, 1, \ldots, n).
\end{aligned}\right\} \tag{19.7}$$

Their explicit form is

$$m_i \frac{d^2 \xi_i}{dt^2} = \frac{\partial U}{\partial \xi_i}, \quad m_i \frac{d^2 \eta_i}{dt^2} = \frac{\partial U}{\partial \eta_i}, \quad m_i \frac{d^2 \zeta_i}{dt^2} = \frac{\partial U}{\partial \zeta_i} \tag{19.8}$$

$$\left.\begin{aligned}
A_i \frac{dp_i}{dt} - (B_i - C_i) q_i r_i &= \left(\frac{\partial U}{\partial \psi_i} - \frac{\partial U}{\partial \Phi_i}\cos\Theta_i\right)\frac{\sin\Phi_i}{\sin\Theta_i} + \frac{\partial U}{\partial \Theta_i}\cos\Phi_i\\
B_i \frac{dq_i}{dt} - (C_i - A_i) r_i p_i &= \left(\frac{\partial U}{\partial \psi_i} - \frac{\partial U}{\partial \Phi_i}\cos\Theta_i\right)\frac{\cos\Phi_i}{\sin\Theta_i} - \frac{\partial U}{\partial \Theta_i}\sin\Phi_i\\
C_i \frac{dr_i}{dt} - (A_i - B_i) p_i q_i &= \frac{\partial U}{\partial \Phi_i}
\end{aligned}\right\} \tag{19.9}$$

$$\left.\begin{aligned}
p_i &= \frac{d\psi_i}{dt}\sin\Phi_i\sin\Theta_i + \frac{d\Theta_i}{dt}\cos\Phi_i\\
q_i &= \frac{d\psi_i}{dt}\cos\Phi_i\sin\Theta_i - \frac{d\Theta_i}{dt}\sin\Phi_i\\
r_i &= \frac{d\psi_i}{dt}\cos\Theta_i \qquad\qquad + \frac{d\Phi_i}{dt} \qquad (i = 0, 1, \ldots, n)
\end{aligned}\right\} \tag{19.10}$$

The three systems of equations (19.8), (19.9) and (19.10) constitute a system of order $12(n+1)$, the integration of which determines the translational as well as the rotational motions of each body of the system. In general, however, these equations cannot be integrated.

19.3. The first integrals. The above equations possess first integrals analogous to the classical integrals of the $(n+1)$-body problem. They follow from the same principles as the classical integrals.

In fact, the force function U of the system is independent of the coordinate trihedral $O\,\xi\,\eta\,\zeta$, and therefore it remains invariant for any infinitesimal transformation of the coordinate system. Therefore an infinitesimal translation parallel to each of the coordinate axes as well as an infinitesimal rotation of the system about each axis leaves the function U invariant. As a result we obtain the following relations

$$\sum_{i=0}^{n} \frac{\partial U}{\partial \xi_i} = 0, \quad \sum_{i=0}^{n} \frac{\partial U}{\partial \eta_i} = 0, \quad \sum_{i=0}^{n} \frac{\partial U}{\partial \zeta_i} = 0 \tag{19.11}$$

$$\left.\begin{aligned}
&\sum_{i=0}^{n}\Big(\eta_i \frac{\partial U}{\partial \zeta_i} - \zeta_i \frac{\partial U}{\partial \eta_i} + \frac{\sin\psi_i}{\sin\Theta_i}\frac{\partial U}{\partial \Phi_i} - \frac{\sin\psi_i\cos\Theta_i}{\sin\Theta_i}\frac{\partial U}{\partial \psi_i} + \\
&\qquad + \cos\psi_i \frac{\partial U}{\partial \Theta_i}\Big) = 0 \\
&\sum_{i=0}^{n}\Big(\zeta_i \frac{\partial U}{\partial \xi_i} - \xi_i \frac{\partial U}{\partial \zeta_i} - \frac{\cos\psi_i}{\sin\Theta_i}\frac{\partial U}{\partial \Phi_i} + \frac{\cos\psi_i\cos\Theta_i}{\sin\Theta_i}\frac{\partial U}{\partial \psi_i} + \\
&\qquad + \sin\psi_i \frac{\partial U}{\partial \Theta_i}\Big) = 0 \\
&\sum_{i=0}^{n}\Big(\xi_i \frac{\partial U}{\partial \eta_i} - \eta_i \frac{\partial U}{\partial \xi_i} + \frac{\partial U}{\partial \psi_i}\Big) = 0
\end{aligned}\right\} \tag{19.12}$$

As a consequence of relations (19.11) we obtain six integrals regarding the motion of the mass center of the whole system,

$$\sum_{i=0}^{n} m_i \frac{d\xi_i}{dt} = a_1, \qquad \sum_{i=0}^{n} m_i \xi_i = a_1 t + b_1$$

$$\sum_{i=0}^{n} m_i \frac{d\eta_i}{dt} = a_2, \qquad \sum_{i=0}^{n} m_i \eta_i = a_2 t + b_2$$

$$\sum_{i=0}^{n} m_i \frac{d\zeta_i}{dt} = a_3, \qquad \sum_{i=0}^{n} m_i \zeta_i = a_3 t + b_3$$

a_j, b_j $(j = 1, 2, 3)$ being constants of integration.

As a consequence of relations (19.12) follow the angular momentum integrals

$$\sum_{i=0}^{n} [m_i(\eta_i \dot{\zeta}_i - \zeta_i \dot{\eta}_i) + A_i p_i a_{11}^{(i)} + B_i q_i a_{12}^{(i)} + C_i r_i a_{13}^{(i)}] = c_1$$

$$\sum_{i=0}^{n} [m_i(\zeta_i \dot{\xi}_i - \xi_i \dot{\zeta}_i) + A_i p_i a_{21}^{(i)} + B_i q_i a_{22}^{(i)} + C_i r_i a_{23}^{(i)}] = c_2$$

$$\sum_{i=0}^{n} [m_i(\xi_i \dot{\eta}_i - \eta_i \dot{\xi}_i) + A_i p_i a_{31}^{(i)} + B_i q_i a_{32}^{(i)} + C_i r_i a_{33}^{(i)}] = c_3$$

c_j $(j = 1, 2, 3)$ being constants of integration.

Finally, the energy integral exists as a consequence of the principle of the conservation of energy

$$T = U + h$$

If the form and mass distribution of the bodies M_i are arbitrary, there are no other integrals besides the ten derived above. However, in certain particular cases the equations of motion may possess some additional integrals. For example, if the bodies M_i $(i=0, 1, \ldots, k, k \leqq n)$ are spheres and have spherical density distribution, then obviously $A_i = B_i = C_i$ $(i = 0, 1, \ldots, k)$, and the force function U does not depend on ψ_i, Φ_i, Θ_i $(i = 0, 1, \ldots, k)$. As a consequence equations (19.9) give additional integrals

$$p_i = p_i^{(0)} = \text{const.}, \qquad q_i = q_i^{(0)} = \text{const.}, \qquad r_i = r_i^{(0)} = \text{const.} \qquad (i = 0, 1, \ldots, k)$$

If the bodies M_i $(i = 0, 1, \ldots, k;\ k \leqq n)$ are bodies of revolution, having axial-symmetric density distribution, then, obviously, $A_i = B_i$ $(i = 0, 1, \ldots, k)$, and the force function U does not depend on Φ_i $(i = 0, 1, \ldots, k)$. Equations (19.9) now provide the integrals

$$r_i = r_i^{(0)} = \text{const.} \qquad (i = 0, 1, \ldots, k)$$

By means of the above indicated integrals the order of the system of differential equations can be lowered by the same number of units.

19.4. Separation of the system of differential equations of motion. Consider now the problem of decomposing the complete system of differential equations of motion describing the combined translational and rotational motions of the bodies into two independent or partially independent motions, each of which determines separately the translational or the rotational motions of the bodies.

Complete separation is possible only in the rather trivial case where all bodies are spheres with spherical density distribution. In such a case the mutual force function of any two bodies M_i and M_j is independent of the EULER angles ψ_i, Φ_i, Θ_i; ψ_j, Φ_j, Θ_j and is given by the formula

$$U_{ij}^{(0)} = f \frac{m_i m_j}{R_{ij}} \qquad (i \neq j)$$

where R_{ij} is the distance between the centers of M_i and M_j. Then the total force function

$$U^{(0)} = f \sum_{i,j} \frac{m_i m_j}{R_{ij}} \qquad (i \neq j)$$

depends only upon the coordinates of the mass centers of the spheres, and system (19.8) determines the translations of these centers.

System (19.9) then decomposes into $n+1$ mutually independent systems which upon integration give

$$p_i = p_i^{(0)}, \qquad q_i = q_i^{(0)}, \qquad r_i = r_i^{(0)} \qquad (i = 0, 1, \ldots, n)$$

Thus each sphere rotates with constant angular velocity about a fixed axis passing through its center. In all other cases systems (19.8), (19.9) and (19.10) must be integrated simultaneously.

Consider again the general case of arbitrary bodies, and let us investigate to what extent the complete equations can be separated into equations for the translations and rotations about their mass centers.

For this purpose let us start from the well known series expression for the mutual force function of the bodies assuming that the mass distribution of each body is symmetric with respect to three mutually orthogonal planes intersecting at the mass center.

The first few terms of this series are

$$U_{ij} = f\frac{m_i m_j}{R_{ij}} + f m_j \frac{A_i + B_i + C_i - 3I_i^{(ij)}}{2R_{ij}^3} + \\ + f m_i \frac{A_j + B_j + C_j - 3I_j^{(ij)}}{2R_{ij}^3} + \cdots \tag{19.13}$$

where $I_i^{(ij)}$ and $I_j^{(ij)}$ are the moments of inertia of the bodies M_i and M_j relative to the straight line through their mass centers G_i and G_j, i.e.

$$\begin{aligned} I_i^{(ij)} &= A_i\alpha_{ij}^2 + B_i\beta_{ij}^2 + C_i\gamma_{ij}^2, \\ I_j^{(ij)} &= A_j\alpha_{ji}^2 + B_j\beta_{ji}^2 + C_j\gamma_{ji}^2 \end{aligned} \tag{19.14}$$

α_{ij}, β_{ij}, γ_{ij} being the cosines of the angles made by $\overrightarrow{G_i G_j}$ with the central principal inertia axes of the body M_i, and α_{ji}, β_{ji}, γ_{ji} are analogous quantities for the body M_j.

Explicit expressions for α_{ij}, β_{ij}, γ_{ij} are

$$\left.\begin{aligned} \alpha_{ij} &= (\cos\psi_i\cos\Phi_i - \sin\psi_i\sin\Phi_i\cos\Theta_i)\frac{\xi_j - \xi_i}{R_{ij}} + \\ &\quad + (\sin\psi_i\cos\Phi_i + \cos\psi_i\sin\Phi_i\cos\Theta_i)\frac{\eta_j - \eta_i}{R_{ij}} + \\ &\quad + \sin\Phi_i\sin\Theta_i\frac{\zeta_j - \zeta_i}{R_{ij}} \\ \beta_{ij} &= -(\cos\psi_i\sin\Phi_i - \sin\psi_i\cos\Phi_i\cos\Theta_i)\frac{\xi_j - \xi_i}{R_{ij}} + \\ &\quad + (-\sin\psi_i\sin\Phi_i + \cos\psi_i\cos\Phi_i\cos\Theta_i)\frac{\eta_j - \eta_i}{R_{ij}} + \\ &\quad + \cos\Phi_i\sin\Theta_i\frac{\zeta_j - \zeta_i}{R_{ij}} \\ \gamma_{ij} &= \sin\psi_i\sin\Theta_i\frac{\xi_j - \xi_i}{R_{ij}} - \cos\psi_i\sin\Theta_i\frac{\eta_j - \eta_i}{R_{ij}} + \cos\Theta_i\frac{\zeta_j - \zeta_i}{R_{ij}} \end{aligned}\right\} \tag{19.15}$$

If third and higher order terms in $\frac{1}{R_{ij}}$ in the expansion (19.13) are neglected, then the force function reduces to that of a system of particles,

$$U \approx U^0 = f\sum_{i,j}\frac{m_i m_j}{R_{ij}}$$

and systems of equations (19.8) and (19.9) assume the form

$$m_i \frac{d^2 \xi_i}{dt^2} = \frac{\partial U^{(0)}}{\partial \xi_i}, \quad m_i \frac{d^2 \eta_i}{dt^2} = \frac{\partial U^{(0)}}{\partial \eta_i}, \quad m_i \frac{d^2 \zeta_i}{dt^2} = \frac{\partial U^{(0)}}{\partial \zeta_i}$$

$$A_i \frac{dp_i}{dt} - (B_i - C_i)\, q_i r_i = 0$$

$$B_i \frac{dq_i}{dt} - (C_i - A_i)\, r_i p_i = 0$$

$$C_i \frac{dr_i}{dt} - (A_i - B_i)\, p_i q_i = 0$$

Thus translations and rotations of the bodies are independent of each other, each body rotating under the influence of no external forces.

If in the expansion of the force function terms up to the third power of $1/R_{ij}$ are retained, then

$$U \approx U^{(0)} + U^{(1)}$$

Hence U depends upon the coordinates of all the mass centers G_i as well as all the EULER angles, and as a consequence systems of equations (19.8) and (19.9) cannot be separated into two independent ones.

19.5. Relative differential equations of motion. Consider the problem of transforming systems (19.8) and (19.9) to a certain coordinate trihedral with the origin at one of the mass centers of the $n+1$ bodies, say G_0.

Let the new coordinate trihedral be $G_0\,x\,y\,z$ with axes parallel to the corresponding axes of the trihedral $O\,\xi\,\eta\,\zeta$. Such a transformation will change equations (19.8) but will leave the equations of system (19.9) unchanged.

Denoting the relative coordinates of G_i by x_i, y_i, z_i, we have that

$$x_i = \xi_i - \xi_0, \quad y_i = \eta_i - \eta_0, \quad z_i = \zeta_i - \zeta_0 \qquad (i = 1, 2, \ldots, n)$$

and

$$\frac{d^2 x_i}{dt^2} = \frac{d^2 \xi_i}{dt^2} - \frac{d^2 \xi_0}{dt^2} = \frac{1}{m_i}\frac{\partial U}{\partial \xi_i} - \frac{1}{m_0}\frac{\partial U}{\partial \xi_0}, \ldots$$

where

$$\frac{\partial U}{\partial \xi_0} = \sum_{j=1}^{n} \frac{\partial U_{j0}}{\partial \xi_0} = -\sum_{j=1}^{n} \frac{\partial U_{j0}}{\partial x_j}$$

and the function U_{j0} depends only on $x_j, y_j, z_j; \psi_j, \Phi_j, \Theta_j; \psi_0, \Phi_0, \Theta_0$.

Hence

$$\frac{1}{m_i}\frac{\partial U}{\partial \xi_i} - \frac{1}{m_0}\frac{\partial U}{\partial \xi_0} = \frac{1}{m_i}\sum_{\substack{j=0\\(j\neq i)}}^{n} \frac{\partial U_{ij}}{\partial x_i} + \frac{1}{m_0}\sum_{j=1}^{n}\frac{\partial U_{j0}}{\partial x}$$

$$= \frac{m_0 + m_i}{m_0 m_i}\frac{\partial U_{i0}}{\partial x_i} + \frac{\partial R_i}{\partial x_i}$$

where

$$R_i = \sum_{\substack{j=1\\(j\neq i)}}^{n} \left\{\frac{1}{m_i} U_{ij} + \frac{1}{m_0}\left(x_i \frac{\partial U_{j0}}{\partial x_j} + y_i \frac{\partial U_{j0}}{\partial y_j} + z_i \frac{\partial U_{j0}}{\partial z_j}\right)\right\}$$

Then equations (19.8) assume the form

$$\left.\begin{aligned}
\frac{d^2 x_i}{dt^2} &= \frac{m_0 + m_i}{m_0 m_i}\frac{\partial U_{i0}}{\partial x_i} + \frac{\partial R_i}{\partial x_i}\\
\frac{d^2 y_i}{dt^2} &= \frac{m_0 + m_i}{m_0 m_i}\frac{\partial U_{i0}}{\partial y_i} + \frac{\partial R_i}{\partial y_i} \qquad (i = 1, 2, \ldots, n)\\
\frac{d^2 z_i}{dt^2} &= \frac{m_0 + m_i}{m_0 m_i}\frac{\partial U_{i0}}{\partial z_i} + \frac{\partial R_i}{\partial z_i}
\end{aligned}\right\} \tag{19.16}$$

Equations (19.9) will preserve their form; however, the EULER angles will henceforth refer to the new coordinate trihedral.

19.6. The two-body problem. Consider now the particular case $n = 1$, i.e. the case of two bodies M_0 and M_1. Then, if we drop the subscript 1 and replace x, y, z by ξ, η, ζ, the differential equations of motion become

$$\left.\begin{aligned}
\ddot{\xi} &= \frac{m_0 + m}{m_0 m}\frac{\partial U}{\partial \xi}\\
\ddot{\eta} &= \frac{m_0 + m}{m_0 m}\frac{\partial U}{\partial \eta}\\
\ddot{\zeta} &= \frac{m_0 + m}{m_0 m}\frac{\partial U}{\partial \zeta}
\end{aligned}\right\} \tag{19.17}$$

$$\left.\begin{aligned}
&A_0 \dot{p}_0 - (B_0 - C_0)\, q_0 r_0\\
&\qquad = \frac{\sin\Phi_0}{\sin\Theta_0}\left(\frac{\partial U}{\partial \psi_0} - \cos\Theta_0 \frac{\partial U}{\partial \Phi_0}\right) + \cos\Phi_0 \frac{\partial U}{\partial \Theta_0}\\
&B_0 \dot{q}_0 - (C_0 - A_0)\, r_0 p_0\\
&\qquad = \frac{\cos\Phi_0}{\sin\Theta_0}\left(\frac{\partial U}{\partial \psi_0} - \cos\Theta_0 \frac{\partial U}{\partial \Phi_0}\right) - \sin\Phi_0 \frac{\partial U}{\partial \Theta_0}\\
&C_0 \dot{r}_0 - (A_0 - B_0)\, p_0 q_0\\
&\qquad = \frac{\partial U}{\partial \Phi_0}
\end{aligned}\right\} \tag{19.18}$$

$$\left.\begin{aligned}
A\dot{p} - (B - C)\, q r &= \frac{\sin\Phi}{\sin\Theta}\left(\frac{\partial U}{\partial \psi} - \cos\Theta \frac{\partial U}{\partial \Phi}\right) + \cos\Phi \frac{\partial U}{\partial \Theta}\\
B\dot{q} - (C - A)\, r p &= \frac{\cos\Phi}{\sin\Theta}\left(\frac{\partial U}{\partial \psi} - \cos\Theta \frac{\partial U}{\partial \Phi}\right) - \sin\Phi \frac{\partial U}{\partial \Theta}\\
C\dot{r} - (A - B)\, p q &= \frac{\partial U}{\partial \Phi}
\end{aligned}\right\} \tag{19.19}$$

$$\left.\begin{aligned}
p_0 &= \dot{\psi}_0 \sin\Phi_0 \sin\Theta_0 + \dot{\Theta}_0 \cos\Phi_0\\
q_0 &= \dot{\psi}_0 \cos\Phi_0 \sin\Theta_0 - \dot{\Theta}_0 \sin\Phi_0\\
r_0 &= \dot{\psi}_0 \cos\Theta_0 + \dot{\Phi}_0
\end{aligned}\right\} \tag{19.20}$$

$$\left.\begin{aligned}
p &= \dot{\psi} \sin\Phi \sin\Theta + \dot{\Theta} \cos\Phi\\
q &= \dot{\psi} \cos\Phi \sin\Theta - \dot{\Theta} \sin\Phi\\
r &= \dot{\psi} \cos\Phi + \dot{\Phi}
\end{aligned}\right\} \tag{19.21}$$

where the mutual force function of the bodies M_0 and M,

$$U = f \int\limits_{(M_0)} dm_0 \int\limits_{(M)} \frac{dm}{\Delta} \tag{19.22}$$

is a known function of

$$\xi, \eta, \zeta; \quad \psi_0, \Phi_0, \Theta_0; \quad \psi, \Phi, \Theta \tag{19.23}$$

The systems of equations (19.17) to (19.21) constitute a system of order 18 which for the initial values of the variables (19.23) determines the translation of the mass center G of M as well as the rotations of the bodies M_0 and M relative to a fixed coordinate trihedral in space.

The above systems possess four first integrals which are of the form

$$\left.\begin{aligned}
&\frac{m_0 m}{m_0+m}(\eta\dot{\zeta}-\zeta\dot{\eta}) + A_0 p_0 a_{11}^{(0)} + B_0 q_0 a_{12}^{(0)} + C_0 r_0 a_{13}^{(0)} + \\
&\qquad + A p a_{11} + B q a_{12} + C r a_{13} = c_1 \\
&\frac{m_0 m}{m_0+m}(\zeta\dot{\xi}-\xi\dot{\zeta}) + A_0 p_0 a_{21}^{(0)} + B_0 q_0 a_{22}^{(0)} + C_0 r_0 a_{23}^{(0)} + \\
&\qquad + A p a_{21} + B q a_{22} + C r a_{23} = c_2 \\
&\frac{m_0 m}{m_0+m}(\xi\dot{\eta}-\eta\dot{\xi}) + A_0 p_0 a_{31}^{(0)} + B_0 q_0 a_{32}^{(0)} + C_0 r_0 a_{33}^{(0)} + \\
&\qquad + A p a_{31} + B q a_{32} + C r a_{33} = c_3
\end{aligned}\right\} \tag{19.24}$$

$$\begin{aligned}
&\frac{m_0 m}{m_0+m}(\dot{\xi}^2+\dot{\eta}^2+\dot{\zeta}^2) + A_0 p_0^2 + B_0 q_0^2 + C_0 r_0^2 + A p^2 + \\
&\qquad + B q^2 + C r^2 = 2U + h
\end{aligned} \tag{19.25}$$

where a_{ik} and $a_{ik}^{(0)}$ are the direction cosines of the principal inertia axes of the bodies M and M_0 relative to the fixed coordinate trihedral mentioned above.

By means of integrals (19.24) and (19.25) the order of the systems (19.17) to (19.21) could be lowered by four. However, it is well known that such an operation does not simplify the problem but only complicates all calculations.

If M_0 is a sphere, homogeneous or with a spherical mass distribution, then $A_0 = B_0 = C_0$, and as a consequence the force function U does not depend upon ψ_0, Φ_0, Θ_0. In such a case it follows from equations (19.18) that

$$p_0 = \text{const.}, \qquad q_0 = \text{const.}, \qquad r_0 = \text{const.}$$

i.e. the body M_0 rotates with constant angular velocity about an axis of fixed direction.

The problem of motion reduces to the determination of the combined translational and rotational motions of the body M, i.e. to the integration of the following system of order 12:

$$\left.\begin{aligned}\ddot{\xi} &= \frac{m_0+m}{m_0 m}\frac{\partial U}{\partial \xi}\\ \ddot{\eta} &= \frac{m_0+m}{m_0 m}\frac{\partial U}{\partial \eta}\\ \ddot{\zeta} &= \frac{m_0+m}{m_0 m}\frac{\partial U}{\partial \zeta}\end{aligned}\right\} \tag{19.26}$$

$$\left.\begin{aligned}A\dot{p}-(B-C)\,q\,r &= \frac{\sin\Phi}{\sin\Theta}\left(\frac{\partial U}{\partial\psi}-\cos\Theta\frac{\partial U}{\partial\Phi}\right)+\cos\Phi\frac{\partial U}{\partial\Theta}\\ B\dot{q}-(C-A)\,r\,p &= \frac{\cos\Phi}{\sin\Theta}\left(\frac{\partial U}{\partial\psi}-\cos\Theta\frac{\partial U}{\partial\Phi}\right)-\sin\Phi\frac{\partial U}{\partial\Theta}\\ C\dot{r}-(A-B)\,p\,q &= \frac{\partial U}{\partial\Phi}\end{aligned}\right\} \tag{19.27}$$

$$\left.\begin{aligned}\dot{\psi}\sin\Theta &= p\sin\Phi+q\cos\Phi\\ \dot{\Theta} &= p\cos\Phi-q\sin\Phi\\ \dot{\Phi} &= r-\cot\Theta\,(p\sin\Phi+q\cos\Phi)\end{aligned}\right\} \tag{19.28}$$

The above systems possess the following integrals

$$\left.\begin{aligned}&\frac{m_0 m}{m_0+m}(\eta\dot{\zeta}-\zeta\dot{\eta})+A\,p\,a_{11}+B\,q\,a_{12}+C\,r\,a_{13}=c_1\\ &\frac{m_0 m}{m_0+m}(\zeta\dot{\xi}-\xi\dot{\zeta})+A\,p\,a_{21}+B\,q\,a_{22}+C\,r\,a_{23}=c_2\\ &\frac{m_0 m}{m_0+m}(\xi\dot{\eta}-\eta\dot{\xi})+A\,p\,a_{31}+B\,q\,a_{32}+C\,r\,a_{33}=c_3\\ &\frac{m_0 m}{m_0+m}(\dot{\xi}^2+\dot{\eta}^2+\dot{\zeta}^2)+A\,p^2+B\,q^2+C\,r^2=2U+h\end{aligned}\right\} \tag{19.29}$$

If M is a homogeneous body of rotation or possesses axial symmetry, then $A=B$, and the force function U, being independent of Φ, depends only upon ξ, η, ζ; ψ, Θ. Then the third equation of system (19.27) gives

$$\dot{r}=0,\qquad r=\text{const}.$$

If we eliminate p, q by means of equations (19.28), equations of motion (19.26) and (19.27) assume the form of a system of order 10:

$$\left.\begin{aligned}\ddot{\xi} &= \frac{m_0+m}{m_0 m}\frac{\partial U}{\partial \xi}\\ \ddot{\eta} &= \frac{m_0+m}{m_0 m}\frac{\partial U}{\partial \eta}\\ \ddot{\zeta} &= \frac{m_0+m}{m_0 m}\frac{\partial U}{\partial \zeta}\\ A\frac{d}{dt}(\dot{\psi}\sin^2\Theta)-C\,r\,\dot{\Theta}\sin\Theta &= \frac{\partial U}{\partial\psi}\\ A\ddot{\Theta}+\dot{\psi}\,C\,r\sin\Theta-\dot{\psi}^2A\sin\Theta\cos\Theta &= \frac{\partial U}{\partial\Theta}\end{aligned}\right\} \tag{19.30}$$

Upon the integration of this system the angle Φ is given by a quadrature

$$\Phi = \Phi_0 + r(t - t_0) - \int_{t_0}^{t} \dot{\psi} \cos\Theta \, dt$$

The known first integrals of system (19.30) are

$$\left.\begin{aligned}
&\frac{m_0 m}{m_0 + m}(\eta \dot{\zeta} - \zeta \dot{\eta}) - A \dot{\psi} \sin\psi \sin\Theta \cos\Theta + A \dot{\Theta} \cos\psi + \\
&\qquad + C r \sin\psi \sin\Theta = c_1 \\
&\frac{m_0 m}{m_0 + m}(\zeta \dot{\xi} - \xi \dot{\zeta}) + A \dot{\psi} \cos\psi \sin\Theta \cos\Theta + A \dot{\Theta} \sin\psi - \\
&\qquad - C r \cos\psi \sin\Theta = c_2 \\
&\frac{m_0 m}{m_0 + m}(\xi \dot{\eta} - \eta \dot{\xi}) + A \dot{\psi} \sin^2\Theta + C r \cos\Theta = c_3 \\
&\frac{m_0 m}{m_0 + m}(\dot{\xi}^2 + \dot{\eta}^2 + \dot{\zeta}^2) + \dot{\psi}^2 A \sin^2\Theta + A \dot{\Theta}^2 \\
&\qquad = 2U + h - C r^2
\end{aligned}\right\} \tag{19.31}$$

In what follows we shall apply equations (19.30) to the case where M_0 as before is a sphere while M is a homogeneous rod of length $2l$. We shall call one end of the rod "head" and the other end "tail". The largest axis of the rod passes through its midpoint G in the direction of the head of the rod, so that Θ is the angle between the direction "tail-head" and the positive direction of $G_0 \zeta$.

The moments of inertia of the rod M being

$$A = \frac{m l^2}{3}, \qquad C = 0$$

the equations of motion (19.30) and their first integrals (19.31) assume the form

$$\left.\begin{aligned}
\ddot{\xi} &= \frac{m_0 + m}{m_0 m} \frac{\partial U}{\partial \xi} \\
\ddot{\eta} &= \frac{m_0 + m}{m_0 m} \frac{\partial U}{\partial \eta} \\
\ddot{\zeta} &= \frac{m_0 + m}{m_0 m} \frac{\partial U}{\partial \zeta} \\
\ddot{\psi} \sin^2\Theta + 2\dot{\psi} \dot{\Theta} \sin\Theta \cos\Theta &= \frac{3}{m l^2} \frac{\partial U}{\partial \psi} \\
\ddot{\Theta} - \dot{\psi}^2 \sin\Theta \cos\Theta &= \frac{3}{m l^2} \frac{\partial U}{\partial \Theta}
\end{aligned}\right\} \tag{19.32}$$

and

$$\left.\begin{aligned}
&\frac{m_0}{m_0+m}(\eta\,\dot\zeta-\zeta\,\dot\eta)+\frac{l^2}{3}(-\dot\psi\sin\psi\sin\Theta\cos\Theta+\dot\Theta\cos\Theta)=c_1\\
&\frac{m_0}{m_0+m}(\zeta\,\dot\xi-\xi\,\dot\zeta)+\frac{l^2}{3}(\dot\psi\cos\psi\sin\Theta\cos\Theta+\dot\Theta\sin\Theta)=c_2\\
&\frac{m_0}{m_0+m}(\xi\,\dot\eta-\eta\,\dot\xi)+\frac{l^2}{3}\dot\psi\sin^2\Theta=c_3\\
&\frac{m_0}{m_0+m}(\dot\xi^2+\dot\eta^2+\dot\zeta^2)+\frac{l^2}{3}(\dot\psi^2\sin^2\Theta+\dot\Theta^2)=\frac{2}{m}U+h
\end{aligned}\right\}\qquad(19.33)$$

where the new constants include the factor $1/m$.

The force function U is given by the integral

$$U=\frac{f\,m_0\,m}{2l}\int\limits_{-l}^{l}\frac{ds}{\Delta}$$

where

$$\Delta^2=r^2+s^2-2r\,s\,\nu,\qquad r^2=\xi^2+\eta^2+\zeta^2$$

and

$$\nu=\cos\alpha=a_{13}\frac{\xi}{r}+a_{23}\frac{\eta}{r}+a_{33}\frac{\zeta}{r}$$

$$a_{13}=\sin\psi\sin\Theta,\qquad a_{23}=-\cos\psi\sin\Theta,\qquad a_{33}=\cos\Theta$$

Evaluating the integral we obtain

$$U=\frac{f\,m_0\,m}{2l}\ln\frac{r_1+l-r\,\nu}{r_2-l-r\,\nu}\qquad(19.34)$$

where

$$r_1=\sqrt{l^2+r^2-2r\,l\,\nu}$$
$$r_2=\sqrt{l^2+r^2+2r\,l\,\nu}$$

are the distances from G_0 to the "tail" and the "head" of the rod respectively.

Using the above expression for U and calculating the partial derivatives of U with respect to ξ, η, ζ; ψ, Θ we obtain the equations of motion in the following form:

$$\left.\begin{aligned}
\ddot\xi&=\frac{\mu\,\xi}{r^3}R+\frac{\mu\,l^2}{r^4}\left(a_{13}-\frac{\xi}{r}\nu\right)H\\
\ddot\eta&=\frac{\mu\,\eta}{r^3}R+\frac{\mu\,l^2}{r^4}\left(a_{23}-\frac{\eta}{r}\nu\right)H\\
\ddot\zeta&=\frac{\mu\,\zeta}{r^3}R+\frac{\mu\,l^2}{r^4}\left(a_{33}-\frac{\zeta}{r}\nu\right)H
\end{aligned}\right\}\qquad(19.35)$$

$$\left.\begin{aligned}
&\ddot\psi\sin\Theta+2\dot\psi\,\dot\Theta\cos\Theta=\frac{3f\,m_0}{r^4}(\xi\cos\psi+\eta\sin\psi)\,H\\
&\ddot\Theta-\dot\psi^2\sin\Theta\cos\Theta=\frac{3f\,m_0}{r^4}[(\xi\sin\psi-\eta\cos\psi)\cos\Theta-\zeta\sin\Theta]\,H
\end{aligned}\right\}\qquad(19.36)$$

where

$$\mu = f(m_0 + m)$$

$$R = \frac{r^2}{f m_0 m} \frac{\partial U}{\partial r} = \frac{r^2}{2l}\left[\frac{r - l\nu - r_1\nu}{r_1(r_1 + l - r\nu)} - \frac{r + l\nu - r_2\nu}{r_2(r_2 - l - r\nu)}\right]$$

$$H = \frac{r^3}{f m_0 m l^2} \frac{\partial U}{\partial \nu} = -\frac{r^4}{2l^3}\left[\frac{l + r_1}{r_1(r_1 + l - r\nu)} + \frac{l - r_2}{r_2(r_2 - l - r\nu)}\right]$$

If l is small in comparison with r, then R and H can be expanded in powers of l/r. Putting

$$U = \frac{f m_0 m}{r} F(\sigma), \qquad \sigma = \frac{l}{r}$$

where

$$F(\sigma) = \frac{1}{2\sigma} \ln \frac{\sqrt{1 + \sigma^2 - 2\nu\sigma} + \sigma - \nu}{\sqrt{1 + \sigma^2 + 2\nu\sigma} - \sigma - \nu}$$

we obtain that

$$\sigma F'(\sigma) + F(\sigma) = \frac{1}{2} \frac{1}{\sqrt{1 + \sigma^2 - 2\nu\sigma}} + \frac{1}{2} \frac{1}{\sqrt{1 + \sigma^2 + 2\nu\sigma}}$$

Denote the LEGENDRE polynomials by $X_k(\nu)$. Then

$$\frac{d}{d\sigma}[\sigma F(\sigma)] = \frac{1}{2}\sum_{k=0}^{\infty} \sigma^k X_k(\nu) + \frac{1}{2}\sum_{k=0}^{\infty} \sigma^k X_k(-\nu) = \sum_{k=0}^{\infty} \sigma^{2k} X_{2k}(\nu)$$

and for the functions U, R and H we obtain the expansions

$$U = \frac{f m_0 m}{r} \sum_{k=0}^{\infty} \frac{X_{2k}(\nu)}{2k + 1} \left(\frac{l}{r}\right)^{2k}$$

$$R = -\sum_{k=0}^{\infty} X_{2k}(\nu) \left(\frac{l}{r}\right)^{2k} = -1 - X_2(\nu)\frac{l^2}{r^2} - \cdots$$

$$H = \sum_{k=0}^{\infty} \frac{X'_{2k+2}(\nu)}{2k + 3} \left(\frac{l}{r}\right)^{2k} = \frac{X'_2(\nu)}{3} + \frac{X'_4(\nu)}{5} \frac{l^2}{r^2} + \cdots$$

where X_{2k} is a LEGENDRE polynomial of order $2k$ and X'_{2k+2} the derivative of X_{2k+2} with respect to ν.

19.7. Particular solutions. Equations (19.35) and (19.36) constitute a system of order 10 of which only four integrals (19.33) are known. Consequently in general such a system cannot be integrated.

Therefore we shall look for particular cases where the integration of the above equations is possible. For this purpose let us investigate whether there exist particular solutions having the property that the mass center G of the rod moves in a plane.

Multiplying the three integrals of (19.33) by ξ, η, ζ respectively and adding we obtain

$$C_1 \xi + C_2 \eta + C_3 \zeta = 0 \tag{19.37}$$

where

$$\left.\begin{aligned} C_1 &= c_1 - \frac{l^2}{3}\bar{c}_1 = c_1 - \frac{l^2}{3}(-\dot{\psi}\sin\psi\sin\Theta\cos\Theta + \dot{\Theta}\cos\Theta) \\ C_2 &= c_2 - \frac{l^2}{3}\bar{c}_2 = c_2 - \frac{l^2}{3}(\dot{\psi}\cos\psi\sin\Theta\cos\Theta + \dot{\Theta}\sin\Theta) \\ C_3 &= c_3 - \frac{l^2}{3}\bar{c}_3 = c_3 - \frac{l^2}{3}\dot{\psi}\sin^2\Theta \end{aligned}\right\} \qquad (19.38)$$

In order that the plane (19.37) be invariable it is necessary and sufficient that $\bar{c}_1$, $\bar{c}_2$, $\bar{c}_3$ be constants. For this purpose the EULER angles ψ and Θ must satisfy the equations

$$\begin{aligned} -\dot{\psi}\sin\psi\sin\Theta\cos\Theta + \dot{\Theta}\cos\Theta &= \bar{c}_1 = \text{const.} \\ \dot{\psi}\cos\psi\sin\Theta\cos\Theta + \dot{\Theta}\sin\Theta &= \bar{c}_2 = \text{const.} \\ \dot{\psi}\sin^2\Theta &= \bar{c}_3 = \text{const.} \end{aligned}$$

By virtue of equations (19.32) this implies that

$$\frac{\partial U}{\partial\psi} = 0$$

or by virtue of (19.36) that

$$\sin\Theta(\xi\cos\psi + \eta\sin\psi)\,H = 0 \qquad (19.39)$$

Assume that the last equation is satisfied. Then the mass center G of the rod moves in an invariable plane passing through the origin. Without loss of generality we can assume that this plane coincides with the $G_0\,\xi\,\eta$-plane. In such a case the third equation of system (19.35) must be satisfied for $\zeta = 0$. This leads to the condition

$$H\cos\Theta = 0 \qquad (19.40)$$

which must be consistent with equation (19.39).

Consider the possible cases which may occur. The case $H = 0$ implies $\nu = \cos\alpha = 0$. In such a case

$$r_1 = r_2 = \sqrt{r^2 + l^2}, \qquad R = -\frac{r}{r_1}$$

and the equations of motion decompose into two independent systems of which the system

$$\ddot{\xi} = -\frac{\mu\,\xi}{r^2\,r_1}, \qquad \ddot{\eta} = -\frac{\mu\,\eta}{r^2\,r_1} \qquad (19.41)$$

determines the translation of the mass center G of the rod while the other system

$$\begin{aligned} \ddot{\psi}\sin\Theta + 2\dot{\psi}\,\dot{\Theta}\cos\Theta &= 0 \\ \ddot{\Theta} - \dot{\psi}^2\sin\Theta\cos\Theta &= 0 \end{aligned}$$

determines the rotation of the rod about its center G. The integration of both systems can be reduced to quadratures.

Equations (19.41) can be interpreted as the equations of motion of a particle subject to attraction from the origin, the magnitude of attraction being equal to $\mu/r\sqrt{r^2 + l^2}$. Consequently for small values of l the orbit will deviate slightly from a KEPLER orbit.

For $H \neq 0$ conditions (19.39) and (19.40) imply that

$$\xi \cos\psi + \eta \sin\psi = 0$$

$$\cos\Theta = 0$$

Hence

$$\sin\Theta = \pm 1, \qquad \cos(\psi - \nu) = 0, \qquad \dot{\psi} = \dot{\nu}$$

Then, however, the condition

$$\dot{\psi} \sin^2\Theta = \text{const.}$$

implies that $\dot{\psi}$ must be constant. Hence also $\dot{\nu} = \text{const.}$, i.e. the orbit of G is a circle or a straight line through the origin.

Introduce now the angle γ between the direction of the rod ("tail-head") and that of the velocity V of translation of the mass center G. Then

$$\cos\gamma = a_{13} \frac{\dot{\xi}}{V} + a_{23} \frac{\dot{\eta}}{V} + a_{33} \frac{\dot{\zeta}}{V} \tag{19.42}$$

Consider now the particular cases for which the orbit of the mass center G is a circle while the angles α and γ remain constant. Then equations (19.33), (19.35) and (19.36) yield three fixed orientations of the rod relative to the orbit of G, known as "arrow"-, "floating bar"-, and "spoke"-type.

Case 1 (arrow-type orientation of the rod):

$$\xi = a \cos nt, \qquad \eta = a \sin nt, \qquad \zeta = 0$$

$$\Theta = \frac{\pi}{2}, \qquad \psi = nt + \pi \quad \text{or} \quad \psi = nt$$

$$r = a, \qquad V = a\,n$$

Hence

$$\nu = \cos\alpha = 0, \qquad \alpha = \frac{\pi}{2}$$

$$\cos\gamma = \pm 1, \qquad \gamma = 0, \pi$$

Further, we have that

$$r_1 = r_2 = \sqrt{a^2 + l^2}, \qquad R = -\frac{a}{r_1}, \qquad H = 0$$

and

$$n^2 = \frac{\mu}{a^2 \sqrt{a^2 + l^2}} < \frac{\mu}{a^3} \tag{19.43}$$

Thus while the mass center G moves along a circle, the rod, lying in the same plane and remaining perpendicular to the radius vector of the circular orbit, has its head directed forward if $\psi = nt + \pi$, $\gamma = 0$ or its tail directed forward if $\psi = nt$, $\gamma = \pi$.

Formula (19.43) shows that the angular as well as the linear velocity assume their largest value for $l = 0$, and they decrease as l increases and assume the value zero for $l = \infty$.

Case 2 (floating bar-type orientation):

$$\xi = a \cos nt, \qquad \eta = a \sin nt, \qquad \zeta = 0$$

$$\Theta = 0 \quad \text{or} \quad \pi, \qquad \psi \text{ undetermined}$$

$$r = a, \qquad V = a\,n$$

Hence

$$\nu = \cos\alpha = 0, \qquad \alpha = \frac{\pi}{2}$$

$$\cos\gamma = 0, \qquad \gamma = \frac{\pi}{2}, \frac{3\pi}{2}$$

while r_1, r_2, R, H and n have the same values as in Case 1.

Thus the mass center G of the rod moves in a circular orbit while the rod itself remains perpendicular to the plane of the motion of G, and is directed upward with the head if $\Theta = 0$, $\gamma = \pi/2$ or with the tail if $\Theta = \pi$, $\gamma = 3\pi/2$.

Case 3 (*spoke-type orientation*):

$$\xi = a\cos nt, \qquad \eta = a\sin nt, \qquad \zeta = 0$$

$$\Theta = \frac{\pi}{2}, \qquad \psi = nt + \frac{3\pi}{2} \quad \text{or} \quad \psi = nt + \frac{\pi}{2}$$

$$r = a, \qquad V = n\,a$$

Hence

$$\nu = \cos\alpha = \mp 1, \qquad \alpha = \pi, 0$$

$$\cos\gamma = 0, \qquad \gamma = \frac{\pi}{2}$$

Further, we find that

$$r_1 = a \pm l, \qquad r_2 = a \mp l, \qquad R = -\frac{a^2}{a^2 - l^2}$$

$$\xi\cos\psi + \eta\sin\psi = 0, \qquad \cos\Theta = 0$$

and

$$n^2 = \frac{\mu}{a(a^2 - l^2)} > \frac{\mu}{a^3} \tag{19.44}$$

Thus the mass center G of the rod moves in a circular orbit while the rod itself, lying in the plane of the orbit, remains perpendicular to the velocity vector $\vec{V}$ of the point G, i.e. it coincides with the radius vector of G and is directed towards the origin with its head if $\psi = nt + \frac{3\pi}{2}$, $\alpha = \pi$, $r_1 = a + l$, $r_2 = a - l$ or with its tail if $\psi = nt + \frac{\pi}{2}$, $\alpha = 0$, $r_1 = a - l$, $r_2 = a + l$.

It is easy to show that besides the circle there are no other planar trajectories of G for which the angle of attack γ satisfies the condition $\cos\gamma = \pm 1$.

Assume now that $\cos\gamma = 0$ for a planar trajectory. Then formula (19.42) implies that

$$a_{13}\dot{\xi} + a_{23}\dot{\eta} = 0$$

This condition holds for $\Theta = 0$ and arbitrary ξ, η satisfying equations (19.41). Therefore there are infinitely many trajectories for which $\cos\gamma = 0$. Also among these are circular trajectories.

Next let us consider the particular cases with a rectilinear trajectory for the mass center G of the rod. By reasons of symmetry it is obvious that such solutions exist and that the rod itself remains perpendicular to the rectilinear trajectory of G or coincides with it. In fact, assume that G moves along a fixed straight line passing through the origin. Then the equations of such a straight line are

$$\frac{\xi}{\xi_0} = \frac{\eta}{\eta_0} = \frac{\zeta}{\zeta_0} = \frac{r}{r_0}$$

where $\xi_0, \eta_0, \zeta_0, r_0$ are the initial values of the coordinates and length of the radius vector of the mass center G.

Equations (19.35) and (19.36) have the following particular solutions:

Case 1: $\xi = \frac{\xi_0}{r_0} r, \ \eta = \frac{\eta_0}{r_0} r, \ \zeta = \frac{\zeta_0}{r_0} r$

where r satisfies the equation

$$\ddot{r} = -\frac{\mu}{r\sqrt{r^2 + l^2}}$$

In such a case

$$H = 0, \qquad R = -\frac{r}{\sqrt{r^2 + l^2}}$$

$$\nu = \cos\alpha = 0, \qquad \cos\gamma = 0$$

i.e. the rod remains perpendicular to the trajectory of its mass center G, and the angle of attack γ is equal to $\pi/2$.

Case 2: $\xi = \frac{\xi_0}{r_0} r, \ \eta = \frac{\eta_0}{r_0} r, \ \zeta = \frac{\zeta_0}{r_0} r$, where r satisfies the equation

$$\ddot{r} = -\frac{\mu}{r^2 - l^2}$$

the angles ψ and Θ remain constant, and the direction cosines a_{13}, a_{23}, a_{33} assume the values

$$a_{13} = \pm\frac{\xi_0}{r_0}, \qquad a_{23} = \pm\frac{\eta_0}{r_0}, \qquad a_{33} = \pm\frac{\zeta_0}{r_0}$$

In such a case we have

$$\xi\cos\psi + \eta\sin\psi = 0$$

$$(\xi\sin\psi - \eta\cos\psi)\cos\Theta - \zeta\sin\Theta = 0$$

$$R = -\frac{r^2}{r^2 - l^2}, \qquad \nu = \cos\alpha = \pm 1$$

so that the rod coincides with the trajectory of its mass center G.

Formula, (19.42) then shows that $\cos\gamma = \pm 1$, independent of the sign of ν. Therefore in the case under consideration the following four subcases are possible:

(i) $\cos\alpha = +1$, $\cos\gamma = +1$; the rod recedes from the origin with its head directed forward;

(ii) $\cos\alpha = +1$, $\cos\gamma = -1$; the rod approaches the origin with its tail directed forward;

(iii) $\cos\alpha = -1$, $\cos\gamma = +1$; the rod recedes from the origin with its tail directed forward;

(iv) $\cos\alpha = -1$, $\cos\gamma = -1$; the rod approaches the origin with its head directed forward.

An investigation of stability by means of Lyapunov V-functions shows that the arrow-type motion of the rod is in general unstable, the floating rod-type motion unconditionally unstable, while the spoke-type motion may be stable, at least in the first approximation.

A particular oscillation of a rod-like satellite about its center of mass is analyzed by numerical integration in a paper by Miss L. M. Domožilova [1].

§ 20. Motion of an artificial Earth satellite

20.1. Introduction. The problem of the motion of celestial bodies about their mass centers has a long history. The following two particular cases of this problem appear in classical celestial mechanics: (i) the physical libration of the Moon, and (ii) the lunar and solar precession of the Earth axis. For further studies of these problems see F. TISSERAND [1], S. D. POISSON [2], E. W. WOOLARD [1]. The basic force in these problems is due to the central NEWTONian gravitational force field subject to which each particle of the body under consideration moves and which generates the gravitational moments. The above particular cases of the general problem carry with them corresponding simplifications. For example, in the theory of the lunar and solar precession of the Earth axis, use is made of the fact that the difference between the equatorial and polar moments of inertia of the Earth is small and the angle of nutation is small, i.e. the unperturbed motion of the Earth is a stationary rotation about its axis of dynamic symmetry, fixed in space. For an artificial Earth satellite such an assumption, in general, has no validity, and therefore the unperturbed motion of an artificial Earth satellite must be considered as the general EULERian motion of a torque-free rigid body.

In a number of papers V. V. BELECKII [3—5], V. A. SARYČEV [1], and F. L. ČERNOUS'KO [1] have investigated the motion of a satellite about its center of mass under the action of gravitational moments, and P. P. VČERAŠNYUK [1] investigated the motion under the attraction of the Earth. If the kinetic energy in the rotational motion of the satellite about its mass center is much greater than the work of the gravitational moments, the satellite possesses dynamic symmetry, and the orbit is circular or nearly circular, then V. V. BELECKII [3, 5] showed that the motion of the satellite can be considered as the result of superposition of certain perturbations upon its unperturbed motion. F. L. ČERNOUS'KO [1] has considered two cases of satellite motion subject to gravitational moments when the existence of a small parameter permits the application of the method of averaging and derivation of an asymptotic solution. Since his results generalize those of V. V. BELECKII we shall follow ČERNOUS'KO's paper.

20.2. Equations of motion. Consider the motion of a satellite in a central gravitational force field. Within the accuracy of the square of the ratio of the linear dimensions of the satellite to those of the orbit it can be assumed that the motion of the satellite about its mass center does not affect the motion of the mass center itself.

The center of mass moves in a KEPLER ellipse with eccentricity e and period of revolution T_0. The dependence of the true anomaly ν on the time t is given by the equation

$$\frac{d\nu}{dt} = \frac{\omega_0(1 + e\cos\nu)^2}{(1 - e^2)^{3/2}}, \quad \nu(t + T_0) = \nu(t) + 2\pi, \quad \omega_0 = \frac{2\pi}{T_0} \tag{20.1}$$

Consider three right-hand Cartesian coordinate trihedrals, the origins of which are at the mass center G of the satellite.

The trihedral $G x_1 x_2 x_3$ translates along with the mass center, the x_1-axis being parallel to the radius vector of the orbit perigee, the x_2-axis being parallel to the velocity vector of the mass center at the perigee, and the x_3-axis being parallel to the normal to the orbit plane.

The y_3-axis of the trihedral $G\,y_1\,y_2\,y_3$ is directed along the angular momentum vector $\vec{I}$ of the satellite about the mass center, the y_1-axis is perpendicular to the y_3-axis and lies in the $x_3\,y_3$-plane while the y_2-axis is perpendicular to the y_1-axis and y_3-axis and, consequently, lies in the plane of the orbit $x_1\,x_2$ (Fig. 63).

The transfer from the orbital trihedral $G\,x_1\,x_2\,x_3$ to the trihedral $G\,y_1\,y_2\,y_3$ can be realized by two rotations: through the angle λ about

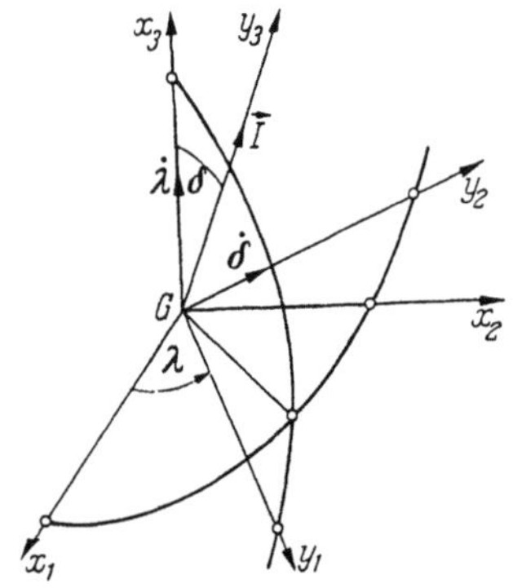

Fig. 63

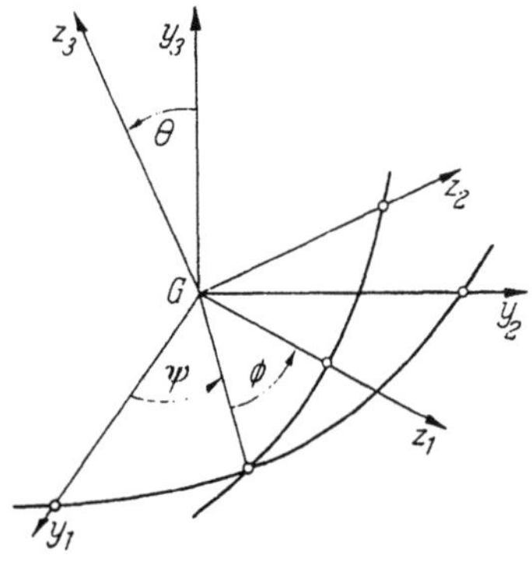

Fig. 64

the x_3-axis and the angle δ about the y_2-axis. The angles λ and δ define the orientation of the angular momentum vector $\vec{I}$ in fixed space.

The axes of the trihedral $G\,z_1\,z_2\,z_3$ by assumption coincide with the principal central axes of inertia of the satellite. Their orientation relative to the trihedral $G\,y_1\,y_2\,y_3$ is defined by the EULER angles Θ, ψ, Φ (Fig. 64) as well as by the direction cosines $\alpha_{ik} = \vec{y}_i \cdot \vec{z}_k$, where $\vec{y}_i$, $\vec{z}_k$ denote the unit vectors along the corresponding axes. By (1.5) the following relations hold between the direction cosines and the EULER angles

$$\left.\begin{aligned}
\alpha_{11} &= \cos\psi\cos\Phi - \cos\Theta\sin\psi\sin\Phi\\
\alpha_{12} &= -\cos\psi\sin\Phi - \cos\Theta\sin\psi\cos\Phi\\
\alpha_{13} &= \sin\Theta\sin\psi\\
\alpha_{21} &= \sin\psi\cos\Phi + \cos\Theta\cos\psi\sin\Phi\\
\alpha_{22} &= -\sin\psi\sin\Phi + \cos\Theta\cos\psi\cos\Phi\\
\alpha_{23} &= -\sin\Theta\cos\psi\\
\alpha_{31} &= \sin\Theta\sin\Phi\\
\alpha_{32} &= \sin\Theta\cos\Phi\\
\alpha_{33} &= \cos\Theta
\end{aligned}\right\} \tag{20.2}$$

For the purpose of deriving the equations of motion of the satellite about its mass center we shall choose for the six required quantities the angular momentum vector $\vec{I}$ and the angles λ, δ, Θ, ψ, Φ. The momentum equation when referred to the trihedral $G y_1 y_2 y_3$ gives us the equations

$$\dot{I} = L_3, \quad \dot{\delta} = \frac{L_1}{I}, \quad \dot{\lambda} = \frac{L_2}{I \sin\delta} \tag{20.3}$$

where L_i $(i = 1, 2, 3)$ are the components of the resulting moment of the external forces about the mass center G referred to the trihedral $G y_1 y_2 y_3$.

The components of the absolute angular velocity $\vec{\omega}$ of the satellite along the axes of the trihedral $G z_1 z_2 z_3$ are

$$\left.\begin{aligned} p &= \dot{\delta}\alpha_{21} + \dot{\lambda}(\alpha_{31}\cos\delta - \alpha_{11}\sin\delta) + \dot{\Theta}\cos\Phi + \dot{\psi}\alpha_{31} \\ q &= \dot{\delta}\alpha_{22} + \dot{\lambda}(\alpha_{32}\cos\delta - \alpha_{12}\sin\delta) - \dot{\Theta}\sin\Phi + \dot{\psi}\alpha_{32} \\ r &= \dot{\delta}\alpha_{23} + \dot{\lambda}(\alpha_{33}\cos\delta - \alpha_{13}\sin\delta) + \dot{\Phi} \qquad\quad + \dot{\psi}\alpha_{33} \end{aligned}\right\} \tag{20.4}$$

On the other hand, the components of the vector $\vec{I}$ along the same axes are

$$\begin{aligned} &I_1 = A p = I\sin\Theta\sin\Phi, \qquad I_2 = B q = I\sin\Theta\cos\Phi, \\ &I_3 = C r = I\cos\Theta \end{aligned} \tag{20.5}$$

Substituting into equations (20.4) for p, q, r; $\dot{\delta}$, $\dot{\lambda}$ and α_{ik} their expressions given by formulas (20.5), (20.3) and (20.2), and solving them for the derivatives of the Euler angles, we obtain

$$\left.\begin{aligned} \dot{\Theta} &= I\left(\frac{1}{A} - \frac{1}{B}\right)\sin\Theta\sin\Phi\cos\Phi + \frac{L_2\cos\psi - L_1\sin\psi}{I} \\ \dot{\psi} &= I\left(\frac{\sin^2\Phi}{A} + \frac{\cos^2\Phi}{B}\right) - \frac{L_1\cos\psi + L_2\sin\psi}{I}\cot\Theta - \\ &\quad - \frac{L_2}{I}\cot\delta \\ \dot{\Phi} &= I\left(\frac{1}{C} - \frac{\sin^2\Phi}{A} - \frac{\cos^2\Phi}{B}\right)\cos\Theta + \frac{L_1\cos\psi + L_2\sin\psi}{I\sin\Theta} \end{aligned}\right\} \tag{20.6}$$

The systems of equations (20.3) and (20.6) are convenient for application of asymptotic methods.

If we use equalities (20.5), (20.3) and (20.6), the kinetic energy T of the satellite moving about its mass center G and its time derivative $\dot{T}$

assume the expressions

$$
\begin{aligned}
T &= \frac{1}{2}(A p^2 + B q^2 + C r^2) \\
&= \frac{I^2}{2}\left[\left(\frac{\sin^2\Phi}{A} + \frac{\cos^2\Phi}{B}\right)\sin^2\Theta + \frac{\cos^2\Phi}{C}\right]
\end{aligned}
\tag{20.7}
$$

$$
\begin{aligned}
\dot{T} = \frac{2T}{I} L_3 + I \sin\Theta \Big[\cos\Theta\left(\frac{\sin^2\Phi}{A} + \frac{\cos^2\Phi}{B} - \frac{1}{C}\right) \times \\
\times (L_2\cos\psi - L_1\sin\psi) + \left(\frac{1}{A} - \frac{1}{B}\right)\sin\Phi\cos\Phi(L_1\cos\psi + L_2\sin\psi)\Big]
\end{aligned}
\tag{20.8}
$$

Assume now that the satellite moves in a NEWTONian central force field. Then, within the accuracy of higher order terms in the ratio of the linear dimensions of the satellite to those of the orbit, the moment of these forces relative to the mass center of the satellite is

$$
\vec{L} = \frac{3\bar{f}}{R^3}\left[(C - B)\gamma_3\gamma_2\vec{z}_1 + (A - C)\gamma_1\gamma_3\vec{z}_2 + (B - A)\gamma_1\gamma_2\vec{z}_3\right] \tag{20.9}
$$

Here γ_i denote the direction cosines of the satellite mass center radius vector $\vec{R}$ (directed from the fixed center of attraction toward the mass center) with the principal central inertia axes $G z_i$, and $\bar{f}$ denotes the gravitational constant.

We shall project equation (20.9) on the $G y_i$-axes. Let β_i denote the direction cosines of the radius vector $\vec{R}$ with these axes. Expressing γ_i in terms of β_i and α_{ik}, and $\bar{f}$, R in terms of ν, e, ω_0 according to the formulas for the elliptic motion, we obtain

$$
\left.
\begin{aligned}
L_1 &= 3\omega_0^2(1 + e\cos\nu)^3(1 - e^2)^{-3}\sum_{j=1}^{3}(\beta_2\beta_j s_{3j} - \beta_3\beta_j s_{2j}) \\
L_2 &= 3\omega_0^2(1 + e\cos\nu)^3(1 - e^2)^{-3}\sum_{j=1}^{3}(\beta_3\beta_j s_{1j} - \beta_1\beta_j s_{3j}) \\
L_3 &= 3\omega_0^2(1 + e\cos\nu)^3(1 - e^2)^{-3}\sum_{j=1}^{3}(\beta_1\beta_j s_{2j} - \beta_2\beta_j s_{1j})
\end{aligned}
\right\}
\tag{20.10}
$$

$$
s_{ij} = A\alpha_{i1}\alpha_{j1} + B\alpha_{i2}\alpha_{j2} + C\alpha_{i3}\alpha_{j3} \tag{20.11}
$$

For the computation of β_i let us note that $\vec{R}$ lies in the orbital plane $x_1 x_2$ and makes the angle ν with the x_1-axis. Therefore (Fig. 63)

$$
\beta_1 = \cos\delta\cos(\nu - \lambda), \quad \beta_2 = \sin(\nu - \lambda), \quad \beta_3 = \sin\delta\cos(\nu - \lambda) \tag{20.12}
$$

20.3. Integration of the equations of motion. In order to solve systems (20.3) and (20.6) F. L. ČERNOUS'KO applies the method of averaging in the form developed by V. M. VOLOSOV [1, 2]. This method generalizes certain methods of N. N. BOGOLYUBOV [1], N. N. BOGOLYUBOV and D. N. ZUBAREV [1], and N. N. BOGOLYUBOV and YU. A. MITROPOL'SKII [1].

VOLOSOV considers a system

$$\dot{x} = \varepsilon X(x, y, t, \varepsilon), \qquad \dot{y} = Y(x, y, t, \varepsilon) \tag{20.13}$$

where x and y are vectors of dimension n and m respectively, and ε is a small parameter. The general solution of the degenerate system

$$\dot{y} = Y(x, y, t, 0), \qquad x = \text{const.} \tag{20.14}$$

is assumed to be known as

$$y = \varphi(x, y_0, t)$$

where $\varphi(x, y_0, t_0) = y_0$. Further it is assumed that the right-hand members in system (20.13) have mean values with respect to t. The asymptotic solution of system (20.13) in the k-th approximation is sought in the form

$$\begin{aligned} x &= \xi + \varepsilon u_1(\xi, \eta, t) + \cdots + \varepsilon^{k-1} u_{k-1}(\xi, \eta, t) \\ y &= \eta + \varepsilon v_1(\xi, \eta, t) + \cdots + \varepsilon^{k-2} v_{k-2}(\xi, \eta, t) \end{aligned} \tag{20.15}$$

where the variables ξ, η satisfy the system of k-th approximation

$$\begin{aligned} \dot{\xi} &= \varepsilon A_1(\xi) + \varepsilon^2 A_2(\xi) + \cdots + \varepsilon^k A_k(\xi) \\ \dot{\eta} &= Y(\xi, \eta, t, 0) + \varepsilon B_1(\xi) + \cdots + \varepsilon^{k-1} B_{k-1}(\xi) \end{aligned} \tag{20.16}$$

The function $A_1(\xi)$ is of the form

$$A_1(\xi) = \lim_{T\to\infty} \frac{1}{T} \int_{t_0}^{t_0+T} X[\xi, \varphi(\xi, y_0, t), t, 0]\, dt \tag{20.17}$$

where the averaging operation is extended along the general solution of the unperturbed system (20.14).

It is assumed that the result of averaging in (20.17) is independent of the initial conditions t_0, y_0. Such an assumption holds for a wide class of cases. Then under certain general restrictions the difference between the k-th approximation (20.15) and the exact solution is of order ε^k for the x-variables, and of order ε^{k-1} for the y-variables when taken over long time intervals of order $1/\varepsilon$. Let us note that system (20.16) is much simpler than the original one in the sense that the equations for the ξ-variables are autonomous and hence can be integrated separately.

In what follows we shall consider two cases.

Case I:

$$A = J_0 + \varepsilon A', \qquad B = J_0 + \varepsilon B', \qquad C = J_0 + \varepsilon C', \tag{20.18}$$

i.e. the principal central moments of inertia of the satellite are nearly equal to each other.

For $\varepsilon = 0$, s_{ij} and L_i given by (20.11) and (20.10) respectively reduce to $s_{ij} = J_0 \delta_{ij}$ (δ_{ij} being the KRONECKER delta symbol) and $L_1 = L_2 = L_3 = 0$. From systems (20.3) and (20.6) it follows that I, δ, λ, Θ become constants, and

$$\psi = \frac{I}{J_0} t + \psi_0 \tag{20.19}$$

i.e. the satellite rotates uniformly about the angular momentum axis which translates in space.

For small $\varepsilon \neq 0$ it follows from (20.11) and (20.10) that

$$s_{ij} = J_0 \delta_{ij} + O(\varepsilon), \qquad L_i = O(\varepsilon)$$

and consequently the system of seven equations (20.1), (20.3), (20.6) is of the type (20.13) in which the role of the "slow" variables (x) is played by I, δ, λ, Θ, Φ, while that of the "rapid" variables (y) by ψ and ν. In order to obtain the first approximation to the solution it is sufficient to average the right-hand sides of equations (20.3) and (20.6), after substituting for ν its expression from equation (20.1) and for ψ the expression (20.19). Then for fixed values of the "slow" variables the right-hand sides of the equations to be averaged will be sums of terms of the form $f_1(\psi) f_2(\nu)$, where the functions f_1, f_2 are periodic in their arguments with periods 2π. In addition, as it may be verified, the FOURIER expansion of $f_1(\psi)$ will not contain harmonics of higher order than three. Therefore, expanding the right-hand sides of equations of systems (20.3) and (20.6) in double FOURIER series in terms of ψ and ν, and substituting for ψ and ν their expressions in terms of the time t, we obtain expansions which will appear in the form of sums of terms of the form

$$C_{mn} \cos\left[m\left(\frac{I}{J_0} t + \psi_0\right)\right] \cos n\, \omega_0 t \quad (m = 0, 1, 2, 3; \quad n = 0, 1, 2, \ldots)$$

and similar sums where one or both cosine terms are replaced by sine terms. Further, assume that none of the equalities

$$I = n J_0 \omega_0, \qquad I = \tfrac{1}{2} n J_0 \omega_0, \qquad I = \tfrac{1}{3} n J_0 \omega_0 \tag{20.20}$$

holds for any natural number n.

Then the result of averaging the right-hand sides is independent of the initial value ψ_0. In such a case averaging with respect to the time t can be replaced by independent averagings with respect to ψ and ν, as functions of t.

If, however, any of the equalities (20.20) is satisfied, then resonance takes place. For such phenomena see F. L. ČERNOUS'KO [2].

By virtue of equation (20.1) the operation of averaging functions of ν with respect to the time t can be replaced by averaging with respect to ν the expression under the sign of the integral on the right, i.e.

$$\frac{1}{T_0}\int_0^{T_0} f(\nu)\,dt = \frac{1}{2\pi}\int_0^{2\pi} \frac{(1-e^2)^{3/2} f(\nu)\,d\nu}{(1+e\cos\nu)^2} \tag{20.21}$$

If we average the right-hand sides of equations (20.3) and (20.6) [observing relations (20.10), (20.11) and (20.2)] first with respect to ψ and afterwards with respect to ν in accordance with formula (20.21), then we obtain in the first approximation the equations

$$\dot I = 0, \quad \dot\delta = 0, \quad \dot\psi = \frac{I}{J_0} + O(\varepsilon)$$

$$\left.\begin{aligned}
\dot\lambda &= \frac{3\omega_0^2\cos\delta}{4I(1-e^2)^{3/2}}\{A+B+C - \\
&\quad - 3[(A\sin^2\Phi + B\cos^2\Phi)\sin^2\Theta + C\cos^2\Theta]\} \\
\dot\Theta &= \left(\frac{1}{A}-\frac{1}{B}\right) I\sin\Theta\sin\Phi\cos\Phi + \\
&\quad + \frac{3\omega_0^2(1-3\cos^2\delta)}{4I(1-e^2)^{3/2}}(A-B)\sin\Theta\sin\Phi\cos\Phi \\
\dot\Phi &= I\cos\Theta\left(\frac{1}{C}-\frac{\sin^2\Phi}{A}-\frac{\cos^2\Phi}{B}\right) - \\
&\quad - \frac{3\omega_0^2(1-3\cos^2\delta)}{4I(1-e^2)^{3/2}}\cos\Theta(A\sin^2\Phi + B\cos^2\Phi - C)
\end{aligned}\right\} \tag{20.22}$$

For further simplification of system (20.22) note that by virtue of (20.18) we have within the accuracy of terms of order ε^2 the equalities

$$\frac{1}{A} = \frac{2}{J_0} - \frac{A}{J_0^2}, \qquad A = 2J_0 - \frac{J_0^2}{A}$$

and two similar equalities for the moments of inertia B and C. Using these approximate equalities we can transform system (20.22), without decreasing the order of accuracy in ε of this system, into the following one

$$\left.\begin{aligned}
&\dot I = 0, \quad \dot\delta = 0, \quad \dot\psi = \omega + O(\varepsilon), \quad \dot\lambda = \frac{3\omega_0^2 F\cos\delta}{4I(1-e^2)^{3/2}} \\
&\dot\Theta = \left(\frac{1}{A}-\frac{1}{B}\right) D\, I\sin\Theta\sin\Phi\cos\Phi, \\
&\dot\Phi = I\cos\Theta\left(\frac{1}{C}-\frac{\sin^2\Phi}{A}-\frac{\cos^2\Phi}{B}\right) D
\end{aligned}\right\} \tag{20.23}$$

where

$$F = \frac{6TJ_0^2}{I^2} - J_0^2\left(\frac{1}{A} + \frac{1}{B} + \frac{1}{C}\right), \quad D = 1 - \frac{3\omega_0^2(1 - 3\cos^2\delta)}{4\omega^2(1 - e^2)^{3/2}},$$

$$\omega = \frac{I}{J_0} \tag{20.24}$$

and T is defined by formula (20.7).

The solution of system (20.23) approximates the exact solution of systems (20.3) and (20.6) in the time interval of order T_0/ε (much larger than the satellite period of revolution) within the accuracy of order ε relative to the "slow" variables I, δ, λ, Θ, Φ, and within the accuracy of order 1 relative to ψ.

The relative motion of the satellite described by system (20.23) decomposes into three parts: a "rapid" motion in terms of the variable ψ and two "slow" motions in terms of the variables Θ, Φ and I, δ, λ.

The "rapid" motion represents the rotation of the satellite with constant angular velocity $\omega = I/J_0$ about the angular momentum vector [by virtue of the first equation of system (20.23)].

The equations for the variables Θ, Φ describe the motion of the angular momentum vector relative to the satellite. It is easy to see that these equations differ from those for Θ, Φ in the case of the torque-free motion of the body [the first and third equations of system (20.6) for $\vec{L} = 0$; case of EULER and POINSOT] only by the constant multiplier D (in our approximation).

Thus the action of gravitational moments merely alters by a multiplicative constant D the velocity of displacement of the vector $\vec{I}$ along its trajectory of motion in the case of EULER and POINSOT. These trajectories are determined by the relations stemming from equalities (20.5) and (20.7),

$$I_1^2 + I_2^2 + I_3^2 = I, \qquad \frac{I_1^2}{A} + \frac{I_2^2}{B} + \frac{I_3^2}{C} = 2T$$

where I and T are constants. A number of such trajectories for fixed T and variable I are shown in Fig. 65, where the arrows give the direction of motion in the case of EULER and POINSOT and by assumption $A > B > C$. The quantity D can have a wide range of variation and become even negative if the direction of motion in Fig. 65 is reversed. As in the case of EULER and POINSOT the permanent axes of rotation are the principal axes of inertia, the rotation about the z_2-axis being unstable while those about the z_1- and z_3-axis are stable.

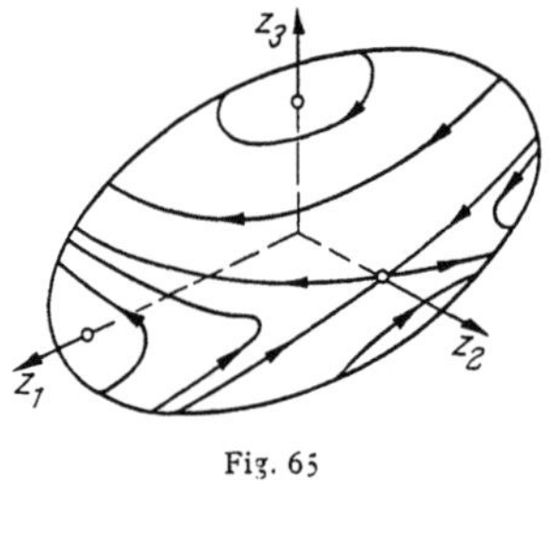

Fig. 65

However, when the condition

$$\omega = \frac{I}{J_0} = \sqrt{\frac{3(1-3\cos^2\delta)}{4(1-e^2)^{3/2}}}\,\omega_0$$

is satisfied, the rotation of the satellite will be stationary for any orientation of the axis of rotation relative to the satellite. In this case the moments of the centrifugal and gravitational forces are in equilibrium in the first approximation, and the vector $\vec{I}$ does not move relative to the body.

The motion of the vector $\vec{I}$, described by the first three equations of system (20.23), is a steady rotation of this vector about the normal to the orbit plane, this normal being at a constant angular distance δ from $\vec{I}$. The angular velocity of rotation $\dot{\lambda}$ is a quantity of order $\varepsilon\,\omega_0^2/\omega$ (since $F \sim \varepsilon\, J_0$) while its sign depends on the nature of the motion of $\vec{I}$ relative to the satellite. If, for example, $\delta < \pi/2$, then in the case of rotation of the satellite about the axis of the largest moment of inertia ($G\,z_1$-axis) ($I^2 = 2TA$) it follows from (20.24) and (20.23) that $F < 0$, $\dot{\lambda} < 0$ (the rotation of $\vec{I}$ is in the direction opposite to the orbital motion) while for $I^2 = 2TC$ (rotation about the $G\,z_3$-axis) $\dot{\lambda} > 0$.

Case II: The moments of inertia are arbitrary, say $A \geqq B \geqq C$. Further, assume that the angular velocity of the satellite in its relative motion is much greater than the angular velocity of its orbital motion, and let $\varepsilon \sim A\,\omega_0/I \ll 1$. Let the unit of time be a quantity of the order of the period of the relative motion. Then $\omega_0 \sim \varepsilon$, and by (20.10) the moments of the gravitational forces $L_i \sim \varepsilon^2$. This case will be realized when the kinetic energy of rotation is large in comparison with the work of the external forces. The asymptotics of such motion for systems of a single degree of freedom have been studied by N. N. Moisefv [1].

The unperturbed motion ($\varepsilon = 0$) will be the Euler and Poinsot motion, the quantities I, δ, λ and T being constant. The function $\psi(t)$ can be expressed in the form $\psi(t) = \psi_1(t) + \psi_2(t)$, where Θ, Φ and ψ_1 are periodic functions of t with the period τ equal to that of the motion of the vector $\vec{I}$ along the closed trajectories in Fig. 65, or they obtain constant increments 2π during the time τ. The second component is of the form $\psi_2(t) = 2\pi\, t/\tau'$. The periods τ and τ' depend on I and T and, in general, are incommensurable.

In the perturbed motion ($\varepsilon \neq 0$) the "slow" variables (x) are I, δ, λ and T while the "rapid" ones are Φ and ψ [Θ can be expressed in terms of T and Φ by means of (20.7)]. The equations of motion and equation (20.1) are of the form

$$\dot{x} = \varepsilon^2 X(x, y, \nu), \quad \dot{y} = Y_0(x, y) + \varepsilon^2 Y_1(x, y, \nu), \quad \dot{\nu} = \varepsilon f(\nu) \quad (20.25)$$

In constructing solutions of type (20.15), (20.16) it is easily seen that $u_1 = v_1 = 0$ while $A_1 = B_1 = A_3 = 0$, $A_2 = M_t\{X\}$. Here M_t denotes the averaging of X with respect to t. The solution for the "slow" variables is sought in the form

$$x = \xi, \quad \dot{\xi} = \varepsilon^2 A_2(\xi, \nu) = \varepsilon^2 M_t\{X(\xi, y, \nu)\} \tag{20.26}$$

neglecting terms of order ε^2 in (20.15) and of order ε^4 in (20.16).

Therefore the error in the asymptotic solution for the "slow" variables will be of order ε^2 in the time-interval of order ε^{-2}, corresponding to the number of revolutions of the satellite in the orbit, the order of which is $\varepsilon^{-1}(\Delta\nu \sim \varepsilon^{-1})$.

In order to construct the averaged system (20.26) it is necessary to average the right-hand sides of the equations of motion (for fixed "slow" variables and ν) along the Euler and Poinsot motion. These right-hand sides are periodic functions of Θ, Φ and ψ with periods 2π, while the periods τ and τ' are incommensurable. Therefore by arguments similar to those in Case *I*, we arrive at the conclusion that averaging with respect to the time t can be accomplished in two independent averaging steps: (i) with respect to Θ, Φ, ψ_1, and (ii) with respect to ψ_2, as functions of the time t. Thus,

$$\begin{aligned} M_t\{f(\Theta, \Phi, \psi)\} &= \frac{1}{\tau\tau'}\int_0^\tau\int_0^{\tau'} f\left[\Theta(t), \Phi(t), \psi_1(t) + \frac{2\pi t'}{\tau'}\right] dt'\, dt \\ &= \frac{1}{2\pi\tau}\int_0^\tau\int_0^{2\pi} f[\Theta(t), \Phi(t), \psi]\, d\psi\, dt \\ &= M_1 M_\psi\{f(\Theta, \Phi, \psi)\} \end{aligned} \tag{20.27}$$

where M_ψ denotes averaging with respect to ψ, and M_1 with respect to Θ and Φ, related by (20.7). Both averagings are to be carried out along the closed trajectories described by the angular momentum vector $\vec{I}$ in the Euler and Poinsot motion (Fig. 65).

Averaging the right-hand sides of equations (20.3) and (20.6) in accordance with the above formula, we obtain in the first approximation the system

$$\left.\begin{aligned} &\dot{I} = 0, \quad \dot{\delta} = -\frac{3\omega_0^2(1 + e\cos\nu)^3}{2(1 - e^2)^3 I}\beta_2\beta_3 N \\ &\dot{\lambda} = \frac{3\omega_0^2(1 + e\cos\nu)^3}{2(1 - e^2)^3 I\sin\delta}\beta_1\beta_3 N \\ &\dot{T} = \frac{3\omega_0^2(1 + e\cos\nu)^3}{2(1 - e^2)^3}(2\beta_3^2 - \beta_1^2 - \beta_2^2) \times \\ &\quad\times\frac{(A - B)(B - C)(C - A)}{A B C I^2} M_1\{I_1, I_2, I_3\} \end{aligned}\right\} \tag{20.28}$$

where

$$N = A + B + C - \\ - 3M_1\{(A\sin^2\Phi + B\cos^2\Phi)\sin^2\Theta + C\cos^2\Theta\} \qquad (20.29)$$

By virtue of the symmetry with respect to the coordinate planes of the trihedral $G z_1 z_2 z_3$ of the trajectories described by the vector $\vec{I}$, it is obvious that $M_1\{I_1, I_2, I_3\} = 0$ and $T = \text{const}$. For nearly equal moments of inertia A, B, C, given by (20.18), N coincides with F, given by (20.24). In the general case, using formulas (20.5) and (20.7), we obtain

$$N = B - 2A - 2C + 6ACTI^{-2} + \\ + 3B(A-B)(B-C)I^{-2}M_1\{q^2\} \qquad (20.30)$$

Substituting into (20.30) for $q(t)$ its expression from the Euler and Poinsot motion and averaging the function q^2 over its period, we obtain for N relative to the trajectories described by the vector $\vec{I}$ around the z_1-axis ($I^2 > 2TB$) the expression

$$N = B + C - 2A + 3\left(\frac{2TA}{I^2} - 1\right)\left[C + (B-C)\frac{K(k) - E(k)}{k^2 K(k)}\right] \qquad (20.31)$$

where $K(k)$ and $E(k)$ are the complete elliptic integrals of the first and second kind respectively, and

$$k^2 = \frac{(B-C)(2TA - I^2)}{(A-B)(I^2 - 2TC)} \qquad (20.32)$$

For the trajectories of the vector $\vec{I}$ enveloping the z_3-axis ($I^2 < 2TB$), the letters A and C in formulas (20.31) and (20.32) must be interchanged.

The quantity N depends on the satellite moments of inertia and the ratio I^2/T which determines the trajectory in Fig. 65 and is constant in the approximation under consideration. If the satellite is rotating about the z_1-axis ($I^2 = 2TA$) or z_3-axis ($I^2 = 2TC$), then it follows from (20.31) that

$$N = B + C - 2A < 0 \quad \text{or} \quad N = B + A - 2C > 0$$

respectively.

In the case of dynamic symmetry ($A = B$) formulas (20.30), (20.7) and (20.23) give

$$N = 6ACTI^{-2} - A - 2C = (A - C)(2 - 3\sin^2\Theta), \\ \Theta = \text{const}. \qquad (20.33)$$

In the approximation under consideration the relative motion of the satellite consists of the Euler and Poinsot motion about the vector $\vec{I}$ (for constant I and T) and the motion of the vector $\vec{I}$ itself in space, described by equations (20.28) in δ and λ. In what follows we shall investigate these equations using the true anomaly ν as the independent variable. Taking equations (20.1), (20.12) and (20.28) into account, we can rewrite the equations for δ and λ in the form

$$\begin{aligned} \frac{d\delta}{d\nu} &= \varkappa(1 + e\cos\nu)\sin\delta\sin(\lambda - \nu)\cos(\lambda - \nu) \\ \frac{d\lambda}{d\nu} &= \varkappa(1 + e\cos\nu)\cos\delta\cos^2(\lambda - \nu) \end{aligned} \tag{20.34}$$

where

$$\varkappa = \frac{3\omega_0 N}{2(1 - e^2)^{3/2} I}$$

It is obvious that the introduced dimensionless quantity $\varkappa$ is of order ε and constant in the approximation under consideration (by virtue of the fact that I, T and N are constants).

In the case of a circular orbit ($e = 0$) equations (20.34) possess the first integral

$$\cos\delta + \tfrac{1}{2}\varkappa\sin^2\delta\cos^2(\lambda - \nu) = \text{const.}$$

and as a consequence their integration can be reduced to quadratures.

It is preferable, however, to apply the asymptotic methods for solving system (20.34). The averaged equations determine the quantities δ, λ to within the accuracy of order ε^2 (or $\varkappa^2$) in the interval of length $\Delta\nu \sim \varepsilon^{-1} \sim \varkappa^{-1}$. Therefore it is sufficient to solve equations (20.34) within this accuracy by finding the asymptotic solution in the second approximation relative to $\varkappa$. System (20.34) being in the standard form (see N. N. Bogolyubov, and Yu. A. Mitropol'skii [1]), its solution in the second approximation is sought in the form

$$\delta = \xi + \varkappa u(\nu, \xi, \eta), \qquad \lambda = \eta + \varkappa v(\nu, \xi, \eta) \tag{20.35}$$

Determining the functions u, v by the known procedure (see N. N. Bogolyubov and Yu. A. Mitropol'skii [1]) we obtain

$$\begin{aligned} u &= \frac{1}{12}\sin\xi[3\cos(2\nu - 2\eta) + 3e\cos(\nu - 2\eta) + e\cos(3\nu - 2\eta)] \\ v &= \frac{1}{12}\cos\xi[3\sin(2\nu - 2\eta) + 6e\sin\nu + 3e\sin(\nu - 2\eta) + \\ &\quad + e\sin(3\nu - 2\eta)] \end{aligned} \tag{20.36}$$

The variables ξ, η satisfy the system of the second approximation

$$\begin{aligned} \frac{d\xi}{d\nu} &= \frac{1}{8}\varkappa^2 e^2 \sin\xi\cos\xi\sin 2\eta \\ \frac{d\eta}{d\nu} &= \frac{1}{2}\varkappa\cos\xi + \frac{1}{16}\varkappa^2(3\cos^2\xi - 1)\left(1 + \frac{2}{3}e^2 + e^2\cos 2\eta\right) \end{aligned} \tag{20.37}$$

Next let us find to within the required accuracy (within the error of order $\varkappa^2$ in the time interval of order $\varkappa^{-1}$) the solution of system (20.37), satisfying the initial conditions $\xi(0) = \xi_0$, $\eta(0) = \eta_0$. It is easy to see that for such a relation in the interval of length $\Delta\nu \sim \varkappa^{-1}$ the estimates

$$\xi - \xi_0 = O(\varkappa), \qquad \eta - \eta_0 - \tfrac{1}{2}\varkappa\nu\cos\xi_0 = O(\varkappa) \tag{20.38}$$

hold, and yield the solution of system (20.37) in the first approximation. Transforming the right-hand sides of equations (20.37) by means of estimates (20.38) and neglecting terms of order $O(\varkappa^3)$ (introducing herewith in the solution an error of order $O(\varkappa^2)$], we obtain

$$\left.\begin{aligned} \frac{d\xi}{d\nu} &= \frac{1}{8}\varkappa^2 e^2 \sin\xi_0 \cos\xi_0 \sin(2\eta_0 + \varkappa\nu\cos\xi_0) \\ \frac{d\eta}{d\nu} &= \frac{1}{2}\varkappa\cos\xi_0 - \frac{1}{2}\varkappa(\xi - \xi_0)\sin\xi_0 + \\ &\quad + \frac{1}{16}\varkappa^2(3\cos^2\xi_0 - 1)\left[1 + \frac{2}{3}e^2 + e^2\cos(2\eta_0 + \varkappa\nu\cos\xi_0)\right] \end{aligned}\right\} \tag{20.39}$$

The solution of system (20.39), satisfying the above initial conditions, is

$$\begin{aligned} \xi &= \xi_0 + \frac{1}{8}\varkappa e^2 \sin\xi_0[\cos 2\eta_0 - \cos(2\eta_0 + \varkappa\nu\cos\xi_0)] \\ \eta &= \eta_0 + \frac{1}{2}\varkappa\nu\cos\xi_0 + \\ &\quad + \frac{1}{16}\varkappa^2\nu\left[(3\cos^2\xi_0 - 1)\left(1 + \frac{2}{3}e^2\right) - e^2\sin^2\xi_0\cos 2\eta_0\right] + \\ &\quad + \frac{1}{8}\varkappa e^2\cos\xi_0[\sin(2\eta_0 + \varkappa\nu\cos\xi_0) - \sin 2\eta_0] \end{aligned} \tag{20.40}$$

Substituting (20.36) and (20.40) into (20.35) we obtain

$$\begin{aligned} \delta &= \xi_0 + \frac{1}{24}\varkappa\sin\xi_0\{6\cos[(2 - \varkappa\cos\xi_0)\nu - 2\eta_0] + \\ &\quad + 6e\cos[(1 - \varkappa\cos\xi_0)\nu - 2\eta_0] + 2e\cos[(3 - \varkappa\cos\xi_0)\nu - 2\eta_0] + \\ &\quad + 3e^2\cos 2\eta_0 - 3e^2\cos(\varkappa\nu\cos\xi_0 + 2\eta_0)\} \\ \lambda &= \eta_0 + \frac{1}{2}\varkappa\nu\cos\xi_0 + \\ &\quad + \frac{1}{16}\varkappa^2\nu\left[(3\cos^2\xi_0 - 1)\left(1 + \frac{2}{3}e^2\right) - e^2\sin^2\xi_0\cos 2\eta_0\right] + \\ &\quad + \frac{1}{24}\varkappa\cos\xi_0\{6\sin[(2 - \varkappa\cos\xi_0)\nu - 2\eta_0] + 12e\sin\nu + \\ &\quad + 6e\sin[(1 - \varkappa\cos\xi_0)\nu - 2\eta_0] + 2e\sin[(3 - \varkappa\cos\xi_0)\nu - 2\eta_0] + \\ &\quad + 3e^2\sin(\varkappa\nu\cos\xi_0 + 2\eta_0) - 3e^2\sin 2\eta_0\} \end{aligned} \tag{20.41}$$

This solution of system (20.34) differs from the exact one by quantities of order $\varkappa^2$ on the interval of length of order $\varkappa^{-1}$. If we limit ourselves to accuracy of order $\varkappa$, then the solution (20.41) reduces to

$$\delta = \xi_0, \qquad \lambda = \eta_0 + \tfrac{1}{2}\varkappa\,\nu\cos\xi_0 \tag{20.42}$$

This solution describes in the first approximation the rotation of the angular momentum vector $\vec{I}$ about the normal to the orbit plane, the velocity of rotation being

$$\dot{\lambda} = \frac{3\omega_0^2 N\cos\delta}{4I(1-e^2)^{3/2}} \tag{20.43}$$

For $A = B$ and $e = 0$, using the expression (20.33) for N, we arrive at the formula derived by V. V. Beleckii [3].

In the first approximation the trajectories of the trace of the angular momentum vector $\vec{I}$ on the unit sphere, fixed in the trihedral $G\,x_1\,x_2\,x_3$, are circles $\delta = \text{const}$. In the second approximation there appear oscillations of the angles δ and λ, and as the mean-value δ (i.e. ξ_0) increases from 0 to $\pi/2$, the amplitude of δ-oscillations increases, while the amplitude of λ-oscillations, as well as the mean angular velocity of rotation, given by (20.43), decreases. For $\delta \neq \pi/2$ the total velocity $\dot{\lambda}$ does not change sign and vanishes only at separate points, at which $\dot{\delta} = 0$ also [see formula (20.34)]. For $\delta \sim \pi/2$ the variations in λ are second order quantities when compared with variations in δ. The trajectories described by the trace of the vector $\vec{I}$ on the unit sphere, fixed relative to the trihedral $G\,x_1\,x_2\,x_3$, are indicated in Fig. 66 for $\varkappa > 0$, with the above effects taken into account. For $\varkappa < 0$, only the direction of motion, indicated by arrows, changes along the trajectories.

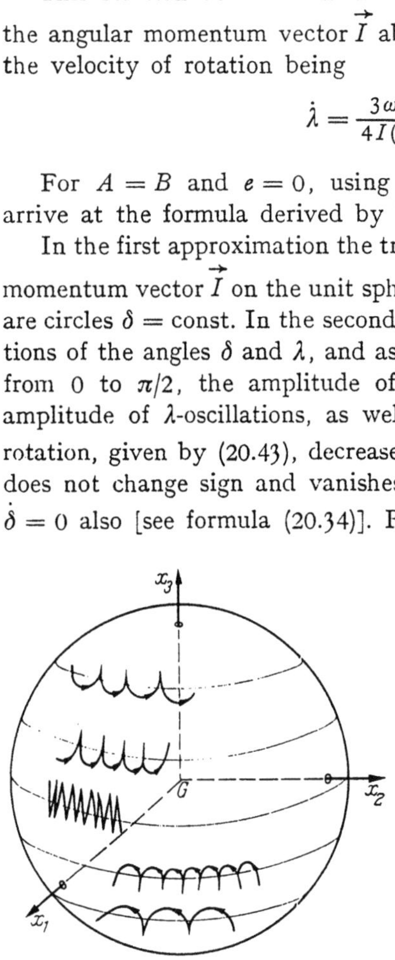

Fig. 66

On a circular orbit ($e = 0$) the oscillations of δ and λ are almost sinusoidal with angular frequency equal to twice the velocity of the orbital motion. The corresponding curves in Fig. 66 are nearly cycloids subject to compression or extension along the axes of the coordinates. In the case of an elliptic orbit the oscillations of δ and λ become more complex: there appear the first and third harmonics as well as a substantial dependence of the form of the oscillations upon the initial value η_0, but the basic properties of the trajectories in Fig. 66 remain unaltered (particularly the existence of cusps directed towards the poles).

The regions of applicability of the asymptotic solutions in Cases I and II apparently intersect: for $\omega \gg \omega_0$ (rapid relative motion) and A, B, C nearly equal, the results obtained in Case I coincide with the first approximation in Case II. The method, however, is applicable for the study of a rapid relative motion of a rigid body subject to gravitational moments of any nature.

For further investigations on the subject see A. I. Rybakov [1], M. S. Yarov [1], M. S. Volkov [1], and N. G. Magnaradze [1].

Bibliography

Abbreviations of names of journals are as in MATHEMATICAL REVIEWS. An (R) after the title of a paper indicates that this paper was written in Russian. Where no translations into English existed, the titles were translated by the author.

AGOSTINELLI, C.: [1] Sopra alcuni integrali particolari delle equazioni del moto di un corpo rigido pesante, intorno a un punto fisso. Acta Pontif. Acad. Sci. Novi Lyncaei **88**, 64—84 (1935). — [2] Sul moto intorno a un punto fisso di un corpo rigido pesante il cui baricentro appartiene all'asse di uno dei piani ciclici dell'ellissoide d'inerzia. Ann. Mat. Pura Appl. (4) **30**, 211—224 (1949). — [3] Sulla stabilità di un particolare moto di precessione regolare di un solido pesante asimmetrico. Ist. Veneto Sci. Lett. Arti. Parte II. Cl. Sci. Mat. Nat. **107**, 193—203 (1949). — [4] Sul moto di un corpo rigido pesante asimmetrico col baricentro appartenente all'asse di uno dei piani ciclici dell'ellissoide d'inerzia. Atti Sem. Mat. e Fis. Univ. Modena **3**, 248—260 (1949).

ALFIERI, L.: [1] Risoluzione di un problema comprendente quello di STAUDE Univ. Roma. Ist. Naz. Alta Mat. Rend. Mat. e Appl. (5) **12**, 367—372 (1953), (1954).

ANČEV, A.: [1] On the stability of permanent rotations of a quasi-symmetric gyrostat (R). Prikl. Mat. Meh. **28**, 164—166 (1964).

ANGELI, E. DE: [1] Su alcuni casi integrabili di movimento di un giroscopio asimmetrico pesante. Atti R. Accad. Naz. Lincei. Rend. Cl. Sci. Fis. Mat. Nat. (6) **25**, 699—703 (1937).

APPELL, P.: [1] Interprétation de la période imaginaire dans un mouvement à la POINSOT. Bull. Soc. Math. France **26**, 98—112 (1898).

APPEL'ROT, G. G.: [1] Concerning Section 1 of the memoir of S. V. KOVALEVSKAYA "Sur le problème de la rotation d'un corps solide autour d'un point fixe" (R), and the appendix to this paper. Mat. Sbornik **16**, 483—507, 592—596 (1892). — [2] Certain supplements to the work of N. B. DELONE "Algebraic integrals of the motion of a heavy rigid body about a fixed point" (R). Trudy Otdel. Fiz. Nauk Obšč. Lyubit. Estestvozn. **6**, 1—11 (1893). — [3] The problem of motion of a heavy rigid body about a fixed point (R). Učenye Zap. Imp. Moskov. Univ. Otdel. Fiz. Mat. Nauk, no. 11, 1—112 (1894). — [4] Certain additions to my paper "The problem of motion of a heavy rigid body about a fixed point" (R). Trudy Otdel. Fiz. Nauk Obšč. Lyubit. Estestvozn. **7**, no. 1, 1—5 (1894). — [5] Remarks on Prof. A. M. LYAPUNOV's communication at the Har'kov Math. Society's session on May 10, 1893 (R). Mat. Sbornik **18**, 723—727 (1896). — [6] The simplest cases of motion of a heavy asymmetric gyroscope of S. V. KOVALEVSKAYA (R). Ibid. **27**, no. 3 262—335 (1909), no. 4 477—561 (1910). — [7] Determination of classes of kinetically symmetric heavy gyroscopes which are able to admit simplified motions, near to the inertial or a certain simplified motion of the LAGRANGE gyroscope (R). Bull. Acad. Sci. URSS. Cl. Sci. Math. Nat. Ser. Phys. **1938**, no. 3, 385—411. — [8] Not entirely symmetric heavy gyroscopes (R). KOVALEVSKAYA Memorial

Collection "Motion of a Rigid Body about a Fixed Point", edited by S. A. Čaplygin and N. I. Mercalov. Moscow-Leningrad: Izdat. Akad. Nauk SSSR 1940, pp. 61—155.

Apyhtin, N. G.: [1] Permanent axes of rotation of a rigid body with a fixed point when the integrals of D. N. Goryačev exist (R). Prikl. Mat. Meh. **27**, 894—898 (1963); transl. as J. Appl. Math. Mech. **27**, 1361—1369 (1964).

Arhangel'skii, Yu. A.: [1] Motion of the fast gyroscope of Goryačev and Čaplygin (R). Izvestiya Akad. Nauk SSSR. Otdel. Tehn. Nauk, no. 7, 122—124 (1957). — [2] On the stability of motion of a heavy solid about a fixed point in a certain special case (R). Prikl. Mat. Meh. **24**, 294—302 (1960); transl. as J. Appl. Math. Mech. **24**, 422—433. — [3] On single-valued integrals in the problem of motion of a solid body in a Newtonian field of forces (R). Ibid. **26**, 568—570 (1962); J. Appl. Math. Mech. **26**, 850—854. — [4] On a theorem of Poincaré concerning the motion of a rigid body in a Newtonian force field (R). Ibid. **26**, 1116—1117 (1962); J. Appl. Math. Mech. **26**, 1693—1696. — [5] On algebraic and single-valued integrals in the problem of motion of a rigid body in the Newtonian force field (R). Ibid. **27**, 697—698 (1963); J. Appl. Math. Mech. **27**, 1059—1061. — [6] On algebraic integrals in the problem of motion of a rigid body in a Newtonian force field (R). Prikl. Mat. Meh. **27**, 171—175 (1963). — [7] On the motion of a fast rotating heavy rigid body about a fixed point (R). Ibid. **27**, 864—877 (1963). — [8] On a motion of a balanced gyroscope in a Newtonian force field (R). Ibid. **27**, 1099—1101 (1963). — [9] On a motion of the Kovalevskaya gyroscope (R). Ibid. **28**, 521—522 (1964). — [10] On new particular solutions of the problem of motion of a heavy rigid body about a fixed point (R). Dokl. Akad. Nauk SSSR **158**, 292—293 (1964). — [11] New particular solutions of the problem of motion of a heavy rigid body about a fixed point (R). Ibid. **159**, 36—38 (1964).

Arnold, R. N., and L. Maunder: [1] Gyrodynamics and its engineering applications. New York: Acad. Press 1961.

Arrighi, G.: [1] Sul giroscopio magnetico. Atti Accad. Naz. Lincei. Memorie. Cl. Sci. Fis. Mat. Nat. (8) **1**, Sez. I, 195—204 (1947).

Bankovskaya, N. V.: [1] Stability of the vertical position of the axis of a heavy gyroscope with harmonic oscillations of the point of support (R). Ukraïn. Mat. Ž. **6**, 418—422 (1954).

Barbarin, P.: [1] Note sur l'herpolhodie. Nouv. Ann. de Math. (3) **4**, 538—556 (1885). The paper was actually written in 1882, but its publication delayed.

Battaglia, L.: [1] Sul principio dell'effetto giroscopico nei solidi autoeccitati. Riv. Mat. Univ. Parma **8**, 73—80 (1957).

Beleckii, V. V.: [1] Integrability of equations of motion of a solid around a fixed point under the action of a central Newtonian field of force (R). Dokl. Akad. Nauk SSSR (N. S.) **113**, 287—290 (1957). — [2] Some problems of motion of a body in a Newtonian force field (R). Prikl. Mat. Meh. **21**, 749—758 (1957). — [3] The motion of an artificial Earth satellite about its center of mass (R). Iskusstv. Sputniki Zeml'i, vyp. 1, pp. 25—43. Moscow: Izdat. Akad. Nauk SSSR 1958. — [4] On satellite librations (R). Ibid. vyp. 3, pp. 13—31, 1959. — [5] Classification of motions of an artificial Earth satellite about the mass center (R). Ibid. vyp. 6, pp. 11—32, 1961.

Bloch, F.: [1] Nuclear induction. Phys. Rev. **70**, 460—473 (1946).

—, W. W. Hansen, and M. Packard: [1] The nuclear induction experiment. Phys. Rev. **70**, 474—485 (1946).

—, and A. Siegert: [1] Magnetic resonance for nonrotating fields. Phys. Rev. **57**, 522—527 (1940).

BOBYLEV, D.: [1] On a certain particular solution of the differential equations of rotation of a heavy rigid body about a fixed point (R). Trudy Otdel. Fiz. Nauk Obšč. Estestvozn. **8**, no. 2 21—25 (1896).

BÖDEWADT, U. T.: [1] Der symmetrische Kreisel bei zeitfester Drehkraft. Math. Z. **55**, 310—320 (1952).

BOGOLYUBOV, N. N.: [1] On certain statistical methods in mathematical physics (R). Kiev: Izdat. Akad. Nauk Ukraïn. SSR 1945.

—, and YU. A. MITROPOL'SKII: [1] Asymptotic methods in the theory of non-linear oscillations (R). Third edition. Moscow: Gos. Izdat. Fiz. Mat. Lit. 1963.

—, and D. N. ZUBAREV: [1] The method of asymptotic approximation for systems with revolving phase and its application to the motion of charged particles in a magnetic field (R). Ukraïn. Mat. Ž. **7**, 5—17 (1955).

BOGOYAVLENSKII, A. A.: [1] On particular cases of motion of a heavy rigid body about a fixed point (R). Prikl. Mat. Meh. **22**, 622—645 (1958); transl. as J. Appl. Math. Mech. **22**, 873—906. — [2] On some special solutions of the problem of motion of a heavy rigid body about a fixed point (R). Ibid. **22**, 738—749 (1958); J. Appl. Math. Mech. **22**, 1049—1064. — [3] A particular solution of the problem of the motion of a gyroscope on gimbals (R). Ibid. **23**, 958—960 (1959); J. Appl. Math. Mech. **23**, 1365—1369. — [4] On the linear integral of a cyclic displacement for a rigid body containing moving masses (R). Prikl. Mat. Meh. **28**, 360—363 (1964). — [5] The linear integral of a cyclic displacement for a gyroscope in a CARDAN suspension (R). Ibid. **28**, 508—510 (1964).

BOTTEMA, O.: [1] The stability of STAUDE's top motion. Koninkl. Nederland. Akad. Wetenschap., Proc. **48**, 316—325 (1945). — [2] On STAUDE's motion in five dimensional space. Ibid. Proc., Ser. A, **60** = Indag. Math. **19**, 248—253 (1957). — [3] Über die Stabilität der Präzessionsbewegung eines symmetrischen Kreisels. Ing.-Arch. **29**, 425—427 (1960).

—, and H. J. E. BETH: [1] EULER's equations for the motion of a rigid body in n-dimensional space. Koninkl. Nederland. Akad. Wetenschap., Proc., Ser. A, **54**, 106—108 (1951). — [2] The stationary motions of a rigid body under no forces in four-dimensional space. Ibid. **54**, 123—129 (1951).

BRAAMS, C. M.: [1] On the influence of friction on the motion of a top. Physica **18**, 503—514 (1952).

BRAUNBEK, W.: [1] Der symmetrische Kreisel mit zeitlich periodischem Richtmoment. Z. angew. Math. Mech. **33**, 174—188 (1953).

BRESSAN, A.: [1] Sulle precessioni d'un corpo rigido costituenti moti di HESS. Rend. Sem. Mat. Univ. Padova **27**, 276—283 (1957).

BRUN, F. DE: [1] Rotation kring fix punkt. Öfvers. Kongl. Svenska Vetenskaps-Akad. Förhandl., Stockholm, **1893**, no. 7 455—468. — [2] Rotation kring fix punkt, II, III. Arkiv Mat. Astr. Fys. **4**, no. 5, 1—4 (1907); **6**, no. 5, 1—10 (1909). — [3] Sur le mouvement d'un solide autor d'un point fixe. Ibid. **6**, no. 9, 1—17 (1909).

BRUNS, H.: [1] Über die Rotation eines starren Körpers. Ber. Verh. Königl. Sächs. Ges. Wiss. Leipzig. Math.-phys. Kl. **37**, 55—59 (1885). — [2] Über die Integrale des Vielkörperproblems. Ibid. **39**, 1—39, 55—82 (1887), and Acta Math. **11**, 25—96 (1887/88).

BULGAKOV, B. V.: [1] Two problems from the theory of motion of a gyroscope (R). Mat. Sbornik **41**, 73—91 (1934). — [2] Quelques problèmes de la stabilité du mouvement dans la théorie des gyroscopes. Učenye Zap. Moskov. Gos. Univ. **24**, 70—82 (1938). — [3] Applied theory of gyroscopes (R). Second edition. Moscow: Gosud. Izdat. Tehn. Teor. Lit. 1955. First edition appeared in 1939. — [4] Applied theory of gyroscopes. Second edition. Transl. from the Russian by J. J. SCHORR-KON. Published for the Nat. Sci. Foundation, Washington, D. C., by the Israel Program for Scientific Translations, Jerusalem, 1960.

BUNATJAN, L. A.: [1] Motion of a free gyroscope on a non-stationary basis with allowance for the masses of the CARDAN suspension rings (R). Dokl. Akad. Nauk SSSR **152**, 305—308 (1963).

BURGATTI, P.: [1] Sugl'integrali primi dell'equazioni del moto d'un corpo pesante intorno a un punto fisso. Ann. Mat. Pura Appl. (3) **12**, 81—100 (1906). — [2] Dimostrazione della non esistenza d'integrali algebrici (oltre i noti) nel problema del moto d'un corpo pesante intorno a un punto fisso. Rend. Circ. Mat. Palermo **29**, 369—377 (1910).

ČAPLYGIN, S. A.: [1] Concerning HESS's loxodromic pendulum (R). Trudy Otdel. Fiz. Nauk Obšč. Lyubit. Estestvozn. **7** (1894). Sobranie Sočin., vol 1, 133—135. Moscow-Leningrad: Gosud. Izdat. Tehn. Teor. Lit. 1948. — [2] Linear particular integrals of the problem of motion of a rigid body supported at one point (R). Ibid. **9**, no. 2, 15—21 (1898). Sobranie Sočin., vol 1, 110—117. — [3] A new case of rotation of a heavy rigid body supported at one point (R). Ibid. **10**, no. 2, 31—34 (1901). Sobranie Sočin., vol. 1, 118—124. — [4] A new particular solution of the problem of rotation of a heavy body about a fixed point (R). Ibid. **12**, no. 1 (1904). Sobranie Sočin., vol. 1, 125—132. — [5] On the case of motion of a heavy rigid body about a fixed point obtained by KOVALEVSKAYA (unpublished manuscript) (R). See L. N. SRETENSKII [2].

CAPRIZ, G.: [1] Precessioni di un giroscopio pesante con armatura asimmetrica. Univ. Roma. Ist. Naz. Alta Mat. Rend. Mat. e Appl. (5) **12**, 229—236 (1953).

CASSINA, U.: [1] Su un teorema di Peano ed il moto del polo. Ist. Lombardo Accad. Sci. Lett. Rend. A **92**, 631—655 (1957/58).

CAYLEY, A.: [1] Reproduction of EULER's memoir of 1758 on the rotation of a solid body. Quart. J. Pure and Appl. Math. **9**, 361—373 (1868).

ČERNOUS'KO, F. L.: [1] On the motion of a satellite about its center of mass under the action of gravitational moments (R). Prikl. Mat. Meh. **27**, 474—483 (1963); transl. as J. Appl. Math. Mech. **27**, 708—722. — [2] Resonance phenomena of the satellite motion relative to its mass center (R). Ž. Vyčisl. Matem. i Matem. Fiziki **3**, no. 3, 528—538 (1963).

ČETAEV, N. G.: [1] On the stability of rotation of a rigid body with a fixed point in LAGRANGE's case (R). Prikl. Mat. Meh. **18**, 123—124 (1954). — [2] On certain problems of stability of motion in mechanics (R). Ibid. **20**, 309—314 (1956). — [3] On a gyroscope mounted in a universal suspension (on gimbals) (R). Ibid. **22**, 379—381 (1958); transl. as J. Appl. Math. Mech. **22**, 521—525. — [4] Stability of motion. Works on analytical mechanics (R). Moscow: Izdat. Akad. Nauk SSSR 1962.

CHERRY, T. M.: [1] On POINCARÉ's theorem of the non-existence of uniform integrals of dynamical equations. Proc. Cambridge Philos. Soc. **22**, 287—294 (1924).

CHRAPAN, J.: [1] The Lagrangian rigid body (Slovak.). Mat.-Fys. Sborn. Slovensk. Adak. Vied Umeni **2**, 23—51 (1952).

CLEBSCH, A.: [1] Zur Theorie der Trägheitsmomente und der Drehung eines festen Körpers um einen Punkt. J. reine angew. Math. **57**, 73—77 (1860).

COLOMBO, G.: [1] Osservazioni sulla stabilità dei moti merostatici di un giroscopio ed applicazioni ad un caso notevoli. Rend. Sem. Mat. Univ. Padova **20**, 59—77 (1951). — [2] Sul moto di due corpi rigidi pesanti collegati in un punto, di cui uno ha un punto fisso. Ibid. **22**, 305—312 (1953).

Copeland, A. H.: [1] Types of motion of the gyroscope. Trans. Amer. Math. Soc. **30**, 737—764 (1928).

Cor, P.: [1] Essai sur l'explication de l'effet gyroscopic. J. École Polytechn. (2) **31**, 41—46 (1933).

Corliss, J. J.: [1] On the unsymmetric top. Acta Math. **59**, 423—441 (1932). — [2] On the unsymmetric top. Ibid. **62** 301—312 (1934).

Čžan, Sy-In: [1] On the stability of motion of a gyroscope (R). Prikl. Mat. Meh. **23**, 604—605 (1959); transl. as J. Appl. Math. Mech. **23**, 860—863.

Darboux, G.: [1] Cours de mécanique par Despeyrous, avec des notes par G. Darboux, 2 vols., Paris: A. Hermann 1884—1886. — [2] Sur le mouvement d'un corps pesant de révolution fixé par un point de son axe. C. R. Acad. Sci. Paris **101**, 115—119 (1885). — [3] Sur le mouvement d'un corps pesant de révolution fixé par un point de son axe. J. Math. Pures Appl. (4) **1**, 403—430 (1885). — [4] Leçons sur la théorie générale des surfaces, 4 volumes, Paris: Gauthier-Villars 1887—1896.

Davidson, M., Editor: [1] The gyroscope and its applications. London: Hutchinson's Scientific and Technical Publications 1946.

Delone, N. B.: [1] Algebraic integrals of the motion of a heavy rigid body about a fixed point (R). St. Petersburg, 1892, 78 pp. and 2 plates. — [2] On the question of geometrical interpretation of integrals given by S. Kovalevskaya in the motion of a heavy rigid body about a fixed point (R). Mat. Sbornik **16**, 346—351 (1892).

Diaz, J. B., and F. T. Metcalf: [1] Upper and lower bounds for the apsidal angle in the theory of the spherical pendulum. Proc. 4th U. S. Congress Appl. Mechanics. Univ. of Calif., Berkeley, 1962. Vol. 1, pp. 127—135. Amer. Soc. Mech. Engrs., New York, 1962. — [2] On a result of Hadamard concerning the sign of the precession of a heavy symmetrical top. Proc. Amer. Math. Soc. **13**, 669—670 (1962). — [3] Upper and lower bounds for the apsidal angle in the theory of the heavy symmetrical top. Arch. Rational Mech. Anal. **16**, 214—229 (1964).

Dobronravov, V. V.: [1] On certain relations in the problem of motion of a rigid body about a fixed point in Euler's case (R). Moskov. Gos. Univ. Učenye Zap. **154**, Mehanika 4, 55—59 (1951). — [2] On a relation in Euler's problem on the motion of a rigid body about a fixed point (R). Dokl. Akad. Nauk SSSR **65**, 143—144 (1949).

Dokševič, A. I.: [1] Some inequalities in the problem of the rotation of a heavy rigid body about a fixed point (R). Izv. Akad. Nauk UzSSR. Ser. Fiz.-Mat. Nauk **1963**, no. 2, 16—23.

Domogarov, A.: [1] On free motion of a gyroscope (R). Diss., St. Petersburg, 1893.

Domožilova, L. M.: [1] Investigation of a special case of rotational motion of a celestial body (R). Vestnik Moskov. Univ. Ser. III. Fiz. Astron. **1961**, no. 2, 82—87.

Drach, J.: [1] Sur le mouvement d'un solide pesant qui a un point fixe (Détermination du groupe de rationalité de l'équation différentielle du problème). C. R. Acad. Sci. Paris **179**, 735—737 (1924).

Draper, C. S.: [1] Mechanization of inertial guidance systems. Gyrodynamics. Symposium Celerina, August 1962. International Union of Theoretical and Applied Mechanics. Edited by H. Ziegler, pp. 92—118. Berlin-Göttingen-Heidelberg: Springer 1963.

—, W. Wrigley, and L. R. Grohe: [1] The floated integrating gyro and its applications to geometrical stabilization problems on moving bases. S. M. F. Fund Paper FF-13, New York: Institute of the Aeronautical Sciences, 1955.

—, —, and J. Hovorka: [1] Inertial guidance. New York: Pergamon Press 1962.

DROFA, V. N.: [1] On the permanent axes of motion of a heavy gyrostat about a fixed point (R). Prikl. Mat. Meh. **25**, 941—945 (1961); transl. as J. Appl. Math. Mech. **25**, 1411—1419.

DUBOŠIN, G. N.: [1] The differential equations of translational-rotational motion of mutually attracting rigid bodies (R). Astron. Ž. **35**, 265—276 (1958). — [2] On a particular case of the problem of translational-rotational motions of two bodies (R). Ibid. **36**, 153—163 (1959). — [3] On the stability of regular motions of artificial celestial bodies (R). Ibid. **36**, 723—733 (1959); transl. as Soviet Astron. AJ **3**, 702—712 (1960). — [4] Sur le mouvement de translation-rotation des corps célestes artificiels. Dynamics and Satellites. Symposium Paris, 1962. International Union of Theoretical and Applied Mechanics. Edited by M. ROY, pp. 14—20. Berlin-Göttingen-Heidelberg: Springer 1963.

DUVAKIN, A. P.: [1] On the stability of motion of a tippe-top on a perfectly rough horizontal plane (R). Inžen. Ž. **2**, 222—230 (1962).

EULER, L.: [1] Decouverte d'un nouveau principe de mécanique. Mém. Acad. Roy. Sci. et Belles-Lettres de Berlin **6**, 1752, 185—217 (1750). Also Opera omnia, Ser. secunda, vol. 5, pp. 81—108, Lausannae, Orell Füsli Turici, 1957. — [2] Du mouvement de rotation des corps solides autour d'un axe variable. Ibid. **14**, 1765, 154—193 (1758). Opera omnia, Ser. secunda, vol. 8 (in press). — [3] Theoria motus corporum solidorum seu rigidorum ex primis nostrae cognitionis principiis stabilita et ad omnes motus, qui in hujusmodi corpora cadere possunt, accomodata. Rostochii et Gryphiswaldiae, 1765. Also Opera omnia, Ser. secunda, vol. 3, 1948, Chapt. I—X, and vol. 4, 1950, Chapt. XI—XIX, Suppl. — [4] Du mouvement d'un corps solide quelconque lorsqu'il tourne autour d'un axe mobile. Mém. Acad. Roy. Sci. et Belles-Lettres de Berlin **16**, 1767, 176—227 (1760). Opera omnia, Ser. secunda, vol. 8 (in press).

FABBRI, M. R.: [1] Sopra una soluzione particolare delle equazioni del moto di un solido pesante intorno ad un punto fisso. Atti R. Accad. Naz. Lincei. Rend. Cl. Sci. Fis. Mat. Nat (6) **19**, 407—415 (1934). — [2] Sopra un particolare movimento di un solido pesante intorno a un punto fisso. Ibid. (6) **19**, 495—502 (1934). — [3] Sui coni di POINSOT in una particolare rotazione dei solidi pesanti. Ibid. (6) **19**, 872—873 (1934).

FEDORČENKO, A. M.: [1] On the method of canonical averaging in the theory of nonlinear oscillations (R). Ukraïn. Mat. Ž. **9**, 220—224 (1957). — [2] On the motion of an asymmetric heavy gyroscope with a vibrating fulcrum (R). Ibid. **10**, 209—218 (1958). — [3] On a dynamical method for increasing stability of a rapidly rotating symmetrical gyroscope (R). Prikl. Mat. Meh. **25**, 938—940 (1961); transl. as J. Appl. Math. Mech. **25**, 1406—1410.

FERRY, E. S.: [1] Applied gyrodynamics. New York: John Wiley and Sons 1932.

FIELD, P.: [1] On the unsymmetrical top. Acta Math. **56**, 355—362 (1931). — [2] On the unsymmetrical top. Ibid. **62**, 313—316 (1934).

FILATOV, A. N.: [1] Generalized LIE series and their applications (R). Taškent: Izdat. Akad. Nauk Uzbek. SSR 1963.

FOUCAULT, L.: [1] Sur une nouvelle démonstration expérimentale du mouvement de la terre fondée sur la fixité du plan de rotation. C. R. Acad. Sci. Paris **35**, 421—424 (1852). — [2] Sur les phénomènes d'orientation des corps tournants entraînés par un axe fixe à la surface de la terre. Ibid. **35**, 424—427 (1852). — [3] Démonstration expérimentale du mouvement de la terre. Additions aux communications faites dans les précédentes séances. Ibid. **35**, 469—470 (1852). — [4] Sur la tendance des rotations au parallélisme. Ibid. **35**, 602 (1852).

FOX, C.: [1] The mechanical determination of position and velocity on the Earth's surface. Proc. Cambridge Philos. Soc. **45**, 311—315 (1949).

FRAHM, W.: [1] Note über die Rotation eines starren Körpers. Math. Ann. **8**, 31—34 (1874).

FRANKE, J. N.: [1] Sur la courbure de l'herpolhodie, C. R. Acad. Sci. Paris **100**, 1573—1576 (1885).

FRENZEL, C.: [1] Neue Lösung des Rotationsproblemes. Z. Math. Phys. **26**, 104—127 (1881).

FRIDLENDER, G. O.: [1] On the precession of a gyroscope under the action of an external moment (R). Inžen. Sbornik **12**, 229—233 (1952).

GEBELEIN, H.: [1] Über den unsymmetrischen schweren Kreisel. Ann. Physik (5) **12**, 889—926 (1932).

GÉRARD, L.: [1] Théorèmes de PUISEUX et de HALPHEN sur le pendule sphérique. Nouv. Ann. de Math. (6), **2**, 147—148 (1927).

GERMAIN, A. DE ST.: [1] Sur l'herpolhodie. C. R. Acad. Sci. Paris **100**, 1126—1128 (1885). — [2] Note sur le pendule sphérique. Bull. Sci. Math. (2) **20**, 114—116 (1896). — [3] Note sur le pendule sphérique. Ibid. (2) **22**, 95—96 (1898).

GOLUBEV, V. V.: [1] Lectures on integration of the equations of motion of a rigid body about a fixed point (R). Moscow: Gosud. Izdat. Tehn. Teor. Lit. 1953. — [2] Lectures on integration of the equations of motion of a rigid body about a fixed point. Transl. from Russian by J. SHORR-KON and published for the Nat. Sci. Foundation by the Israel Program for Sci. Translations. Washington, D. C., Office of Technical Services, U. S. Department of Commerce, 1960.

GOODSTEIN, R.: [1] A perturbation solution of the equations of motion of a gyroscope. J. Appl. Mech. **26**, 349—352 (1959).

GORYAČEV, D. N.: [1] A new particular solution of the problem of motion of a heavy rigid body about a fixed point (R). Trudy Otdel. Fiz. Nauk Obšč. Lyubit. Estestvozn. **9**, 22—24 (1898). — [2] On the motion of a heavy rigid body about a fixed point in the case $A = B = 4C$ (R). Mat. Sbornik **21**, 431—438 (1900). — [3] Certain general integrals in the problem on the motion of a rigid body (R). Warsaw, 1910, 1—22. — [4] New cases of motion of a rigid body about a fixed point (R). Varšav. Univ. Izvest., no. 3, 1—11 (1915). — [5] New case of integrability of the EULER dynamical equations (R). Rostov-Don. Varšav. Univ. Izvest. no. 3, 1—13 (1916).

GRAMMEL, R.: [1] Nutation des unsymmetrischen Kreisels. Phys. Z. **20**, 398—400 (1919). — [2] Über einige Bewegungen des unsymmetrischen schweren Kreisels. Jber. Deutschen Math. Verein **29**, 150—174 (1920). — [3] Die Stabilität der Staudeschen Kreiselbewegungen. Math. Z. **6**, 124—142 (1920). — [4_1] Der Kreisel, seine Theorie und seine Anwendungen. Zweite, neubearbeitete Auflage. Erster Band: Die Theorie des Kreisels. Berlin-Göttingen-Heidelberg: Springer 1950. — [4_2] Der Kreisel, seine Theorie und seine Anwendungen. Zweite, neubearbeitete Auflage. Zweiter Band: Die Anwendungen des Kreisels. Berlin-Göttingen-Heidelberg: Springer 1950. — [5] Zur Berechnung der Poinsotbewegung. Ing.-Arch. **18**, 53—59 (1950). — [6] Die stationären Bewegungen des selbsterregten Kreisels und ihre Stabilität. Ibid. **21**, 149—156 (1953). — [7] Der selbsterregte unsymmetrische Kreisel. Ibid. **22**, 73—97 (1954). — [8] Drehzahlabhängige Selbsterregung des unsymmetrischen Kreisels. Ibid. **29**, 153—159 (1960). — [9] Ein Problem des selbsterregten unsymmetrischen Kreisels. Ann. Mat. Pura Appl. (4) **50**, 187—196 (1960).

—, and H. ZIEGLER: [1] Der schnelle symmetrische Kardankreisel mit Lagerreibung. Z. angew. Math. Mech. **36**, 278—279 (1956).

GRAY, A.: [1] A treatise on gyrostatics and rotational motion. Theory and applications. London: The MacMillan Co. 1918. Reprinted by Dover Publications, New York, 1959.

GREENHILL, A. G.: [1] Solution of the equations of motion of a rigid body movable about a fixed point under the action of no forces. Quart. J. Pure and Appl. Math. **14**, 182—183 (1877). — [2] Solution of EULER's equations of motion by means of elliptic functions. Ibid. **14**, 265—271 (1877). — [3] The applications of elliptic functions. London and New York: The MacMillan Co. 1892. Reprinted by Dover Publications, New York, 1959. — [4] Pseudo-elliptic integrals and their dynamical applications. Proc. London Math. Soc. **25**, 195—304 (1894). — [5] Dynamics of a top. Ibid. **26**, 215—256 (1895). — [6] The associated dynamics of a top and of a body under no forces. Ibid. **27**, 545—612 (1896). — [7] The mathematical theory of the top. Science (N.S.) 973—975 (1901). — [8] The mathematical theory of the top considered historically. Verhandl. 3. intern. Math.-Kongr. Heidelberg, 1904, 100—108. — [9] The mathematical theory of the top. Ann. of Math. (2) **5**, 1—20, 67—98 (1904). — [10] Report on gyroscopic theory. London: Advisory Committee for Aeronautics 1914. — [11] Pseudo-regular precession. Phil. Mag. **14**, 179—193 (1922).

GRIOLI, G.: [1] Esistenza e determinazione delle precessioni regolari dinamicamente possibili per un solido pesante asimmetrico. Ann. Mat. Pura Appl. (4) **26**, 271—281 (1947). — [2] Moto attorno al baricentro di un giroscopio soggetto a forze di potenza nulla. Univ. Roma Ist. Naz. Alta Mat. Rend. Mat. e Appl. (5) **6**, 439—463 (1947). — [3] Questioni di stabilità riguardanti le precessioni regolari del solido pesante asimmetrico. Ann. Scuola Norm. Sup. Pisa (3) **1**, 1947, 43—74 (1949). — [4] Precessioni regolari di un solido pesante asimmetrico. Atti Accad. Naz. Lincei. Rend. Cl. Sci. Fis. Mat. Nat. (8) **4**, 420—423 (1948). — [5] Una proprietà caratteristica delle precessioni regolari del solido pesante asimmetrico. Rend. Sem. Mat. Univ. Padova **19**, 237—248 (1950). — [6] Questioni di dinamica del solido pesante asimmetrico. Ibid. **21**, 256—277 (1952). — [7] Forma intrinseca delle equazioni dinamiche del solido pesante asimmetrico con un punto fisso e ricerca dei moti di precessione. Ann. Univ. Ferrara. Sez. VII (N.S.) **3**, 31—43 (1954). — [8] Movimenti dinamicamente possibili per un solido asimmetrico soggetto a forze di potenza nulla. Atti Accad. Naz. Lincei. Rend. Cl. Sci. Fis. Mat. Nat. (8) **22**, 459—463 (1957). — [9] Sul moto di un corpo rigido asimmetrico soggetto a forze di potenza nulla. Rend. Sem. Mat. Univ. Padova **27**, 90—102 (1957). — [10] On the general theory of asymmetric gyros. Gyrodynamics. Symposium Celerina 1962, International Union of Theoretical and Applied Mechanics. Edited by H. ZIEGLER. Berlin-Göttingen-Heidelberg: Springer 1963, pp. 26—31.

GRISERI, B.: [1] Soluzioni meromorfe delle equazioni del moto di un solido pesante intorno a un punto fisso col baricentro situato sull'asse di una delle sezioni circolari dell'ellissoide d'inerzia relativo al punto fisso. Univ. e Polit. Torino. Rend. Sem. Mat. **9**, 225—243 (1950).

GRÖBNER, W.: [1] Die LIE-Reihen und ihre Anwendungen. Mathem. Monographien, 3. Herausgegeben von W. BLASCHKE. Berlin: VEB Deutscher Verlag der Wiss. 1960.

GULYAEV, M. P.: [1] On a new particular solution of the equations of motion of a heavy rigid body having one fixed point (R). Vestnik Moskov. Univ., Ser. Fiz. Mat. i Estestv. Nauk **1955**, vyp. 2, 15—21. — [2] On dynamically possible, regular precession of a solid body with one fixed point (R). Akad. Nauk Kazah. SSR. Trudy Sektor. Mat. Meh. **1**, 202—208 (1958). — [3] On the stability of rotation of a rigid body with one fixed point in the EULER case (R). Prikl. Mat. Meh. **23**, 406—408 (1959); transl. as J. Appl. Math. Mech. **23**, 579—582. — [4] One version of the regular precession of a solid with a liquid-filled cavity (R). Dokl. Akad. Nauk SSSR **139**, 574—575(1961); transl. as Soviet Phys. Dokl. **6**, 583—584 (1962).

GULYAEV, M. P., and M. OŠIBAEV: [1] On stability of rotation of a heavy solid body with one fixed point in the case of D. N. GORYAČEV and S. A. ČAPLYGIN (R). Akad. Nauk Kazah. SSR, Trudy Sektor. Mat. Meh. **1**, 144—146 (1958).

HADAMARD, J.: [1] Sur la stabilité des rotations dans le mouvement d'un corps solide pesant autour d'un point fixe. Assoc. Française pour l'avancement des sciences. C. R. de la 24me session. Congrès de Bordeaux, **1895**, 1—6. — [2] Sur la précession dans le mouvement d'un corps pesant de révolution fixé par un point de son axe. Bull. Sci. math. (2) **19**, 228—230 (1895).

HALPHEN, G.: [1] Sur le mouvement d'un corps grave de révolution suspendu par un point de son axe. C. R. Acad. Sci. Paris **100**, 1065—1068 (1885). — [2] Traité des fonctions elliptiques, t. 2, Paris: Gauthier-Villars 1888.

HAMEL, G.: [1] Über den allgemeinen schweren Kreisel. Z. angew. Math. Mech. **25/27**, no. 5/6, 159—160 (1947).

HARLAMOVA, E. I.: [1] On a particular solution of the EULER-POISSON equations (R). Prikl. Mat. Meh. **23**, 681—690 (1959); transl. as J. Appl. Math. Mech. **23**, 975—988. — [2] A particular case of the integrability of the EULER-POISSON equations (R). Dokl. Akad. Nauk SSSR **125**, 996—997 (1959); transl. as Soviet Phys. Dokl. **4**, 313—315. — [3] On the motion of a rigid body about a fixed point in the central NEWTONian force field (R). Izvestiya Sibirsk. Otdel. Akad. Nauk SSSR, no. 6 (1959).

HARLAMOVA-ZABELINA, E. I.: [1] Rapid rotation of a solid body about a fixed point with non-holonomic constraints (R). Vestnik Moskov. Univ. Ser. Mat. Meh. Astr. Fiz. Him. **12**, no. 6, 25—34 (1957).

HARLAMOV, P. V.: [1] On linear integrals of the equations of motion of a heavy solid body about a fixed point (R). Dokl. Akad. Nauk SSSR **143**, 805—807 (1962); transl. as Soviet Phys. Dokl. **7**, 308—309 (1962). — [2] On the equations of motion for a heavy body with a fixed point (R). Prikl. Mat. Meh. **27**, 703—707 (1963); transl. as J. Appl. Math. Mech. **27**, 1070—1078 (1964). — [3] Cinematical interpretation of the motion of a body with a fixed point (R). Ibid. **28**, 502—507 (1964).

HARLAMOV, S. A.: [1] The motion of a gyroscope in a CARDAN suspension wiht respect to the axis of natural rotation (R). Dokl. Akad. Nauk SSSR **139**, 327—330 (1961); transl. as Soviet Phys. Dokl. **6**, 580—582 (1962). — [2] Nutational oscillations and drift of a synchronous gyroscope mounted in a CARDAN suspension (R). Dokl. Akad. Nauk SSSR **146**, 550—553 (1962). — [3] On the theory of an astatic gyroscope with an electric drive mounted in a CARDAN suspension (R). Izv. Akad. Nauk SSSR. Meh. i. Mašinostroenie, no. 6, 45—54 (1963).

HAUG, J.: [1] Über die Drehung eines starren Körpers um seinen Schwerpunkt. Pr. Theresien-Gymnasium München, 1902, 17 pp.

HAYWARD, R. B.: [1] On a direct method of estimating velocities, accelerations, and all similar quantities, with respect to axes movable in any manner in space. Trans. Cambridge Philos. Soc. **10**$_1$, 1—20 (1864).

HESS, W.: [1] Über das Rollen einer Fläche zweiten Grades auf einer invariablen Ebene, Diss. München, 1880. — [2] Über das Gyroskop. Math. Ann. **19**, 121—154 (1881). — [3] Über das Problem der Rotation, Ibid. **20**, 461—470 (1882). —

[4] Über das Gyroskop bei allgemeinster Wahl des zur Bewegung anregenden Momentankräftesystems. Ibid. **29**, 500—580 (1887). — [5] Über das Jacobische Theorem von der Ersetzbarkeit einer Lagrangeschen Rotation durch zwei Poinsotsche Rotationen. Z. Math. Phys. **33**, 292—305 (1888). — [6] Über die Eulerschen Bewegungsgleichungen und ihre singulären Lösungen. Programm des Lyceums zu Bamberg, 1889, 60 pp. — [7] Über die Eulerschen Bewegungsgleichungen und über eine neue particuläre Lösung des Problems der Bewegung eines starren Körpers um einen festen Punkt. Math. Ann. **37**, 153—181 (1890).

Hill, E. L.: [1] Rotations of a rigid body about a fixed point. Amer. J. Phys. **13**, 137—140 (1945).

Hugenholtz, N. M.: [1] On tops rising by friction. Physica **18**, 515—527 (1952).

Husson, E.: [1] Recherches des intégrales algébriques dans le mouvement d'un corps solide pesant autour d'un point fixe. C. R. Acad. Sci. Paris **141**, 100—102 (1905). — [2] Sur un théorème de M. Poincaré relativement au mouvement d'un solide pesant. Ibid. **141**, 821—823 (1905). — [3] Recherche des intégrales algébriques dans le mouvement d'un solide pesant autour d'un point fixe. Ann. Fac. Sci. Toulouse (2) **8**, 73—152 (1906). — [4] Sur un théorème de M. Poincaré, relativement au mouvement d'un solide pesant. Acta Math. **31**, 71—88 (1908).

Isaeva, L. S.: [1] On the sufficient conditions of stability of rotation of a tippe-top on a perfectly rough horizontal surface (R). Prikl. Mat. Meh. **23**, 403—406 (1959); transl. as J. Appl. Math. Mech. **23**, 572—578.

Išlinskii, A. Yu.: [1] On relative equilibrium of a physical pendulum with a movable fulcrum (R). Prikl. Mat. Meh. **20**, 297—308 (1956). — [2] On the theory of a horizontal gyrocompass (R). Ibid. **20**, 487—499 (1956). — [3] On the theory of a gyroscopic pendulum (R). Ibid. **21**, 3—14 (1957). — [4] Theory of a double-gyroscopic vertical (R). Ibid. **21**, 175—183 (1957). — [5] On the equations for the problem of determining the position of a moving object by means of gyroscopes and accelerometers (R). Ibid. **21**, 725—739 (1957). — [6] On the theory of complicated systems of gyroscopic stabilization (R). Ibid. **22**, 359—375 (1958); transl. as J. Appl. Math. Mech. **22**, 493—512. — [7] On the autonomous determination of position of a moving object by means of a space gyrocompass, a directional gyroscope and an integrating device (R). Ibid. **23**, 58—63 (1959); J. Appl. Math. Mech. **23**, 75—82. — [8] On the equations of the precessional theory of a gyroscope in the form of equations of motion of the pole in the phase plane (R). Ibid. **23**, 801—809 (1959); J. Appl. Math. Mech. **23**, 1153—1163.

Jacobi, C. G. J.: [1] Sur la rotation d'un corps. J. reine u. angew. Math. **39**, 293—350 (1849). Ges. Werke, vol. 2, pp. 289—352. Berlin: G. Reimer 1882. — [2] Second mémoire sur la rotation d'un corps non soumis à des forces accélératrices. Ibid. vol. 2, pp. 427—467. — [3] Supplément au mémoire précédent. Ibid. vol. 2, pp. 468—476. — [4] Nouvelle théorie de la rotation d'un corps de révolution grave suspendu en un point quelconque de son axe. Ibid. vol. 2, pp. 477—492. — [5] Sur la rotation d'un corps de révolution grave autour d'un point quelconque de son axe. Ibid. vol. 2, pp. 493—510.

Jonquières, E. de: [1] Au sujet de certaines circonstances qui se présentent dans le mouvement de la toupie. C. R. Acad. Sci. Paris **102**, 1519—1522 (1886).

Keis, I. A.: [1] On the existence of certain integrals of the equations of motion of a gyrostat, fixed at one point (R). Vestnik Moskov. Univ. **1963**, Ser. I Mat. Meh., no. 6, 55—62. — [2] On the motion of a gyrostat, fastened at a fixed point (R). Ibid. **1964**, no. 1, 76—79. — [3] On the algebraic integrals in the problem of motion of a heavy gyrostat, fastened at a point (R). Prikl. Mat. Meh. **28**, 516—520 (1964).

KELVIN, Lord (Wm. THOMSON) and P. G. TAIT: [1] Treatise on Natural Philosophy. Part 1. Cambridge: University Press 1912.

KIRCHHOFF, G.: [1] Vorlesungen über Mechanik, 4. Aufl., Leipzig: B. G. Teubner 1897.

KIRCHNER, F.: [1] Einfache Resonanzversuche am symmetrischen Kreisel. Z. angew. Math. Phys. **6**, 355—361 (1955).

KLEIN, F.: [1] Sur le mouvement d'un corps grave de révolution suspendu par un point de son axe. Nouv. Ann. de Math. (3) **15**, 218—222 (1896). — [2] On the stability of a sleeping top. Bull. Amer. Math. Soc. (2) **3**, 129—132 (1897). — [3] The mathematical theory of the top. Princeton Lectures. New York: Charles Scribner's Sons 1897.

—, and A. SOMMERFELD: [1] Über die Theorie des Kreisels. Heft 1: Die kinematischen und kinetischen Grundlagen der Theorie. Leipzig: B. G. Teubner 1897, pp. 1—196. Heft 2: Durchführung der Theorie im Falle des schweren symmetrischen Kreisels, 1898, pp. 197—512. Heft 3: Die störenden Einflüsse. Astronomische und geophysikalische Anwendungen, 1903, pp. 513—759. Heft 4: Die technischen Anwendungen, 1910, pp. 760—960.

KLIMOV, D. M.: [1] On the motion of a gyroscope with universal (CARDAN's) suspension and a nonaxially placed rotor (R). Dokl. Akad. Nauk SSSR **124**, 537—539 (1959); transl. as Soviet Phys. Dokl. **124**, no. 4, 80—82 (1959).

KOBB, G.: [1] Sur le mouvement d'un point matériel sur une surface de révolution. Acta Math. **10**, 89—108 (1887). — [2] Sur le problème de la rotation d'un corps autour d'un point fixe. Bull. Soc. Math. France **23**, 210—215 (1895). — [3] Sur les solutions périodiques du problème de la rotation d'un corps autour d'un point fixe. Ann. Fac. Sci. Toulouse (2) **1**, 5—30 (1899).

KOENIGS, G.: [1] Sur les solutions périodiques du problème du mouvement d'un corps pesant quelconque, suspendu par un de ses points. C. R. Acad. Sci. Paris **122**, 1048—1049 (1896).

KOHN, W.: [1] Contour integration in the theory of the spherical pendulum and the heavy symmetrical top. Trans. Amer. Math. Soc. **59**, 107—131 (1946).

KOLESNIKOV, N. N.: [1] On the stability of a free gyrostat (R). Prikl. Mat. Meh. **27**, 699—702 (1963); transl. as J. Appl. Math. Mech. **27**, 1063—1069 (1964).

KOLOSOV, G. V.: [1] Über einen Fall der Bewegung eines allgemeinen Kreisels, dessen Unterstützungspunkt sich in der horizontalen Ebene frei bewegen kann. Nachr. Königl. Ges. Wiss. Göttingen. Math.-phys. Kl., **1898**, 80—83. — [2] On a case of motion of a heavy rigid body supported at a sharp point on a smooth plane (R). Trudy Otdel. Fiz. Nauk Obšč. Lyubit. Estestvozn. **9**, no. 2, 11—12 (1898). — [3] On a case of motion of a rigid body. Mess. of Math. (2) **30**, 174—177 (1901). — [4] On some property of the KOVALEVSKAYA problem on the rotation of a heavy rigid body about a fixed point (R). Trudy Otdel. Fiz. Nauk Obšč. Lyubit Estestvozn. **11**, no. 1, 5—13 (1902). — [5] Sur le cas de M. GORIATSCHOFF de la rotation d'un corps pesant autour d'un point fixe. Rend. Circ. Mat. Palermo **16**, 346—348 (1902). — [6] Über eine Eigenschaft der Differentialgleichungen der Rotation eines schweren Körpers um einen festen Punkt im Falle von Frau KOWALEVSKI. Math. Ann. **56**, 265—272 (1903). — [7] On the GORIATSCHOFF's case of rotation of a heavy body about a fixed point. Mess. of Math. (2) **32**, 84—88 (1903).

KONDURAR', V. T.: [1] Problem of motion of two ellipsoids under the action of mutual attraction. Part 1. Problem of two spheroids with coinciding planes of equatorial sections (R). Astron. Ž. **13**, 563—588 (1936). — [2] Part 2. Problem of two spheroids with mutually perpendicular axes of rotation (R). Moskov. Gos. Univ. Trudy Gos. Astron. Inst. Šternberga **9**, vyp. 2, 307—368 (1939). — [3] Part 3. On stability of circular motions in the problem of two ellipsoids with common plane of symmetry (R). Ibid. **21**, 115—134 (1952). — [4] Part 4. On periodic solutions near to circular ones in the problem of motion of two spheroids with mutually perpendicular axes of rotation (R). Ibid. **21**, 135—158 (1952). — [5] Part 5. Periodic solutions near to circular ones in the problem of motion of two spheroids with mutually perpendicular axes of rotation (R). Ibid. **24**, 155—198 (1954). — [6] Particular solutions of the general problem on the translational-rotational motion of a spheroid attracted by a sphere (R). Astron. Ž. **36**, 890—901 (1959); transl. as Soviet Astron. AJ **3**, 863—875 (1960). — [7] The translational-rotational motion of a satellite under the attraction of a planet and the sun (R). Ibid. **38**, 969—981 (1961); Soviet Astron. AJ **6**, 739—748 (1962).

KOOY, J. M. J., and J. W. H. UYTENBOGAART: [1] Ballistics of the future, with special reference to the dynamical and physical theory of the rocket weapons. New York and London: McGraw-Hill Co. 1946.

KOŠLYAKOV, V. N.: [1] On deviations of the gyrovertical under variable velocity of the proper rotation of the gyroscope (R). Inžen. Sbornik **6**, 185—196 (1950). — [2] On certain particular cases of integration of the dynamical equations of EULER connected with the motion of a gyroscope in a resisting medium (R). Prikl. Mat. Meh. **17**, 137—148 (1953). — [3] The theory of a gyrocompass (R). Ibid. **23**, 810—817 (1959); transl. as J. Appl. Math. Mech. **23**, 1164—1173. — [4] On the reduction of the equations of motion of a gyrohorizoncompass (R). Ibid. **25**, 801—805 (1961); J. Appl. Math. Mech. **25**, 1205—1211 (1962). — [5] On the stability of motion of a horizontal gyrocompass in the presence of dissipative forces (R). Ibid. **26**, 412—417 (1962); J. Appl. Math. Mech. **26**, 615—622 (1962). — [6] On GECKELER's equations in the theory of gyrocompasses (R). Prikl. Mat. Meh. **28**, 708—715 (1964).

—, and V. F. LIAŠENKO: [1] On a certain integral in the theory of a horizontal gyrocompass (R). Prikl. Mat. Meh. **27**, 10—15 (1963); transl. as J. Appl. Math. Mech. **27**, 10—17.

KÖTTER, F.: [1] Über das KOWALEVSKIsche Rotationsproblem. Jber. Deutschen Math. Verein. **1**, 65—68 (1890/91). — [2] Sur le cas traité par Mme. KOWALEVSKI de rotation d'un corps solide autour d'un point fixe. Acta Math. **17**, 209—264 (1893). — [3] Ein Beweis des JACOBIschen Theorems von der Zusammensetzbarkeit einer Kreiselbewegung aus den Inversionen zweier POINSOT-Bewegungen. Ber. Berliner Math. Ges. **1**, 11—12 (1902).

KOVALEVSKAYA, S. V. (KOWALEVSKY, S., also KOWALEVSKI, S.): [1] Sur le problème de la rotation d'un corps solide autour d'un point fixe. Acta Math. **12**, 177—232 (1889). — [2] Sur une propriété du système d'équations différentielles qui définit la rotation d'un corps solide autour d'un point fixe. Ibid. **14**, 81—93 (1890). — [3] Mémoire sur un cas particulier du problème de la rotation d'un corps pesant autour d'un point fixe, où l'intégration s'effectue à l'aide de fonctions ultraelliptiques du temps. Académie des Sciences de l'Institut de France. Mémoires présentés par divers savants, vol. **31**, pp. 1—62 (1890) Paris. — [4] The problem of motion of a rigid body about a fixed point (R). Transl. of Ref. [1] from French by P. YA. POLUBARINOVA-KOČINA. KOVALEVSKAYA Memorial Collection, "Motion of a Rigid Body about a Fixed Point", edited by S. A. ČAPLYGIN and N. I. MERCALOV. Moscow-Leningrad: Izdat. Akad. Nauk SSSR 1940, pp. 11—49. — [5] On a certain property of a system of differential equations determining the motion of a rigid body about a fixed point (R). Transl. of Ref. [2] from French by P. Ya. Polubarinova-Kočina. Kovalevskaya Memorial Collection, pp. 50—60. — [6] Scientific Works (R). Edited by P. Ya. Polubarinova-Kočina, Moscow: Izdat. Akad. Nauk SSSR 1948. Ref. [4], pp. 153—220, and Ref. [5], pp. 221—234.

Kowalewski, N.: [1] Eine neue partikuläre Lösung der Differentialgleichungen der Bewegung eines schweren starren Körpers um einen festen Punkt. Math. Ann. **65**, 528—537 (1908).

Krementulo, V. V.: [1] Investigation of the stability of a gyroscope taking into account the dry friction on the axis of the inner Cardan ring (gimbal) (R). Prikl. Mat. Meh. **23**, 968—970 (1959); transl. as J. Appl. Math. Mech. **23**, 1382—1386. — [2] Stability of a gyroscope having a vertical axis of the outer ring with dry friction in the gimbal axes taken into account (R). Ibid. **24**, 568—571 (1960); J. Appl. Math. Mech. **24**, 843—849. — [3] Application of the second method of Lyapunov in the investigation of the steady motion of a gyroscope when elastic properties of the rotor axis are taken into account (R). Ibid. **25**, 579—582 (1961); J. Appl. Math. Mech. **25**, 859—865.

Kreyszig, E.: [1] Über den allgemeinen Integralsinus $\mathrm{Si}(z, \alpha)$. Acta Math. **84/85**, 117—181 (1950/51). — [2] Der allgemeine Integralcosinus $\mathrm{Ci}(z, \alpha)$. Ibid. **89/90**, 107—131 (1953).

Krotova, P. G.: [1] Integration of the equations of motion of an asymmetric gyroscope (R). Izv. Sibirsk. Otdel. Akad. Nauk SSSR **1958**, no. 7, 75—86.

Krylov, A. N., and Yu. A. Krutkov: [1] General theory of gyroscopes and certain technical applications of them (R). Leningrad: Izdat. Akad. Nauk SSSR 1932.

Kuzmak, G. E.: [1] The motion of an axially symmetric solid in space about a stationary point under the action of moments slowly varying with time (R). Dokl. Akad. Nauk SSSR **132**, 549—552 (1960); transl. as Soviet Phys. Dokl. **5**, 526—529.

Kuz'min, P. A.: [1] Supplement to V. A. Steklov's case of motion of a heavy rigid body about a fixed point (R). Prikl. Mat. Meh. **16**, 243—245 (1952).

Kuznecov, L. I.: [1] On the motion of a gyroscope in a resisting medium by taking into account friction in the suspension (R). Vestnik Leningrad. Univ. Ser. Mat. Meh. **1958**, no. 4, 151—155. — [2] Equations of the precessional theory of gyroscopes. AIAA J. **1**, 271—274 (1963).

Kuzovkov, N. T.: [1] Uniaxial force gyrostabilizer with an oblique mounting of the gimbal axis (R). Izv. Akad. Nauk SSSR. Otd. Tehn. Nauk **1958**, no. 11, 70—74.

Lacour, E.: [1] Formules elliptiques pour l'étude des mouvements de Poinsot. Ann. Ecole Norm. Sup. (3) **17**, 283—294 (1900).

Lagerborg, N.: [1] Sur le problème du mouvement d'un corps solide autour d'un point fixe. Bull. Soc. Math. France **18**, 118—122 (1890).

Lagrange, J. L.: [1] Méchanique Analitique, Paris: Veuve Desaint 1788, 2 parts or Oeuvres, vols. 11 and 12. Paris: Gauthier-Villars 1889.

Lazzarino, O.: [1] Sull'equivalenza fra le equazioni differenziali di Hess-Schiff e quelle di Euler-Poisson nella teoria dei giroscopi asimmetrici pesanti. Atti Accad. Naz. Lincei. Rend. Cl. Sci. Fis. Mat. Nat. (5) $\mathbf{28}_1$, 325—331, 341—346 (1919). — [2] Sopra alcuni casi singolari nella teoria dei giroscopi asimmetrici pesanti. Ibid. (5) $\mathbf{28}_2$, 9—14, 259—263, 329—333 (1919). — [3] Sul moto dei giroscopi asimmetrici pesanti nel caso in cui l'invariante principale S e costantemente nullo. Ibid. (5) $\mathbf{28}_2$, 422—426, 489—493 (1919). — [4] Teoria sintetica dell'equivalenza fra i sistemi di equazioni differenziali di Euler-Poisson e di Hess-Schiff nella teoria dei giroscopi rigidi pesanti e studio dei casi singolari per i quali l'equivalenza non susiste I, II. Ann. Scuola Norm. Sup. Pisa (2) **7**, 91—107, 109—136 (1938).

LEE, R.: [1] The motion of a self-excited rigid body. Ph. D. Thesis, Department of Mathematics, Univ. of British Columbia, Vancouver, B. C., 1964, vi + 126 pp. (mimeographed).

LEIMANIS, E.: [1] Some recent advances in the dynamics of rigid bodies and celestial mechanics. Surveys in Applied Mathematics, vol. 2, pp. 1—108. New York: John Wiley and Sons, Inc. 1958.

LEIPHOLZ, H.: [1] Ein Beitrag zu dem Problem des Kreisels mit drehzahlabhängiger Selbsterregung. Z. angew. Math. Mech. **32**, 255—285 (1963). — [2] Über die POINSOT-Bewegung eines gyroscopischen Systems, das aus zwei starren Körpern besteht. Gyrodynamics. Symposium Celerina 1962. International Union of Theoretical and Applied Mechanics. Edited by H. ZIEGLER, pp. 32—44. Berlin-Göttingen-Heidelberg: Springer 1963.

LETOV, A. M.: [1] On the theory of gyrosemicompasses (R). Inžen. Sbornik **13**, 123—130 (1952).

LEVI-CIVITA, T.: [1] Sul moto di un corpo rigido intorno ad un punto fisso. Atti R. Accad. Lincei. Rend. Cl. Sci. Fis. Mat. Nat. (5) $\mathbf{5}_2$, 3—9, 122—127 (1896). — [2] Sui moti stazionari di un corpo rigido nel caso della KOWALEVSKY. Ibid. (5) **10**, 338—346, 429—434, 461—466 (1901). — [3] Sur la résolution qualitative du problème restreint des trois corps. Acta Math. **30**, 305—327 (1906).

LEVY, P.: [1] Explication élémentaire de l'effect gyroscopic. Bull. Sci. Math. (2) **65**, 9—20 (1941).

LIEBMANN, H.: [1] Klassifikation der Kreiselprobleme nach der Art der zugehörigen Parametergruppe. Math. Ann. **50**, 51—67 (1898).

LINDEMANN, F.: [1] Über die Drehung eines starren Körpers um seinen Schwerpunkt. S.-B. Math.-phys. Kl. Königl. Bayer. Akad. Wiss. München **28**, 181—202 (1898).

LINDSKOG, N.: [1] Über die Drehung eines starren Körpers, auf den keine Kräfte wirken, um einen festen Punkt. Nova Acta Reg. Soc. Sci. Upsal. (3) **12**, 24 pp. (1885).

LINNIK, YU. V., and V. S. NOVOSELOV: [1] Random disturbances of the regular precession of a gyroscope (R). Prikl. Mat. Meh. **17**, 361—368 (1953).

LIOUVILLE, R.: [1] Sur la rotation des solides. C. R. Acad. Sci. Paris **120**, 903—905 (1895). — [2] Sur le mouvement d'un corps solide pesant suspendu par l'un de ses points. Acta Math. **20**, 239—284 (1897).

LOTTNER, E.: [1] Reduktion eines schweren, um einen festen Punkt rotierenden Revolutionskörpers auf die elliptischen Transcendenten. J. reine u. angew. Math. **50**, 111—125 (1855). — [2] Note de l'éditeur. Appendix to C. G. J. JACOBI [5].

LYAPUNOV, A. M.: [1] On a certain property of the differential equations of the problem of motion of a heavy rigid body, having a fixed point (R). Soobšč. Har'kovsk. Mat. Obšč. **4**, no. 3, 123—140 (1894). Sobranie Sočin., vol. 1, pp. 402—417. Moscow: Izdat. Akad. Nauk SSSR 1954. KOVALEVSKAYA, S. V. Scientific Works, pp. 286—304, Moscow: Izdat. Akad. Nauk SSSR 1948.

MAC CULLAGH, J.: [1] On the rotation of a solid body. Proc. Royal Irish Acad. Dublin, **2**, 520—526, 542—545 (1840/44); **3**, 370—371 (1845/47).

MACMILLAN, W. D.: [1] Dynamics of rigid bodies. New York: McGraw-Hill Co. 1936 or New York: Dover Publications 1960.

MAGNARADZE, N. G.: [1] Translational-rotational motion of a cosmic body relative to the Earth (R). Probl. Dviženiya Iskusstv. Nebesn. Tel. Moscow: Akad. Nauk SSSR 1963, pp. 278—292.

MAGNUS, K.: [1] Beiträge zur Dynamik des kräftefreien kardanisch gelagerten Kreisels. Z. angew. Math. Mech. **35**, 23—34 (1955). — [2] On the stability of motion of a heavy symmetrical gyroscope on gimbals (R). Prikl. Mat. Meh. **22**, 173—178 (1958); transl. as J. Appl. Math. Mech. **22**, 237—243. — [3] Der schwere symmetrische Kreisel in kardanischer Lagerung. Ing.-Arch. **28**, 184—198 (1959). — [4] Ergebnisse und Probleme der allgemeinen Kreiseltheorie. Gyrodynamics. Symposium Celerian 1962. International Union of Theoretical and Applied Mechanics. Edited by H. ZIEGLER, pp. 4—25. Berlin-Göttingen-Heidelberg: Springer 1963.

MALKIN, I. G.: [1] Some problems in the theory of nonlinear oscillations (R). Moscow: Gosud. Izdat. Tehn. Teor. Lit. 1956. — [2] Some problems in the theory of nonlinear oscillations. AEC-TR-3766, Book 1 and 2. Office of Technical Services, Washington 25, D. C.

MANACORDA, T.: [1] Sul moto di un solido asimmetrico attorno ad un punto fisso. Atti Accad. Naz. Lincei. Rend. Cl. Sci. Fis. Mat. Nat. (8) **6**, 711—714 (1949). — [2] Sul moto di un solido planare intorno ad un punto fisso. Riv. Mat. Univ. Parma **1**, 383—391 (1950). — [3] Sul moto di un solido attorno ad un punto fisso. Ricerca Sci. **20**, 487—490 (1950).

MANARINI, M.: [1] Sopra un teorema di STAUDE-VAN DER WOUDE relativo al moto di un corpo pesante intorno ad un punto fisso. Atti R. Accad. Naz. Lincei. Rend. Cl. Sci. Fis. Mat. Nat. (6) **14**, 572—577 (1931). — [2] Alcune notevoli proprietà del centro d'inerzia relative alle rotazioni permanenti nella dinamica del corpo rigido con un punto fisso. Boll. Unione Mat. Ital. (3) **3**, 214—219 (1948). — [3] Sul centro d'inerzia e sul cono d'inerzia relativi ad un sistema rigido e loro applicazioni alle rotazioni permanenti di moti per inerzia e dei giroscopi. Giorn. Mat. Battaglini (5) **6** (86), 66—110 (1958).

MANNHEIM, A.: [1] Sur l'herpolhodie. C. R. Acad. Sci. Paris **100**, 963—966 (1885). — [2] Sur la polhodie. Ibid. **100**, 938—940 (1885).

MARCOLONGO, R.: [1] Sopra due moti di POINSOT concordanti. Ann. Mat. Pura Appl. (2) **22**, 157—174 (1894). — [2] Sur une propriété de deux mouvements à la POINSOT concordants. J. de Sciencias Mat. Astr. **13**, 17—21 (1896). — [3] Osservazioni intorno alla nota del Sig. KOLOSSOFF "Sur le cas de M. GORIATSCHOFF de la rotation d'un corps pesant autour d'un point fixe". Rend. Circ. Mat. Palermo **16**, 349—357 (1902). — [4] Teoria del gyroscopio simmetrico pesante. Ann. Mat. Pura Appl. (3) **7**, 99—128 (1902). — [5] Sul moto di un corpo pesante intorno a un punto fisso. Atti R. Accad. Naz. Lincei. Rend. Cl. Sci. Fis. Mat. Nat. (5) **17_2**, 698—705 (1908).

MAXWELL, J. C.: [1] On a dynamical top, for exhibiting the phenomena of the motion of a system of invariable form about a fixed point, with some suggestions as to the Earth's motion. Trans. Roy. Soc. Edinburgh **21**, 559—570 (1857). Sci. Papers, vol. 1, pp. 248—262, Cambridge: Univ. Press 1890.

MAYER, A.: [1] Symmetrische Lösung der Aufgabe, die Rotation eines starren Körpers, dessen Winkelgeschwindigkeiten bereits gefunden wurden, vollständig zu bestimmen. Ber. Verh. Königl. Sächs. Ges. Wiss. Leipzig. Math.-phys. Kl. **54**, 53—62 (1902).

MERCALOV, N. I.: [1] The problem of motion of a rigid body, having a fixed point, with $A = B = 4C$ and the area integral $\neq 0$ (R). Bull. Acad. Sci. URSS. Cl. Sci. Techn. (Izv. Akad. Nauk SSSR, Otd. Tehn. Nauk) **1946**, 697—701.

MERKIN, D. R.: [1] Gyroscopic systems (R). Moscow: Gosud. Izdat. Tehn. Teor. Lit. 1956. — [2] On the stability of motion of a gyroframe (R). Prikl. Mat. Meh. **25**, 983—991 (1961), transl. as J. Appl. Math. Mech. **25**, 1482—1493. — [3] On the theory of self-exciting gyroscopes. Izv. Vysš. Učebn. Zaved. Priborostroenie **1959**, no. 3, 21—24. Translation from Russian, OTS: 60-11614, 5 pp. Office of Technical Services, Washington 25, D. C.

Metelicyn, I. I.: [1] On gyroscopic stabilization (R). Dokl. Akad. Nauk SSSR (N.S.) **86**, 31—34 (1952).

Métral, A.: [1] Sur la précession dans le phénomène gyroscopic. C. R. Acad. Sci. Paris **197**, 1582—1584 (1933).

Mettler, E.: [1] Periodische und asymptotische Bewegungen des unsymmetrischen schweren Kreisels. Math. Z. **43**, 59—100 (1938).

Mlodzeevskii, B. K.: [1] On permanent axes in the motion of a heavy rigid body about a fixed point (R). Trudy Otdel. Fiz. Nauk Obšč. Lyubit. Estestvozn. **7**, no. 1, 46—48 (1894). — [2] On a particular case of motion of a heavy rigid body about a fixed point (R). Mat. Sbornik **18**, 76—85 (1896).

—, and P. A. Nekrasov: [1] On conditions for the existence of asymptotic periodic motions in Hess's problem (R). Trudy Otdel. Fiz. Nauk Obšč. Lyubit. Estestvozn. **6**, no. 1, 43—52 (1893).

Moiseev, N. N.: [1] Asymptotics of fast rotations (R). Ž. Vyčisl. Mat. i Mat. Fiziki **3**, no. 1, 145—158 (1963).

Morozov, V. M.: [1] A case of stability of the unsteady motion of a top on a plane (R). Vestnik Moskov. Univ. Ser. I Mat. Meh. **1964**, no. 3, 70—74.

Myasnikov, P. V.: [1] On a new method of singling out integrable cases of motion from the general problem of rotation of a heavy rigid body about a fixed point; a new particular case of motion (R). Vestnik Moskov. Univ. Ser. Fiz. Mat. Estestv. Nauk **9**, no. 3, 47—55 (1954). — [2] A new method of finding a class of integrable cases of motion from the general problem of rotation of a heavy rigid body about a fixed point (R). Moscow Gos. Univ. Učen. Zap. **172**, Meh. 5, 143—162 (1954).

Nadile, A.: [1] Sul moto intorno a un punto fisso di un corpo rigido pesante il cui baricentro appartiene a uno dei piani principali di inerzia. Matematiche Catania **5**, 68—82 (1950).

Nadolschi, V. L.: [1] Sur un nouveau cas intégrable de mouvement d'un corps solide autour d'un point fixe. Ann. Sci. Univ. Jassy. Sect. I, **30**, 43—74 (1944/47), (1948).

Nekrasov, P. A.: [1] On the problem of motion of a heavy rigid body about a fixed point (R). Mat. Sbornik **16**, 508—517 (1892). — [2] On the motion of a heavy rigid body about a fixed point (R). Trudy Otdel. Fiz. Nauk Obšč. Lyubit. Estestvozn. **5**, no. 2, 17—37 (1893). — [3] Appendix to the paper "On the motion of a heavy rigid body about a fixed point" (R). Ibid. **6**, no. 1, 18—20 (1893). — [4] Analytic investigation of a certain case of motion of a heavy rigid body about a fixed point (R). Mat. Sbornik **18**, 161—274 (1895). — [5] Recherches analytiques sur un cas de rotation d'un solide pesant autour d'un point fixe. Math. Ann. **47**, 445—530 (1896).

Nikolai, E. L.: [1] On the motion of a balanced gyroscope in a Cardan suspension (R). Prikl. Mat. Meh. **3**, 3—34 (1939); also in Trudy po mehanike (Works on mechanics), pp. 455—496. Moscow: Gosud. Izdat. Tehn. Teor. Lit. 1955. — [2] Gyroscope in a Cardan suspension (R). Moscow-Leningrad: Gosud. Izdat. Tehn. Teor. Lit. 1944. Second edit. Moscow: Izdat. Nauka 1964.

Novoselov, V. S.: [1] Motion of stabilized gyroscopic systems on a moving base (R). Prikl. Mat. Meh. **23**, 964—967 (1959); transl. as J. Appl. Math. Mech. **23**, 1375—1381.

O'Brien, S., and J. L. Synge: [1] The instability of the tippe-top explained by sliding friction. Proc. Roy. Irish Acad. Sect. A **56**, 23—35 (1954).

Olsson, O.: [1] Ett integrabelt enskildt fall af fasta kroppars rotation kring en fast punkt under tyngdkraftens inverkan. Arkiv Mat. Astr. Fys. **4**, no. 7, 1—32 (1908).

Oseen, C. W.: [1] Über eine in der Theorie des Kreisels auftretende Familie von Flächen sechsten Grades I, II. Arkiv Mat. Astr. Fys. **6**, no. 28, 1—32 (1910); **7**, no. 28, 1—36 (1912).

Osgood, W. F.: [1] On the gyroscope. Trans. Amer. Math. Soc. **23**, 240—264 (1923).

Padova, E.: [1] Sulla rotazione di un corpo di rivoluzione pesante che gira attorno ad un punto del suo asse di simmetria. Atti R. Accad. Sci. Torino **19**, 1007—1016 (1884). — [2] Proprietà del moto di un corpo di rivoluzione soggetto a forze che hanno la funzione potenziale $H\cos^2\Theta$. Atti R. Accad. Lincei. Rend. Cl. Sci. Fis. Mat. Nat. (4) **2**, 135—140, 168—174 (1886). — [3] Del moto di rotazione dei corpi rigidi. Ibid. (5) $\mathbf{3}_1$, 161—165 (1894). — [4] Del moto di un corpo di rivoluzione attorno ad un punto del suo asse. Ibid. (5), $\mathbf{4}_2$, 198—202 (1895).

Paladini, B.: [1] Sul moto di rotazione di un corpo rigido attorno ad un punto fisso. Ann. R. Scuola Norm. Sup. Pisa. Sci. Fis. Mat. **5**, 943—944 (1888).

Parkin, D. G.: [1] The inverting top. Math. Gaz. **40**, 260—265 (1956). — [2] The rising of tops with rounded pegs. Physica **24**, 313—330 (1958).

Peano, G.: [1] Sopra lo spostamento del polo sulla terra. Atti. R. Accad. Sci. Torino **30**, 515—523 (1895). — [2] Sul moto del polo terrestre. Ibid. **30**, 845—852 (1895). — [3] Sul moto del polo terrestre. Atti R. Accad. Lincei. Rend. Cl. Sci. Mat. Nat. (5) $\mathbf{5}_1$, 163—168 (1896).

Peretti, G.: [1] On a curious gyroscopic effect. Atti Sem. Mat. Fis. Univ. Modena **2**, 75—86 (1948).

Petrus, A.: [1] Beiträge zur Theorie der Herpolhodie Poinsot's Diss. Halle, 1902.

Pignedoli, A.: [1] Sul moto di un corpo rigido pesante intorno ad un punto fisso, prossimo al baricentro o poco differente della struttura giroscopica. Atti Soc. Nat. Mat. Modena (6) **76**, 115—143 (1945).

Plymale, B. T., and R. Goodstein: [1] Nutation of a free gyro subjected to an impulse. J. Appl. Mech. **22**, 365—366 (1955).

Poincaré, H.: [1] Sur le problème des trois corps et les équations de la dynamique. Acta Math. **13**, 1—271 (1890). Oeuvres, vol. 7, pp. 262—479, Paris: Gauthier-Villars 1952. — [2] Les méthodes nouvelles de la mécanique céléste, vol. 1, Paris: Gauthier-Villars 1892.

Poinsot, L.: [1] Théorie nouvelle de la rotation des corps, Paris, 1834, republished in 1851 in J. Math. Pures Appl. (1) **16**, 289—336 (1851), and in 1854 as an appendix to the Connaissance des Temps.

Poisson, S. D.: [1] Sur un cas particulier du mouvement de rotation des corps pesans. J. Ecole Polyt. **16**, 247—262 (1813). — [2] Mémoire sur le mouvement de la terre autour de son centre de gravité. Mémoires de l'Acad. des Sci. de l'Institut de France, vol. **7**, 199—265 (1827), vol. **9**, 309—310 (1830). — [3] Traité de mécanique, 2 vols., Paris: Bachelier 1811, second edition 1833.

Polubarinova-Kočina, P. Ya.: [1] On single-valued solutions and algebraic integrals in the problem of rotation of a heavy rigid body about a fixed point (R). Kovalevskaya Memorial Collection "Motion of a Rigid Body about a Fixed Point", pp. 157—186. Moscow-Leningrad: Izdat. Akad. Nauk SSSR 1940.

Poritsky, H.: [1] Topics in gyroscopic motion. J. Appl. Mech. **75**, 1—8 (1953).

POŽARICKII, G. K.: [1] On the stability of permanent rotations of a rigid body with a fixed point under the action of a NEWTONian central force field (R). Prikl. Mat. Meh. **23**, 792—793 (1959); transl. as J. Appl. Math. Mech. **23**, 1134—1137.

PUGLISI, M.: [1] Sul movimento di un punto pesante sopra superficie di rivoluzione. Rend. Circ. Matem. Palermo **12**, 312—321 (1898). — [2] Sul movimento di un punto non soggetto ad alcuna forza sopra un toro. Ibid. **14**, 180—191 (1900).

PUISEUX, V.: [1] Note sur le mouvement d'un point pesant sur une sphère. J. Math. Pures Appl. **7**, 517—520 (1842).

QUATELA, F.: [1] Extensione di una dimostrazione del BURGATTI nel problema del giroscopio pesante. Giorn. Mat. Battaglini **67**, 14—17 (1929).

QUILGHINI, D.: [1] Sul principio dell'effetto giroscopico. Ricerche Mat. **7**, 205—231 (1958).

RAUBER, A.: [1] Über die Lösung der Differentialgleichungen, welche die Bewegung des dreiachsigen Kreisels um einen festen Punkt beschreiben. Diss. München, 1913, 53 pages.

RAWLINGS, A. L.: [1] The theory of the gyroscopic compass and its deviations. Second edition, New York: Macmillan 1944.

RESAL, H.: [1] Développements sur un point de la théorie de la rotation des corps solides. J. Ecole Polyt. **53**, 17—30 (1883). — [2] Note sur la courbure de l'herpolhodie. Ibid. **55**, 275—285 (1885).

REŠETNIKOV, V. L.: [1] Some questions in the dynamics of a CARDAN suspension of a triaxial stabilized platform (R). Prikladna Meh. **7**, 189—196 (1961).

REVEILLE, J.: [1] Examen critique du mémoire de POINSOT sur la "Théorie nouvelle de la rotation des corps". J. Ecole Polyt. (2) **20**, 107—114 (1919).

RICHARDSON, K. I. T.: [1] The gyroscope applied. New York: Philosophical Library 1954.

RIONERO, S.: [1] Sul principio dell'effetto giroscopico. Ricerche Mat. **7**, 14—20 (1958). — [2] Sul principio dell'effetto giroscopico nel caso sferico. Ibid. **7**, 281—290 (1958).

ROITENBERG, L. YA.: [1] On the theory of gyroscope follow-up in the presence of random interference (R). Dokl. Akad. Nauk SSSR **142**, 1050—1053 (1962); transl. as Soviet Phys. Dokl. **7**, 110—113 (1962).

ROITENBERG, YA. N.: [1] The accelerated placing of a gyroscopic compass in a meridian (R). Prikl. Mat. Meh. **23**, 961—963 (1959); transl. as J. Appl. Math. Mech. **23**, 1370—1374. — [2] On the placing of a gyroscopic compass into a meridian during the acceleration of the rotors of the gyroscopes (R). Ibid. **24**, 88—92 (1960); J. Appl. Math. Mech. **24**, 113—119. — [3] On the motion of gyroscopic devices under the influence of random forces (R). Ibid. **24**, 463—472 (1960); J. Appl. Math. Mech. **24**, 677—690. — [4] Motion of a gyroscopic compass under the action of random forces (R). Dokl. Akad. Nauk SSSR **133**, 1045—1048 (1960); transl. as Soviet Phys. Dokl. **5**, 718—722 (1961).

ROUTH, E. J.: [1] On a theorem of JACOBI in dynamics. Quart. J. Pure and Appl. Math. **23**, 34—45 (1889). — [2_1] The elementary part of a treatise on the dynamics of a system of rigid bodies, being part I of a treatise on the whole subject, sixth edition. London and New York: The Macmillan Co. 1897. Reprinted by the Dover Publications, New York, 1955. — [2_2] The advanced part of a treatise on the dynamics of a system of rigid bodies, being part II of a treatise on the whole subject, fifth edition. London and New York: The Macmillan Co. 1892. Reprinted by the Dover Publications, New York 1955. — [3_1] Die Dynamik der Systeme starrer Körper in zwei Bänden. Autorisierte deutsche Ausgabe von A. Schepp. Mit einem Vorwort von F. Klein. Erster Band: Die Elemente. Leipzig: B. G. Teubner 1898. — [3_2] Die Dynamik der Systeme starrer Körper in zwei Bänden, Autorisierte deutsche Ausgabe von A. Schepp. Mit Anmerkungen von F. Klein. Zweiter Band: Die höhere Dynamik. Leipzig: B. G. Teubner 1898.

Rumyancev, V. V.: [1] On the stability of rotation of a heavy rigid body with a fixed point in the case of S. V. Kovalevskaya (R). Prikl. Mat. Meh. **18**, 457—458 (1954). — [2] Stability of permanent rotations of a heavy rigid body (R). Ibid. **20**, 51—66 (1956). — [3] On stability of permanent rotations of a solid body around a fixed point (R). Ibid. **21**, 339—346 (1957). — [4] On the problem of motion of a heavy solid having a fixed point (R). Dokl. Akad. Nauk SSSR (N.S.) **116**, 185—188 (1957). — [5] On the stability of motion of a gyroscope on gimbals (R). Prikl. Mat. Meh. **22**, 374—378 (1958); transl. as J. Appl. Math. Mech. **22**, 513—520. — [6] On the stability of motion of a gyroscope on gimbals (R). Ibid. **22**, 499—503 (1958); J. Appl. Math. Mech. **22**, 694—700. — [7] On the stability of motion of gyrostats (R). Ibid. **25**, 9—16 (1961); J. Appl. Math. Mech. **25**, 9—19. — [8] On the stability of motion of certain types of gyrostats (R). Ibid. **25**, 778—784 (1961); J. Appl. Math. Mech. **25**, 1158—1169.

Rybakov, A. I.: [1] Numerical investigation of the rotational motion of a celestial body about its center of inertia (R). Soobšč. Gosud. Astron. Inst. Šternberga, no. 112, 33—41 (1961).

Saint-Guilhem, P.: [1] Nouvelle détermination synthétique du mouvement d'un corps solide autour d'un point fixe. J. Math. Pures Appl. (1) **19**, 356—365 (1854).

Saryčev, V. A.: [1] Influence of Earth oblateness on the rotational motion of an artificial satellite (R). Iskusstv. Sputniki Zeml'i, vyp. 6, pp. 3—10. Moscow: Izdat. Akad. Nauk SSSR 1961.

Sauvage, L.: [1] Sur les solutions régulières d'un système d'équations différentielles I, II, III. Ann. Sci. Ecole Norm. Sup. (3) **3**, 391—404 (1886); **5**, 7—22 (1888); **6**, 157—182 (1889).

Savet, P., Editor: [1] Gyroscopes: Theory and design. With applications to instrumentation, guidance, and control. New York, Toronto, London: McGraw-Hill Book Co. 1961.

Sbrana, F.: [1] Un osservazione sul moto di un solido pesante attorno ad un punto. Boll. Unione Mat. Ital. **13**, 269—271 (1934).

Scarborough, J. B.: [1] The gyroscope, theory and applications. New York: Interscience Publishers 1958.

Schallreuter, W.: [1] Der Spielkreisel mit Selbstaufrichtung. Wiss. Z. Ernst-Moritz-Arndt Univ. Greifswald. Math. Nat. Reihe **8**, 43—46 (1958/59).

Schmidt, G.: [1] Einfache Ableitung der Euler'schen Bewegungsgleichungen. S.-B. Königl. Böhmischen Ges. Wiss. Prag **1878**, 79—81.

Schuh, F.: [1] Motion of an excentrically loaded sphere on a horizontal plane, in connection with the magic top "tippe-top". Nederl. Akad. Wetensch. Proc. Ser. A 56 = Indagationes Math. **15**, 423—432 (1953).

Schuler, M.: [1] Kreisellehre, being 6th chapter of Müller-Pouillet's book Lehrbuch der Physik, 11th edition, 1. Teil, 1. Band, 1927. — [2] Einführung in die Mechanik, Teil 2, 1951.

Siff, E. J., and C. L. Emmerich: [1] An engineering approach to gyroscopic instruments. New York: Speller 1960.

Šiff, P. A.: [1] On the equations of motion of a heavy rigid body having a fixed point (R). Mat. Sbornik **24**, 169—177 (1903).

SIGNORINI, A.: [1] Complementi alla dinamica dei giroscopi e equazioni del problema completo della balistica esterna. Atti Accad. Naz. Lincei. Mem. Cl. Sci. Fis. Mat. Nat. (8) **1**, (Sect. 1), no. 1, 1—41 (1946). — [2] Sopra un particolare tipo di moti rigidi sferici. Rivista Mat. Univ. Parma **3**, 307—316 (1952).

SILVESTRI, C.: [1] Sulla stabilità dei moti stazionari di sistemi con n gradi di libertà. Atti R. Accad. Lincei. Rend. Cl. Sci. Fis., Mat. Nat. (5) $\mathbf{21}_2$, 508—512 (1912). — [2] Sui moti stazionarii nel caso della KOWALEWSKI. Ibid. (5) $\mathbf{21}_2$, 601—606 (1912). — [3] Sui moti stazionarii nel caso della KOWALEWSKI. Ibid. (5) $\mathbf{22}_1$, 18—23 (1913).

SIMONART, F.: [1] Sur le mouvement d'un corps solide autour d'un point fixe. Ann. Soc. Sci. Bruxelles. Sect. A **42**, 96—101 (1922/23).

SIMONI, F. DE: [1] Sull'integrazione del moto di un solido intorno ad un punto fisso col metodo di HAMILTON-JACOBI. Boll. Unione Mat. Ital. (2) **1**, 122—128 (1939). — [2] Sul moto dei corpi rigidi con sospensione quasibaricentrica. Ann. Scuola Norm. Sup. Pisa. Sci. Fis. Mat. (2) **11**, 197—210 (1942).

SKIMEL', V. N.: [1] On problems of stability of motion of a heavy rigid body about a fixed point (R). Prikl. Mat. Meh. **20**, 130—132 (1956). — [2] Certain problems of motion and stability of a heavy gyroscope (R). Trudy Kazansk. Aviac. Inst. **38**, 103—129 (1958).

SLOMYANSKII, G. A.: [1] On integration of the equations of motion of a symmetric astatic gyroscope (R). Prikl. Mat. Meh. **17**, 411—422 (1953).

SLUDSKII, F.: [1] On the motion of the mass center of a heavy rigid body rotating about a fixed point (R). Mat. Sbornik **17**, 791—796 (1895).

SÖDERBLOM, A.: [1] Über die Drehung eines Rotationskörpers um einen festen Punkt. Nova Acta Reg. Soc. Sci. Upsal. (3) **12**, 92 pp. (1884).

SOMMERFELD, A.: [1] Bemerkungen zum HESS'schen Falle der Kreiselbewegung. Nachr. Königl. Ges. Wiss. Göttingen, Math.-phys. Kl., **1898**, 83—86.

SOMOV, O. I.: [1] Démonstration des formules de M. JACOBI relatives à la théorie de la rotation d'un corps solide. J. reine u. angew. Math. **42**, 95—116 (1851). — [2] Solution rigoureuse du problème de la rotation autour d'un point fixe d'un corps solide pesant, lorsque ce corps a deux moments d'inertie principaux égaux et que le point fixe est situé sur l'axe, auquel repond le troisième moment. Bull. Cl. Phys.-Math. Acad. Imp. Sci. St. Petersbourg **14**, 113—135 (1856).

SPARRE, M. DE: [1] Sur l'herpolhodie de POINSOT. C. R. Acad. Sci. Paris **99**, 906—909 (1884). — [2] Sur l'herpolhodie dans le cas d'une surface du second degré quelconque. Ibid. **101**, 370—373 (1885). — [3] Sur le mouvement d'un solide autour d'un point fixe et sur le pendule conique. Ann. Soc. Sci. Bruxelles. Sect. B. **9**, 49—94 (1885).

SPONDER, E.: [1] Eine genäherte Behandlung des schweren symmetrischen Kreisels in nicht-EULERschen Koordinaten. Z. angew. Math. Phys. **6**, 462—478 (1955).

SRETENSKII, L. N.: [1] On the works of S. A. ČAPLYGIN concerning theoretical mechanics (R). S. A. ČAPLYGIN, Sobranie Sočin., vol. 3, pp. 366—376. Moscow-Leningrad: Gosud. Izdat. Tehn. Teor. Lit. 1950. — [2] Motion of the GORYAČEV-ČAPLYGIN gyroscope (R). Izv. Akad. Nauk SSSR. Otdel. Tehn. Nauk, **1953**, no. 1, 109—119. — [3] On certain cases of motion of a heavy rigid body with a gyroscope (R). Vestnik Moskov. Univ., Ser. I Mat. Meh. **1963**, no. 3, 60—71.

STÄCKEL, P.: [1] Über die reduzierten Differentialgleichungen des schweren unsymmetrischen Kreisels. Nachr. Königl. Ges. Wiss. Göttingen, Math.-Phys. Kl., **1908**, 272—274. — [2] Ausgezeichnete Bewegungen des schweren unsymmetrischen Kreisels. Math. Ann. **65**, 538—555 (1908). — [3] Die reduzierten Differentialgleichungen der Bewegung des schweren unsymmetrischen

Kreisels. Ibid. **67**, 399—432 (1909). — [4] Ausgezeichnete Kreiselbewegungen. Jber. Deutschen Math. Verein. **18**, 120—129 (1909).

STANGE, K.: [1] Über die Bewegung des stabilen schweren symmetrischen Kreisels bei kleinen Neigungswinkeln seiner Achse. Ing.-Arch. **16**, 121—134 (1947). — [2] Über die Bewegung eines stabilen schweren symmetrischen Kreisels bei kleinen Störungen des Längsschwunges. Ibid. **16**, 343—356 (1948).

STANKEVIČ, I. V.: [1] Symmetric coordinates for the motion of a rigid body about a fixed point (R). Trudy Moskov. Stanko-Instrum. Inst. **1**, 135—152 (1937). — [2] Rotation of a heavy rigid body about a fixed point. Application of the JACOBI method (R). Ibid. **2**, 17—42 (1938). — [3] Application of the JACOBI method to the investigation of the motion of a heavy rigid body about a fixed point (R). Ibid. **5**, 5—26 (1939). — [4] Rotation of a heavy rigid body about a fixed point. Application of the JACOBI method (R). Ibid. **6**, 5—56 (1940).

STAUDE, O.: [1] Über die Bewegung eines schweren Punktes auf einer Rotationsfläche. Acta Math. **11**, 303—332 (1888). — [2] Über permanente Rotationsaxen bei der Bewegung eines schweren Körpers um einen festen Punkt. J. reine u. angew. Math. **113**, 318—334 (1894).

STEKLOV, V. A.: [1] A certain case of motion of a heavy rigid body having a fixed point (R). Trudy Otd. Fiz. Nauk Obšč. Lyubit. Estestvozn. **8**, no. 2, 19—21 (1896). — [2] A new particular solution of the differential equations of motion of a heavy rigid body having a fixed point (R). Ibid. **10**, no. 1, 1—3 (1899). — [3] Sur le mouvement d'un corps solide ayant une cavité de forme ellipsoidale remplie par un liquide incompressible et sur les variations des latitudes. Ann. Fac. Sci. Toulouse (3) **1**, 145—256 (1909).

STOEWA, P.: [1] Die Stabilität der STAUDEschen Kreiselbewegungen. Jahresber. Phil. Fak. Univ. Leipzig 1922, II. Halbjahr, pp. 131—132.

STOPPELLI, F.: [1] Sul principio dell'effetto giroscopico. Giorn. Mat. Battaglini (4) **4**, (80), 14—38 (1951). — [2] Sui fenomeni giroscopici in un solido qualsiasi. Rend. Sem. Mat. Univ. Padova **21**, 25—43 (1952).

SUSLOV, G. K.: [1] Rotation of a heavy rigid body about a fixed pole (case of S. V. KOVALEVSKAYA) (R). Trudy Otdel. Fiz. Nauk Obšč. Lyubit. Estestvozn. **7**, no. 2, 1—3 (1895). — [2] Über eine continuierliche Gruppe von DARBOUXschen Rotationen. Jber. Deutschen Math. Verein. **4**, 141—144 (1894/95). — [3] Pseudoregular precessions (R). Trudy Otdel. Fiz. Nauk Obšč. Lyubit. Estestvozn. **10**, no. 2, 30—35 (1900).

SVEŠNIKOV, A. A.: [1] On the motion of a gyroscopic pendulum with random displacements of its point of suspension (R). Prikl. Mat. Meh. **26**, 402—411 (1962); transl. as J. Appl. Math. Mech. **26**, 601—614.

SVETLOV, A. V.: [1] On the rotation of a gyroscope in a resisting medium (R). Prikl. Mat. Meh. **1**, 371—376 (1938).

SYLVESTER, J. J.: [1] On the motion of a rigid body acted on by no external forces. Philos. Trans. Roy. Soc. London **156**, 757—780 (1866); also Collected Math. Papers, vol. 2, pp. 577—601, Cambridge: Cambridge Univ. Press 1908.

SYNGE, J. L.: [1] Apsidal angles for symmetrical dynamical systems with two degrees of freedom. Bull. Amer. Math. Soc. **54**, 111—118 (1948). — [2] On a case of instability produced by rotation. Philos. Mag. (7) **43**, 724—728 (1952).

TABAROVSKII, A. M.: [1] On the stability of motion of a heavy gyroscope on gimbals (R). Prikl. Mat. Meh. **24**, 572—574 (1960); transl. as J. Appl. Math. Mech. **24**, 850—855. — [2] On the motion and stability of a gyroscope on gimbals in the NEWTONian central force field (R). Ibid. **25**, 259—264 (1961); J. Appl. Math. Mech. **25**, 382—389.

TALLQVIST, H.: [1] Über die Drehung eines starren Körpers um einen festen Punkt. Acta Soc. Sci. Fenn. **50**, no. 14, 1—27 (1926).

TANNENBERG, W. DE: [1] Sur le mouvement d'un corps solide pesant autour d'un point fixe; cas particulier signalé par Mme KOWALEVSKI, Gounouilhou, Bordeaux, 1898. — [2] Sur la forme donneé par KIRCHHOFF aux équations différentielles du mouvement d'un corps solide. Procès Verbaux des Séances de la Soc. Sci. Phys. Nat. de Bordeaux **1897/98**, 226—228. — [3] Sur le mouvement d'un corps solide pesant autour d'un point fixe. Cas particulier signalé par Mme KOWALEVSKI. Mém. Soc. Sci. Phys. Nat. de Bordeaux (5) **3**, 309—315 (1899).

TEDONE, O.: [1] Sopra i casi in cui il problema del moto di un corpo rigido si riduce alle quadrature. Nuovo Cimento (4) **1**, 220—229, 269—279, 353—365 (1895).

TISSERAND, F.: [1] Traité de mécanique céleste, vol. 2, pp. 371—475. Paris: Gauthier-Villars 1891.

TONDL, A.: [1] The stability of motion of a rotor with unsymmetrical shaft on an elastically supported mass foundation. Ing.-Arch. **29**, 410—418 (1960).

TOROPOVA, V. A.: [1] On integration of the equations of rotation of a heavy solid body around a fixed point in the general case of GORYAČEV-ČAPLYGIN (R). Izv. Akad. Nauk UzSSR. Ser. Fiz. Mat. **1958**, no. 2, 77—85.

UDESCHINI, P.: [1] Giroscopio rotolante sopra un piano orizzontale in presenza di un polo magnetico. Atti Accad. Sci. Torino. Cl. Sci. Fis. Mat. Nat. **77**, 384—398 (1942).

URZI, G.: [1] Un caso particolare del problema della rotazione di un corpo solido di rivoluzione sospeso per un punto del suo asse di simmetria. Giorn. Mat. Battaglini **36**, 185—207 (1898).

VACCA, M. T.: [1] Su alcuni moti, relativi allo spostamento del polo terrestre, dovuti a correnti cicliche marine. Univ. e Politec. Torino. Rend. Sem. Mat. **16**, 413—427 (1956—57). — [2] Su un problema piu generale di quello di DE BRUN per il moto di un corpo rigido intorno a un punto fisso. Bull. Unione Mat. Ital. (3) **10**, 52—58 (1955).

VČERAŠNYUK, P. P.: [1] On the motion of an artificial Earth satellite relative to its center of mass (R). Ukraïn. Mat. Ž. **15**, 305—309 (1963).

VOLKOV, M. S.: [1] Series which represent the periodic translational-rotational motion of a satellite in the gravitational field of a sphere (R). Byull. Inst. Teor. Astr. Akad. Nauk SSSR **9**, no. 2, 120—143 (1963).

VOLOSOV, V. M.: [1] The method of averaging (R). Dokl. Akad. Nauk SSSR **137**, 21—24 (1961); transl. as Soviet Math. Dokl. **2**, 221—224. — [2] Higher approximations in averaging (R). Ibid. **137**, 1022—1025 (1961); Soviet Math. Dokl. **2**, 382—385.

VOLTERRA, V.: [1] Sur la théorie des variations des latitudes. Acta Math. **22**, 201—358 (1899).

VORONEC, P. V.: [1] Geometric investigation of EULER's case of rotation of a rigid body about a fixed point (R). Izv. Kievsk. Univ. **1898**, 1—113.

WAELSCH, E.: [1] Zweibein und Kreisel. Z. angew. Math. Mech. **5**, 331—337 (1925).

WANGSNESS, R. K., and F. BLOCH: [1] The dynamical theory of nuclear induction. Phys. Rev. **89**, 728—739 (1953).

WEIDENHAMMER, F.: [1] Erzwungene Präzession des symmetrischen Kreisels bei lotrechter Erschütterung. Z. angew. Math. Mech. **38**, 480—483 (1958).

WEIERSTRASS, K.: [1] Math. Werke, vol. 5, Chap. 1, p. 13. Berlin: Mayer and Müller 1915.

WEINSTEIN, A.: [1] The spherical pendulum and complex integration. Amer. Math. Monthly **49**, 521—523 (1942).

WIEBELITZ, R.: [1] Zur Theorie der erzwungenen Schwingungen des symmetrischen Kreisels. Z. angew. Math. Phys. **6**, 362—377 (1955).

WOOLARD, E. W.: [1] Theory of the rotation of the Earth around its center of mass. U. S. Naval Observatory Astronomical Papers, vol. XV, Part. 1. U. S. Government Printing Office, Washington 1953.

WOUDE, W. VAN DER: [1] Über die STAUDEschen Kreiselbewegungen. Math. Z. **16**, 170—172 (1923).

YAROV, M. S.: [1] One method of investigating the translational and rotational motion of satellites and planets which possess an axis of symmetry (R). Byull. Inst. Teor. Astr. Akad. Nauk SSSR **8**, 9, 647—659 (1962).

ŽBANOV, YU. K.: [1] Investigation of free vibrations in an autonomous system of coordinates for a moving object (R). Prikl. Mat. Meh. **24**, 1024—1029 (1960); transl. J. Appl. Math. Mech. **24**, 1558—1565. — [2] A horizontal gyrocompass on vibrating elastic foundation (R). Ibid. **25**, 933—937 (1961); transl. as J. Appl. Math. Mech. **25**, 1397—1405.

ZEULI, T.: [1] Sul moto di un corpo rigido pesante asimmetrico col baricentro appartenente all'asse di una delle sezioni circolari dell'ellissoide d'inerzia. Univ. e Polit. Torino. Rend. Sem. Mat. **9**, 263—295 (1950).

ZIEGLER, H.: [1] Der symmetrische Kardankreisel unter einem Moment an der Achse des äußeren Rings. Z. angew. Math. Phys. **7**, 253—256 (1956). — [2] Gyrodynamics. Symposium Celerina, August 20—23, 1962. International Union of Theoretical and Applied Mechanics. Edited by H. ZIEGLER. Berlin-Göttingen-Heidelberg: Springer 1963.

ŽUKOVSKII, N. E.: [1] HESS's loxodromic pendulum (R). Trudy Otdel. Fiz. Nauk Obšč. Lyubit. Estestvozn. **5**, no. 2, 37—45 (1893). Sobranie Sočin., vol. 1, pp. 257—274. Moscow-Leningrad: Izdat. Tehn. Teor. Lit. 1948. — [2] Geometrische Interpretation des HESS'schen Falles der Bewegung eines schweren starren Körpers um einen festen Punkt. Jber. Deutschen Math. Verein. **3**, 1894, 62—70 (1892/93). — [3] Geometric interpretation of the case considered by KOVALEVSKAYA of the motion of a heavy rigid body about a fixed point (R). Mat. Sbornik **19**, 45—93 (1896). Sobranie Sočin., vol. 1, pp. 294—338. — [4] Geometrische Interpretation des von Sophie KOWALEVSKI behandelten Falles der Bewegung eines schweren starren Körpers um einen festen Punkt. Jber. Deutschen Math. Verein. **4**, 1897, 144—150 (1894/95). — [5] Model of the HESS pendulum (R). Trudy Otdel. Fiz. Nauk Obšč. Lyubit. Estestvozn. **10**, no. 1, (1899). Sobranie Sočin., vol. 1, 358—363.

Author Index

MIX
Papier aus verantwortungsvollen Quellen
Paper from responsible sources
FSC® C105338
FSC
www.fsc.org

If you have any concerns about our products,
you can contact us on
ProductSafety@springernature.com

In case Publisher is established outside the EU,
the EU authorized representative is:
Springer Nature Customer Service Center GmbH
Europaplatz 3, 69115 Heidelberg, Germany

Printed by Libri Plureos GmbH
in Hamburg, Germany